ENERGY AND PROCESS OPTIMIZATION FOR THE PROCESS INDUSTRIES

ENERGY AND PROCESS OPTIMIZATION FOR THE PROCESS INDUSTRIES

FRANK (XIN X.) ZHU

AIChE®

WILEY

For general information on our other products and services or for technical support, please contact our
Customer Care Department within the United States at (800) 762-2974, outside the United States
at (317) 572-3993 or fax (317) 572-4002.

Wiley also publishes its books in a variety of electronic formats. Some content that appears in print may
not be available in electronic formats. For more information about Wiley products, visit our web site at
www.wiley.com.

Library of Congress Cataloging-in-Publication Data:

Zhu, Frank Xin X.
    Energy optimization for the process industries / Frank Xin X. Zhu. – First
edition.
        pages cm
    Includes index.
    ISBN 978-1-118-10116-2 (hardback)
    1. Process engineering. 2. Manufacturing processes–Cost control. 3.   ·
Manufacturing processes–Environmental aspects. 4. Energy conservation.   I.
Title.
    TS176.Z53 2013
    658.5–dc23

                                                    2013020443

Printed in the United States of America

10   9   8   7   6   5   4   3   2   1

To Jane, Kathy, and Joshua

These three remain: faith, hope, and love.
The greatest of these is love.
1 Corinthians 13:13

CONTENTS

PREFACE

In recent years, there has been an increased emphasis on industrial energy optimization. However, there are no dedicated books available to discuss basic concepts, provide practical methods, and explain industrial application procedures. This book is written to fill this gap with the following people in mind: managers, engineers, and operators working in the process industries. The book is aimed at providing practical tools to people who face challenges and wish to find opportunities for improved processing energy efficiency. I hope that this book is able to convey concepts, theories, and methods in a straightforward and practical manner.

With these objectives in mind, the focal discussions in this book center around five kinds of energy improvement opportunities. The first is minimizing heat losses via diligence. In reality, steam generated in the boiler house is distributed through an extensive network of steam pipelines to end users. The losses in steam distribution can be 10–20% of fuel fired in boilers. Hence, the net boiler efficiency could be 10–20% lower from the user's point of view.

The second is operation improvement opportunities, which occur due to the age of processes, the nature of operation variations. This usually involves establishing the best operation practices and optimizing process conditions. The third opportunity comes from improved heat recovery within and across process units, which requires design changes to process flow schemes and heat exchange schemes. The fourth is the use of state-of-the-art processes and equipment technology for enhanced processing efficiency. The fifth and final opportunity comes from better operation and planning of the energy supply system. In this book, these opportunities will be discussed and the methods for opportunity identification, assessment, and implementation will be introduced.

As the book covers a wide range of topics, I have attempted to organize the materials in such a way that aids the reader to locate the relevant materials quickly, to be able to understand them readily, and to apply them in the right context. Furthermore, the structure of the book is carefully designed to help readers avoid losing sight of the forest for the trees. The book starts with a provision of an overall context of the process energy optimization, followed by concepts and theory to gain a basic understanding of the energy metrics, gradually transitions to practical assessment methods from equipment- to system-based evaluations, and culminates in establishing an effective energy management system to sustain the benefits. Therefore, the features of material organizations need to be explained:

- An overview of process energy optimization is provided in Chapter 1. Basic concepts for process energy efficiency are introduced in Chapters 2–4 in Part 1. These concepts include energy intensity for determining process-specific energy use, energy benchmarking for setting the energy baseline and identifying the improvement gap, and key energy indicators for determining what operating parameters to monitor and what are their operating targets.

- Energy assessment methods are presented in Part 2. Chapter 5 focuses on reliable and efficient operation for process-fired heaters, while Chapter 6 discusses process energy loss analysis. Chapters 7 and 8 are dedicated to heat exchanger performance assessment as well as fouling mitigation. Methods for heat recovery targeting and retrofit design are explained in Chapters 9 and 10, while process integration methods are illustrated in Chapter 11.

- Part 3 is dedicated to process assessment. The concept of operating window for fractionation is introduced in Chapter 12 where the calculation methods for determining the operating window are explained. Fractionation system assessment and optimization are discussed in Chapters 13 and 14.

- The steam and power system must supply energy in an efficient manner if one wishes to achieve high energy efficiency for an overall processing site. Thus, methods for steam and power system assessment and optimization are provided in Part 4. Steam and power system modeling is explained in Chapter 15. Chapter 16 covers steam and power balances. Chapter 17 discusses practical steam pricing methods. Chapter 18 focuses on steam system benchmarking. By putting the models and opportunities together, Chapter 19 discusses how to build mathematical models for steam and power optimization.

- Finally, Part 5 is dedicated to techno-economical analysis of energy modifications as well as establishing an effective energy management system. To avoid bad investment, true benefits must be determined by considering outside system battery limit conditions and process variations, which are discussed in Chapters 20 and 21. The goal of the capital project evaluation is to achieve minimum investment cost. The key to achieving this goal is to explore alternative design options for each improvement idea and find economical solutions to overcome process/equipment limits. Detailed discussions are given in Chapter 22. The last chapter, Chapter 23, condenses the ideas presented in the other chapters by explaining how to establish an effective energy management system to sustain the benefits gained from implementation through a case study.

It is my sincere hope that readers will find the methods and techniques discussed herein useful for analysis, optimization, engineering design, and monitoring, which are required to identify, assess, implement, and sustain energy improvement opportunities. More importantly, I hope that this book can help readers build mental models in terms of key parameters and their limits and interactions. You can then revisit these methods whenever you need them.

Clearly, it was not a small effort to write this book; but it was the strong need of practical methods for helping people to improving industrial energy efficiency that spurred me to writing. In this endeavor, I owe an enormous debt of gratitude to many colleagues at UOP and Honeywell for their generous support to this effort. First of all, I would like to mention Geoff Miller, vice president of UOP, who has provided encouragement and support. I am very grateful to many colleagues for constructive suggestions and comments on the materials contained in this book, and I apologize if any names are unmentioned. I would especially like to thank John Petri for his critical readings of Chapter 4, Darren Le Geyt and Dennis Clary for Chapter 5, Phil Daly and Lillian Huppler for Chapter 6, Zhanping (Ping) Xu for Chapter 12, and Chuck Welch for Chapter 15; their comments have improved these chapters. Tom King provided meticulous line-by-line reading of the entire first draft and identified pedagogical lapses, typos, better expressions, and better sources of information. My sincere gratitude also goes to Charles Griswold, Margaret (Peg) Stine, and Mark James for their review of the book. I would like to thank all of my colleagues for their help with the book and my debt to them is very great, but I would like to stress that any deficiencies are my responsibility. This book reflects my own opinions and not that of UOP and Honeywell.

I would also like to thank my co-publishers, AIChE and John Wiley, for their help. Special thanks go to Steve Smith at AIChE and Michael Leventhal at John Wiley for guidance. The copyediting and typesetting by Vibhu Dubey at Thomson Digital is very helpful in polishing the book.

Finally, I am truly grateful to my family: my wife Jane and my children Kathy and Joshua, for their understanding, unwavering support, and generosity of spirit in tolerating the absentee *paterfamilias* during the writing of this book. Jane, my beloved wife, produced beautiful drawings for many figures in the book with her graphic design skills and Kathy helped to polish this book with her linguistic skills. Your contributions to this book and to my life are deeply appreciated.

FRANK ZHU

Long Grove, Illinois, USA
May 7, 2013

PART 1

BASIC CONCEPTS AND THEORY

1

OVERVIEW OF THIS BOOK

1.1 INTRODUCTION

Energy management is a buzzword nowadays. What is the objective of energy management in the process industry? It is not simply energy minimization. The ultimate goal of energy management is to control energy usage in the most efficient manner to make production more economical and efficient. To achieve this goal, energy use must be optimized with the same rigor as how product yields and process safety are managed.

The time of "let the plant engineers do their technical work" is long gone. The reduction of the technical workforce due to automation and technology advances has also increased the level of responsibility on business management of plant operations, often resulting in fewer workers taking on more tasks. Furthermore, it is often the case that plant managers and engineers are ill-prepared to take on widespread responsibilities, particularly when working under time pressures. This in turn results in their devoting less time on plant operation and equipment reliability and maintenance. Therefore, the current challenge for energy optimization is: How can we develop effective enablers to support engineers and management?

In addition, plant management and engineers are presented with modern management concepts and techniques. Not all these methods are easily translatable or applicable to any given company. Even if implemented, some of these methods require tailor-made revisions to fit into specific applications. The challenge here becomes: Which methods should be selected and how to implement them for specific circumstances?

Energy and Process Optimization for the Process Industries, First Edition. Frank (Xin X.) Zhu.
© 2014 by the American Institute of Chemical Engineers, Inc. Published 2014 by John Wiley & Sons, Inc.

This reminds me of a project I led a few years into the new millennium. My company took on a project to provide technical support to a large oil refining plant and I was tasked with leading a team of engineers to spearhead this effort. When I met with the general manager of the refinery plant, his words were brief. "My plant spends huge amounts of money on operating costs, in the order of hundreds of million dollars per year." The general manager started after a quick introduction. "I know someone out there can help my plant to cut down the energy cost by more than 10%. I hope it is you." These simple words from the general manager became a strong motivation like a heavy weight on my shoulder. I took the challenge and worked with the team and the plant staff to achieve the goal. By the end of the fifth year, a survey team from corporate management came on site. After reviewing the data and various utility costs, the team issued the statement that the plant had achieved the corporate goal of saving 10% energy costs. Our efforts were successful and the results were recognized by the plant and corporate management.

Over time, I applied the methods and tools I had developed over the course of my career to other projects I was staffed on in the past 10 years. The theory and practice of these methods and experience has become the foundation of this book. The book will present the core of a systematic approach covering energy optimization strategy, solution methodology, supporting structure, and assessment methods. In short, it will describe what it takes to make sizable reductions in energy operating costs for process plants and how to sustain energy-saving benefits. The benefits of this effective approach include identification of large energy-saving projects via applying assessment methods, capturing hidden opportunities in process operation via use of key energy indicators, closing of various loose ends in steam system and off-site utilities via good steam balances, optimizing utility system operation via setting up appropriate steam prices, and maintaining continuous improvement via regular review and performance matrices.

The concepts, methods, and tools presented in this book provide a glimpse of recent advances in energy utilization techniques based on simultaneous optimization of process and energy considerations. The case studies show that very substantial improvements in energy utilization can be made by applying these methods and tools not only in new investment projects but also in existing plants.

1.2   WHO IS THIS BOOK WRITTEN FOR?

This book is written with the following people in mind: managers, engineers, and operators working in the process industries who face challenges and wish to find opportunities for improved processing energy efficiency and are searching for tools for better energy management.

It is my hope that readers are able to take away methods and techniques for analysis, optimization, engineering design, and monitoring, which are required to identify, assess, implement, and sustain energy improvement opportunities. The analysis methods are used for energy benchmarking and gap assessment, while optimization methods are used for operation improvement, heat integration, process changes, and utility system optimization. Engineering methods are applied for developing energy revamp projects, while monitoring methods are used for establishing energy management systems. More importantly, I would like to help readers to build mental models for critical equipment and processes in terms of key parameters and their limits and interactions. You can then revisit these models whenever you need them.

1.3 FIVE WAYS TO IMPROVE ENERGY EFFICIENCY

The five ways in which improved energy efficiency can be achieved within plant processes are highlighted below and will be discussed in detail in this book:

- Minimizing wastes and losses
- Optimizing process operation
- Achieving better heat recovery
- Determining process changes
- Optimizing energy supply system

1.3.1 Minimize Waste and Losses

In reality, steam generated in the boiler house is distributed through an extensive network of steam pipelines to end users. The losses in steam distribution can be 10–20% of fuel fired in boilers. Hence, the net boiler efficiency could be 10–20% lower from the user's point of view.

The losses do not necessarily attribute to a single cause but are the result of a combination of various causes. It is common to observe the major steam loss caused by steam trap failure and condensate discharge problems. Steam loss could also occur due to poor insulation of steam pipes, leaks through flanges and valve seals, opened bypass and/or bleeder valves, and so on. Simple measures such as maintenance of steam traps and monitoring of steam distribution to determine if steam generated is in accordance with steam consumed can lead to significant cost-saving benefits.

Apart from distribution losses, other forms of energy losses could occur due to poor insulation, condensate loss to drainage, pressure loss from steam letdown through valves, pump spill backs, and so on. To detect losses, you must know how much energy is generated versus how much is used in individual processes. The benchmarking method in Chapter 3 could be used to determine the overall gap of the energy performance, and individual losses are identified using different methods. Process energy losses can be detected using the energy loss assessment methods discussed in Chapter 8, while identification of steam losses and the ways to overcome the losses in the steam system are discussed in Chapter 18.

1.3.2   Optimizing Process Operation

The most important step in developing an energy management solution to optimize a process is to be able to measure what process performance looks like against a reasonable set of benchmarks. This involves capturing energy data related to the process and organizing it in a way that allows operations to quickly identify where the big energy consumers are and how well they are doing against a consumption target that reflects the current operations. Only then is it possible to do some analysis to determine the cause of deviations from target and take appropriate remedial action. For this purpose, the concept of key energy indicators is introduced in Chapter 4.

The operation performance gaps are mainly caused by operation variability. Two kinds of operation variability are common in the industry. The first is the so-called operation inconsistency, which is mainly caused by different operation policy and practices applied due to different experience from operators. The second operation inefficiency refers to the kind of operation that is consistent but nonoptimal. This occurs when there are no tools available to indicate to the shift operators the optimal method to run the process and equipment when conditions of feeds and product yields vary.

Once operational gaps are identified, assessment methods (Chapters 5–8 for energy operation, Chapters 12 and 13 for process operation, and Chapter 16 for utility system operation) are then applied to identify root causes—potential causes include inefficient process operation, insufficient maintenance, inadequate operating practices, procedures, and control, inefficient energy system design, and outdated technology. Assessment results are translated into specific corrective actions to achieve targets via either manual adjustments, the best practices, or by automatic control systems. Finally, the results are tracked to measure the improvements and benefits achieved.

1.3.3   Achieving Better Heat Recovery

Using monitoring and optimization tools to improve energy efficiency usually results in pushing the process up against multiple physical constraints. To reach the next level of energy efficiency requires capital cost modifications to increase heat recovery within and across process units. One of the key values of implementing operational solutions first is that it can clearly highlight where the physical constraints exist to the process.

Once specific process units have been identified for improved heat integration, pinch technology can be applied to efficiently screen potential modification options, which is explained in Chapter 9. Practical assessment (Chapter 10) is required, which considers not only the value and cost of improved heat recovery but also the impact in terms of operating flexibility, especially with respect to start-up, shutdown, maintenance, control, and safety.

1.3.4   Determining Process Changes

Improved heat recovery is the most common type of capital projects implemented to improve energy efficiency. However, the use of advanced process/equipment technology may provide significant opportunities. Many of these areas make use of advanced process technology, such as enhanced heat exchangers, high-capacity fractionator internals, dividing wall columns, new reactor internals, power recovery turbines, improved catalysts, and other design features.

There are a variety of advanced technologies that can be applied, all of which vary in terms of implementation cost and return on investment. Careful evaluation of each of these solutions is required to select only the best opportunities that provide the highest return on the capital employed. Chapter 11 provides directions and principles for making process changes.

1.3.5 Optimizing Energy Supply System

In addition to using energy more efficiently in the process, another common strategy is to produce energy more efficiently. Many plants have their own on-site power plants that primarily exist to provide steam and power to the process units, but may also supply electricity to the grid when electricity price is high.

Energy supply optimization is achieved by optimizing the configuration and operating profiles of the boilers and turbines to meet energy demand while taking into account tiered pricing for power and natural gas, power contracts to the grid while meeting environmental limits on NO_x and CO_2 emissions. Energy supply optimization is discussed in Chapter 19.

1.4 FOUR KEY ELEMENTS FOR CONTINUOUS IMPROVEMENT

An effective energy optimization consists of four key elements: target setting, measuring, gap identification, and implementation. Achieving continuous energy improvement occurs only when all these four elements are working in good order as shown in Figure 1.1.

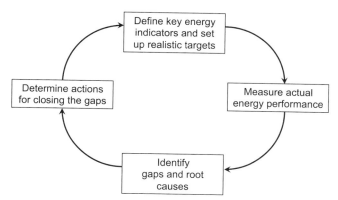

FIGURE 1.1. Four elements of energy management system.

The energy targeting implies setting up a base line energy performance against which actual energy performance can be compared. The base line energy performance should take into account the production rate and processing severity. The ratio of actual performance and base line performance is the energy performance indicator for a process area and an overall plant. The base line energy performance becomes the energy guideline or target for operation. For the energy target to be practical, it must be achievable based on equipment integrity, technology capability, availability of required tools, and skills.

1.5 PROMOTING IMPROVEMENT IDEAS IN THE ORGANIZATION

As a technical manager or process engineer or operator, you may have already acquired some good ideas for improving your plant and process unit. However, it is not an easy feat to persuade the technical committee to consider your ideas and then proceed to accept and eventually implement them. I have observed many good ideas that have died in the infancy stage because they could not pass the evaluation gates. Such failure is commonly due to a lack of techno-economic assessment and communications. Remember, it is always necessary to sell your ideas to key stakeholders.

First, you need to develop technical and economic merits to build a business case. Therefore, it is imperative that you determine the benefit of your ideas, that is, what is the value to the stakeholders, in the very early stages. Next, you should identify, with the help of process specialists, what it takes to implement the idea. You need to do the necessary homework to come up with rough estimates of the capital cost required to deliver the benefit for your ideas.

If the benefit outweighs the cost significantly, it is then necessary to elicit comments and feedback from technical specialists in the areas of operation, engineering, maintenance, and control. Their feedback will provide additional insights for the feasibility of implementing your ideas. Several review meetings may be required during idea development and assessment. Try to limit the scope of these meetings with highly selective attendees because a focused meeting could allow in-depth discussions leading to idea expansion and improvements. In the end, a thorough safety review is essential.

Once you pass reviews based on technical merits, you need to sell your ideas to get buy-in from management. Although management expresses a strong voice for supporting energy efficiency improvement, management will not provide a blank check. You should remember the fact that the business objective of your plant is to produce desirable products and realize targeted economic margins. To successfully convince management, you need to connect your ideas with key business drivers.

In the chapters that follow, all the essential tools will be provided in a clear, step-by-step manner together with application examples. My hope is that by applying the methods in your work—one step at a time, whether you are a manager, an engineer, or an operator—it will enable you to discover improvement ideas, to asses them, and then finally to prioritize them in a good order. Once all these boxes are checked, you will have a good chance to communicate and implement your ideas successfully within your organization.

2

THEORY OF ENERGY INTENSITY

Management's vision and intent is not good enough to achieve energy improvements. Technical concepts and targets must be used as the basis for measuring and improving process energy efficiency. Energy intensity is one of the key technical concepts as it lays down the foundation for process energy benchmarking.

2.1 INTRODUCTION

In some industrial plants, energy optimization work falls into no-man's land. If you ask process engineers, supervisors, and operators, they will tell you that they have done everything they can in making their process units energy efficient. It is understandable that technical people feel proud of themselves in trying to do their job right. If you ask plant managers, they may tell you everything is in good order.

The truth of the matter is that there is large room for energy efficiency improvement. To find out the truth, you may ask a few questions: What metrics are applied to measure the process energy efficiency? What energy indicators are defined for the key equipment? How do you set up targets for these indicators?

The answers to these three questions will show if the plant management only stays in good intention but without proper measures in place. If no energy metrics are used to measure performance level and no indicators are applied for major equipment and no targets are employed for identifying improvements, the energy management program is only on the basis of good intent. It is possible to get people motivated with

Energy and Process Optimization for the Process Industries, First Edition. Frank (Xin X.) Zhu.
© 2014 by the American Institute of Chemical Engineers, Inc. Published 2014 by John Wiley & Sons, Inc.

good intention. However, the motivation will decline gradually if people do not know what to do and have no directions.

To overcome this shortfall, two key concepts are introduced, namely, energy intensity and key energy indicators. The concept of energy intensity sets the basis for measuring energy performance, while the concept of key energy indicators provides guidance for what to do and how. Both energy intensity and key indicators are the cornerstones of an effective and sustainable energy management system. Energy intensity is introduced in this chapter, while example calculations for energy intensity are given in Chapter 3. The concept of key indicators will be discussed in Chapter 4.

2.2 DEFINITION OF PROCESS ENERGY INTENSITY

Meaning must transfer to correct concepts and then concepts must be expressed in mathematical forms for the meaning to be precise and measurable for industrial applications. Adjectives like excellent, good, and bad, have no quantifiable values for technical applications because they cannot be measured. Thus, we need a clear definition of mere linguistic terms from management intent to make sustainable energy perform-ance improvement. In other words, we need to have an operational definition of process energy performance that everyone can agree on and relate to and act upon.

Let us start with the specific question: how to define energy performance for a process? People might think of energy efficiency first. Although energy efficiency is a good measure as everyone knows what it is about, it does not relate energy use to process feed rate and yields, and thus it is hard to connect the concept of energy efficiency to plant managers and engineers.

To overcome this shortcoming, the concept of energy intensity is adopted, which connects process energy use and production activity. The energy intensity originated from Schipper et al. (1992a, 1992b), who attempted to address the intensity of energy use by coupling energy use and economic activity through the energy use history in five nations: the United States, Norway, Denmark, Germany, and Japan. The concept of energy intensity allows them to better examine the trends that prevailed during both increasing and decreasing energy prices.

By definition, energy intensity (I) is described by

$$I = \frac{\text{Energy use}}{\text{Activity}} = \frac{E}{A}. \tag{2.1}$$

Total energy use (E) becomes the numerator, while common measure of activity (A) is the denominator. For example, commonly used measures of activity are vehicle miles for passenger cars in transportation, kWh of electricity produced in the power industry, and unit of production for the process industry, respectively.

Physical unit of production can be t/h or m^3/h of total feed (or product). Thus, industrial energy intensity can be defined as

$$I = \frac{\text{Quantity of energy}}{\text{Quantity of feed or product}}. \tag{2.2}$$

Energy intensity defined in equation (2.2) directly connects energy use to production as it puts production as the basis (denominator). In this way, energy use is measured on the basis of production, which is in the right direction of thought: a process is meant to produce products supported by energy. For a given process, energy intensity has a strong correlation with energy efficiency. Directionally, efficiency improvements in processes and equipment can contribute to observed changes in energy intensity.

Therefore, we can come to agree that energy intensity is a more general concept for measuring of process energy efficiency indirectly.

Before adopting the concept of energy intensity, you may ask the question: Which one, feed rate or product rate, should be used as the measure of activity? For plants with a single most desirable product, the measure of activity should be product. For plants making multiple products, it is better to use feed rate as the measure of activity. The explanation is that a process may produce multiple products and some products are more desirable than others in terms of market value. Furthermore, some products require more energy to make than others. Thus, it could be very difficult to differentiate products for energy use. If we simply add all products together for the sum to appear in the denominator in equation (2.2), we encounter a problem, which is the dissimilarity in product as discussed. However, if feed is used in the denominator, the dissimilarity problem is nonexistent for cases with single feed, and the dissimilarity is much less a concern for multiple feed cases than for multiple products because, in general, feeds are much similar in composition than products.

The above discussions lead us to define the process energy intensity on the feed basis as

$$I_{process} = \frac{\text{Quantity of energy}}{\text{Quantity of feed}} = \frac{E}{F}. \tag{2.3}$$

It is straightforward to calculate the energy intensity for a process using equation (2.3) where E is the total net energy use and F is the total fresh feed entering the process. Net energy use is the difference of total energy use and total energy generation. Process energy use mainly includes fuel fired in furnaces, steam consumed in column stripping and reboiling as well as steam turbines as process drivers, and electricity for motors. Process energy generation mainly comes from process steam generation, and power generation. In many cases, a process makes fuel gas and/or fuel oil, which are exported to other processes for firing or sold to markets. This type of fuel is not counted as energy generation as it is regarded as a part of product slates.

2.3 THE CONCEPT OF FUEL EQUIVALENT (FE)

There is an issue yet to be resolved for the energy intensity defined in equation (2.3). The energy use (E) for a process consists of fuel, steam, and electricity. They are not additive because they are different in energy forms and quality. However, if these energy forms can be traced back to fuel fired at the source of generation, which is the meaning of fuel equivalent (FE), they can be compared on the same basis, which is fuel. In other words, they can be added or subtracted after being converted to their fuel equivalent. For simplicity of discussions, definitions of FE for different energy forms are given here, while examples of FE calculations are provided in Chapter 3.

In general, FE can be defined as the amount of fuel fired (Q_{fuel}) at the source to make a certain amount of energy utility (G_i):

$$FE_i = \frac{Q_{fuel}}{G_i}, \quad i \in (fuel, steam, power). \tag{2.4}$$

In most cases, Q_{fuel} is calculated based on the lower heating value of fuel. G_i is quantified in different units according to specifications in the marketplace, namely, Btu/h for fuel, lb/h for steam, and kWh for power. Thus, specific FE factors can be developed as follows based on this general definition of fuel equivalent. Energy are required for making boiler feed water (BFW), condensate and cooling water. The FE factors for these utilities will be discussed in Chapter 3.

2.3.1 FE Factors for Fuel

By default, fuel is the energy source. No matter what different fuels are used, tracing back to itself makes "fuel equivalent for fuel" equal to unity:

$$FE_{fuel} = \frac{Q_{fuel@source}}{G_{fuel}} \equiv 1 \text{ Btu/Btu}. \tag{2.5}$$

2.3.2 FE Factors for Steam

A typical process plant has multiple steam headers, typically designated as high, medium, and low pressure. In some cases, very high pressure steam is generated in boilers, which is mainly used for power generation. For calculating the fuel equivalent of steam, a top–down approach is adopted starting from steam generators. The total FE for each steam header is the summation of all FEs entering the steam header via different steam flow paths, which include steam generated from on-purpose boilers and waste heat boilers, steam from turbine exhaust, steam from pressure letdown valves, and so on. The FE for each steam header is the total FE divided by the amount of steam generated from this header:

$$FE_i = \frac{\text{Total FE consumed}}{\text{Total steam generated}}\bigg|_i, \quad \text{header } i \in (HP, MP, LP) \text{ kBtu/lb}. \tag{2.6}$$

2.3.3 FE Factors for Power

For power, FE_{power} is expressed as

$$FE_{power} = \frac{Q_{fuel}(Btu/h)}{Q_{power}(Btu/h)} = \frac{1}{\eta_{cycle}} \text{ Btu/Btu}, \tag{2.7}$$

where η_{cycle} is the cycle efficiency of power generation and Q_{power} represents the amount of heat content associated with power in unit of Btu/h.

By using the conversion factor of $1\,kW = 3414\,Btu/h$, equation (2.7) is converted to

$$FE_{power} = \frac{1}{\eta_{cycle}}(Btu/Btu) \times 3414(Btu/kWh) = \frac{3414}{\eta_{cycle}}\,Btu/kWh. \qquad (2.8)$$

Equation (2.8) can be generally applied to different scenarios for power supply such as power import, on-site power generation from back pressure and condensing steam turbines as well as from gas turbines, which are discussed in detail in Chapter 3.

2.3.4 Energy Intensity Based on FE

By converting different energy forms to their fuel equivalent, process energy intensity in equation (2.3) can be revised to give

$$I_{process} = \frac{FE}{F}\,Btu/unit\ of\ feed, \qquad (2.9)$$

where FE is the total fuel equivalent as a summation of individual fuel equivalent for different energy forms across the process battery limit.

2.4 ENERGY INTENSITY FOR A TOTAL SITE

The structure of energy intensity indicators can be organized in a hierarchal manner. That is, intensity indicators are developed for processes first and toward a total site. One may question why the concept of energy intensity does not apply to process sections (e.g., reaction section, product fractionation section) of a process. The reason is that there is strong heat integration between sections of a process unit, and thus energy intensity for sections cannot fairly represent section energy performance. Energy transfer across process units could occur, but the chance is much slim compared with between-process sections. In case of heat transfer between processes, some adjustment must be made to account for it.

To calculate the energy intensity index for the whole site, *aggregate* energy intensity could be defined simply as the ratio of total energy in fuel equivalent divided by total activity:

$$I_{site} = \frac{\sum_i FE_i}{\sum_i F_i}, \qquad (2.10)$$

where FE_i is the total fuel equivalent for process i.

However, there is a problem here with this simple aggregate approach: Although energy in fuel equivalent is additive, feeds (F) are not because processes usually have different feeds with very different compositions. In other words, the problem with equation (2.10) is the dissimilarity in feeds, which cannot be added without treatment.

To overcome this dissimilarity problem in feeds, we could think of a reference site with energy intensity for each process known in prior. Thus, the total amount of energy use could be calculated for the reference site, as the summation of the energy intensity for the reference processes. Let us derive the mathematical expressions along this line of thought.

When the energy intensity for a reference process is known or can be calculated, applying equation (2.9) gives the energy use for a reference process as

$$FE_{i,ref} = I_{i,ref}F_{i,ref}. \tag{2.11}$$

Since FE is additive, the total energy use for the reference site is

$$FE_{site,ref} = \sum_i FE_{i,ref} = \sum_i I_{i,ref}F_{i,ref}. \tag{2.12}$$

Then, an intensity index for the site of interest can be defined as the ratio of actual and reference energy use:

$$I_{site} = \frac{FE_{site,actual}}{FE_{site,ref}}. \tag{2.13}$$

$FE_{site,actual}$ can be readily calculated from individual energy users consisting of fuel, power, steam, BFW and cooling water accross the site battery limit.

You may ask a critical question: A real process could differ from the reference process in terms of feed rates and process conditions. How can we deal with these differences in the energy intensity index calculations? This question can be addressed by defining the intensity as a function of three major factors:

$$I_{i,ref} = f_{i,ref}(design, conditions, weather), \tag{2.14}$$

where design, conditions, and weather reflect the actual process. In this way, equation (2.14) describes the energy performance for the reference processes with the same attributes as the actual processes, but the energy intensity could be different. This is because the energy intensity in equation (2.14) for reference processes is developed based on peers' performance, while the energy intensity for actual processes is calculated based on real data.

The simplest form is a linear expression. For example, if two operating parameters are considered, the linear form becomes

$$I_{i,ref} = a + bx_1 + cx_2 + d(T_{ambient}), \tag{2.15}$$

where a is a structural term that reflects the design performance, while b and c are the sensitivity factors for x_1 (process condition 1) and x_2 (process condition 2), respectively; d is the correction factor for weather; and $T_{ambient}$ is the ambient temperature in local area.

2.5 CONCLUDING REMARKS

The decline in energy intensity is a proxy for efficiency improvement; however, energy intensity reflects production and hence is much more universal and communicable across the process industry.

Clearly, structural and operational changes for efficiency improvements in processes and equipment can contribute to reduction in process energy intensity in a big way. A state-of-the-art process gives low energy intensity by design. However, it could end up with high operating energy intensity if the process is poorly operated. On the other hand, a poorly designed process could achieve its best potential if it is operated with diligence. However, good operation could reach the design limit because the performance is handicapped due to inherently inefficient design. To improve the process beyond this design limitation, structural changes must be made.

NOMENCLATURE

A activity such as processing feed or making products
E energy use
F feed rate
FE   fuel equivalent; amount of fuel at the source to make a unit of energy utility (power, steam)
G energy utility
I energy intensity; energy units/feed or product unit
Q heat content
T temperature

Greek Letters

η_{cycle} power generation efficiency; the ratio of the amount of fuel to make a unit of power

Subscript

ref   reference process or total site

REFERENCES

Schipper L, Howarth RB, Carlassare E (1992a) Energy intensity, sector activity, and structural changes in the Norwegian economy, *Energy: The International Journal*, **17**, 215–233.

Schipper L, Meyers S, Howarth RB, Steiner RL (1992b) *Energy Efficiency and Human Activity: Past Trends, Future Prospects*, pp. 250–285, Cambridge University Press, Cambridge.

3

BENCHMARKING ENERGY INTENSITY

Energy benchmarking defines an intensity measure of process energy performance. It can be used to determine the baseline of energy performance to compare with peers and measure the effects by operation and process changes.

3.1 INTRODUCTION

When you are given a task to improve energy performance for the total site or process unit, your immediate response would be: Where should I start? The answer is to know where the process unit stands in energy performance. In other words, you need to determine both the current energy use and energy consumption target. Only then is it possible to establish the baseline and to know how well the process unit is doing by comparing current performance against the target. We call the exercise of establishing a baseline as energy benchmarking.

The most important result of energy benchmarking is the indication of energy intensity for individual processes. If a performance target can be defined based on a corporate target, industrial peer performance, or the best technology performance for each process, then the benchmarking audit can determine the process energy performance overall in comparison with targeted performance. In general, benchmarking assessment can determine several scenarios:

- *The Need of Having an Overall Energy Optimization Effort:* If large gaps are available for majority of the process units, this could imply that there are many

Energy and Process Optimization for the Process Industries, First Edition. Frank (Xin X.) Zhu.
© 2014 by the American Institute of Chemical Engineers, Inc. Published 2014 by John Wiley & Sons, Inc.

opportunities available and require consorted effort across the site. A dedicated energy team may need to be established to coordinate the overall effort in identifying and capturing the opportunities.

- *Areas for Focus:* Some of the process units are identified with large performance gaps and these processes can be selected as focus areas. This allows us to effectively concentrate efforts on areas with the greatest potential for improvement. Specialists may need to be assembled to form a project team.

- *Update Targets:* If all major process units are under good performance relative to the targets, the plant may concentrate its efforts on continuous improvements via setting more aggressive targets.

3.2   DATA EXTRACTION FROM HISTORIAN

For the purpose of energy benchmarking of a process unit, the important thing is to identify the main energy consumers and give a reasonable estimate for missing data. Going overboard to collect miniature details and chase utmost precision should be avoided at this stage. Doing so may actually waste effort because such fine details are most likely not needed in the benchmarking calculations and will not make a reasonable impact on energy optimization.

Table 3.1 gives an example on the relevant data needed for energy benchmarking. Although all the data look familiar in the table, you may question the need for including the fuel generated in the unit as part of energy generation. As a general guideline, the fuel produced by a process unit, in the forms of fuel gas, LPG, and fuel oil, is treated as part of products from the process and thus should not be included in the energy balance for the unit.

To have a clear view of energy flows into and out of the process, we derive Figure 3.1 based on data in Table 3.1. The left-hand side of the figure shows the energy input to the process. At the same time, exothermic reaction provides additional heat to the process. On the right-hand side of the figure, energy leaves the process, which includes heat exported and lost. In addition, the raw feed and boiler feed water (BFW) carry a certain amount of energy into the process based on an assumed reference temperature of 100 °F. A different reference temperature could be used and fuel equivalent calculations should be conducted based on the chosen reference temperature. A reference temperature is selected based on the consideration that any heat below the reference temperature is not economically viable to recover.

3.3   CONVERT ALL ENERGY USAGE TO FUEL EQUIVALENT

In the industry, steam is measured in mass flow, fuel in volumetric flow, and electricity in electrical current. To compare them on the same basis, all the energy

TABLE 3.1. Example Data Set for Energy Use and Generation

Item	Normal Load (kW)	Electric Power (kW)	HP Steam at 610 psig (klb/h)	MP Steam at 150 psig (klb/h)	LP Steam at 60 psig (klb/h)	Condensate at 3 psig (klb/h)	Fuel Fired Duty (MMBtul/h)
Steam import from steam headers			−188.6	0	0		
Pumps and drivers							
Charge pumps		−2101.0					
Lean amine pumps		−53.3					
Rich amine pumps	−798.8			−18.9		18.9	
Stripper bottoms pumps		−101.8					
Stripper overhead pumps		−68.4					
Product fractionator bottoms pumps	−512.2			−12.1		12.1	
Diesel product pumps		−43.0					
Kerosene product pumps		−28.5					
Product fractionator overhead pumps		−89.6					
Wash water pumps		−80.4					
RX air cooler		−175.2					
Separation coolers		−405.8					
Water coolers							
Heavy naphtha product	6.0						
Kerosene product	4.5						
Diesel product	11.4						
Reaction effluent	27.5						
Fractionator kerosene pump-around	39.4						
H2 make-up gas	30.0						
H2 make-up compressor intercoolers	6.4						
Lean amine	1.4						

Air coolers					
Reaction effluent	192.0				
Debutanizer column overhead	21.3				
Fractionation column overhead	73.3				
Naphtha splitter overhead	12.9				
Compressors					
Recycle gas compressor	−2610.0	−86.6	86.6		
Make-up H$_2$ compressor	−5936.0	−70.5		70.5	
Fired heaters					
Charge heater for train 1	−29.5				−53.5
Charge heater for train 2	−29.5				−55.9
Product fractionator feed heater	−95.0				−111.8
DeC4 Col reboiler	−80.0				−94.1
Diesel stripper reboiler	−12.0				−21.8
Steam generation					
Product fractionator bottom steam gen			16.50		
Energy export					
Condensate return to boiler house				129.4	
Condensate lost				−10.0	
MP steam goes to steam header			50.0		
Energy entering into the process	**3147.0**	**188.6**			
Energy exporting outside the process			**50.0**	**129.4**	**337.1**

Note : A positive value indicates quantity produced. A negative value (−) indicates quantity consumed.

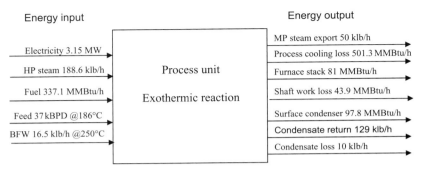

FIGURE 3.1. Energy flows into and out of the process unit.

use and generation need to be traced back to fuel fired at the source of energy generation to obtain the fuel equivalent (FE), which is a cardinal rule for energy balance calculations. The following illustrates how to conduct FE calculations based on Figure 3.1.

Assumptions: First, assumptions for related fuel equivalent factors need to be made and the basis for deriving these assumptions will be explained later. Assumed FE factors are as follows:

- FE for purchased power = 9.09 MMBtu/MWh
- FE for high-pressure (HP) steam = 1550 Btu/lb
- FE for medium-pressure (MP) steam = 1310 Btu/lb
- FE for condensate = 94.6 Btu/lb
- FE for BFW @250 °F = 177 Btu/lb

Convert Energy Inputs and Outputs to Fuel Equivalent:

- FE for power = 3.15 MW × 9.09 MMBtu/MWh = 28.6 MMBtu/h
- FE for HP steam = 188.6 klb/h × 1.55 MMBtu/klb = 292.3 MMBtu/h
- FE for fuel fired = 337.1 MMBtu/h
- FE for MP steam export = 50 klb/h × 1.31 MMBtu/klb = 65.5 MMBtu/h
- FE for Condensate return = 129.4 klb/h × 94.6 Btu/lb × 10^3 lb/klb × 1 MMBtu/ 10^6 Btu = 12.2 MMBtu/h
- FE for Condensate loss = 10 klb/h × 94.6 Btu/lb × 10^3 lb/klb × 1 MMBtu/ 10^6 Btu = 0.9 MMBtu/h

To reveal the significance of FE calculations, let us assume a process receives 20 klb/h of HP steam in which 10 klb/h comes from a boiler with efficiency of 75% and another 10 klb/h from a boiler with efficiency of 85%. Obviously, the fuel required or fuel equivalent for the same amount of HP steam, that is, 10 klb/h, by the two boilers is very different: The fuel equivalent from the boiler with 85% efficiency is 15.35 MMBtu/h, resulting in FE factor of 1.535 MMBtu/klb. The fuel equivalent for the boiler with 75% efficiency is 16.38 MMBtu/h giving the FE factor of 1.638 MMBtu/klb. We can think of another example of power generation on site by a combined cycle (gas and steam turbines) cogeneration facility versus a coal-fired steam turbine power plant. The fuel equivalent for the same amount of power from these two sources can be very different. Therefore, we cannot overstate the importance for tracing any energy back to fuel equivalent.

3.4   ENERGY BALANCE

After converting all energy forms to fuel equivalent, these energy forms are leveled on the equal basis and thus we are ready to conduct energy balance. For a chemical process, energy balance is defined as

$$\text{Energy supply} + \text{Heat of Reaction} = \text{Energy export} + \text{Energy loss.} \qquad (3.1)$$

The sum of energy supply and energy generation (heat of reaction) makes total energy input, while both energy export and energy loss forms total energy output. Energy supply implies the energy coming into the process battery limit. Energy generation for a chemical process implies heat of reaction. If a reaction is exothermic, the term of energy generation takes a positive sign because it contributes to total energy input. An endothermic reaction takes a negative sign because it takes energy away from energy supply and needs energy input to make up the difference. Energy export denotes the energy leaving out of the process that is used by other processes. Energy loss indicates the energy flows leaving out of the process but lost to the environment.

After obtaining fuel equivalent values for all energy flows, we can convert Figure 3.1 to Figure 3.2, which gives a visualized energy balance around the process unit including energy supply, energy generation by heat of reaction, and energy export and losses. The heat of exothermic reaction is calculated as 141 MMBtu/h for

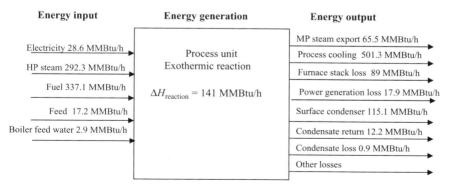

FIGURE 3.2. Energy balance in a visualized form.

TABLE 3.2. Tabulated Energy Balance for the Example

Energy Input	FE (MMBtu/h) (above 100 °F)	Energy Output	FE (MMBtu/h) (above 100 °F)
Power	28.6	Energy export	
Fuel	337.1	MP export	65.5
HP steam	292.3	Condensate return @ 141 °F	12.2
Heat of reaction	141.0	Total	77.7
Feed @ 170 °F	17.2	Energy lost	FE MMBtu/h (above 100 °F)
Boiler feed water @ 250 °F	2.9	Power gen losses	17.9
Total	819.2	Air coolers	352.4
		Water coolers	148.9
	FE MMBtu/h	Furnaces stack loss	89.0
Net energy input	741.5	Pumps and motors (mechanical loss)	2.86
	FE kBtu/bbl	Surface condensers	115.1
Specific energy use	480.9	Condensate loss	0.9
		Unaccounted losses	14.6
		Total	741.5

Balance check: Energy input − Energy output = 819.2 − (77.7 + 741.5) = 0.

this example based on the feed composition and reaction conditions. Heat content of the feed and boiler feed water above 100 °F are treated as energy input. At the same time, the figure shows energy output including energy export and energy losses. It can be observed that only energy flows entering and leaving the process battery limit are addressed in the energy balance described in Figure 3.2.

The detailed energy balance is given in Table 3.2. The total energy input is 819.2 MMBtu/h for the process unit currently operated. The heat of exothermic reaction contributes positively to the total energy input. Fuel fired in process heaters is 337.1 MMBtu/h, which is the most dominant accounting for about 40% of total energy input. The second most dominant energy use is the process shaftwork demand. HP steam of 292.3 MMBtu/h is used for steam turbines as process drivers, while purchased electricity of 28.6 MMBtu/h is for running motors. The total fuel equivalent for meeting the process shaftwork demand is 321 MMBtu/h (28.6 + 292.3), which accounts for another 40% of total energy input. Heat of reaction contributes a significant portion of the energy input at 17%. The remaining minor contributions to the energy input come from feed and boiler feed water.

Energy output is grouped into two categories, namely, energy export and energy losses. Energy export includes any energy flow going out of the process and being used for a meaningful purpose. In the example, the energy export is 77.7 MMBtu/h, which includes MP steam to the steam header and condensate return to the boilers. It could also include hot products directly sent to downstream processes as feeds, which is not present in this example.

Energy losses are mainly caused by process water and air cooling. To derive the fuel equivalent, a process cooling duty is divided by the boiler efficiency (85% for this example) assuming low-temperature heat available in process cooling could be used for boiler feed water preheating. Total cooling duty accounts for 68% of total energy losses. Therefore, one critical area for improving process energy efficiency is to identify opportunities to reduce heat losses in process cooling although the heat is usually available at low temperatures.

Fuel equivalent for purchased power is assumed to be 9090 Btu/kWh compared with the normal conversation factor of 3414 Btu/kWh. This assumption implies power generation loss of 5676 (= 9090 − 3414) Btu for each kWh imported. Thus, power generation loss is 17.9 MMBtu/h for 3.15 purchased. The rationale for this assumption will be discussed later with the FE calculation given in equation (3.7).

Furnace stack loss is calculated based on actual heater efficiency. For this example, 55% furnace efficiency is assumed for the charge heater and the diesel stripper heater, which have a radiant section only. A furnace efficiency of 85% is used for the product fractionator heater and the debutanizer reboiler heater, which have both radiant and convection sections.

The mechanical losses for pumps and motors are calculated based on motor efficiency, which is assumed at 90% for this example.

The net energy input is expressed as

$$\text{Net energy input} = \text{Energy input} - \text{Energy export}. \qquad (3.2)$$

For the example in question, net energy input $= 819.2 - 77.7 = 741.5$ MMBtu/h.
Let us define specific energy use the same as the energy intensity:

$$\text{Specific energy} = \frac{\text{Net energy input}}{\text{Feed rate}}. \qquad (3.3)$$

Applying equation (3.3) yields
Specific energy $=$ 741.5 MMBtu/h $\times$ 1000 kBtu/MMBtu/37,000 bbl/day $\times$ 24 h/day $=$ 480.9 kBtu/bbl, where 37,000 bbl/day is the process feed rate.

As manifested in Chapter 2, specific energy use is a very insightful concept as it represents the *energy intensity of production* indicated by the amount of energy required for processing one unit of feed.

3.5 FUEL EQUIVALENT FOR STEAM AND POWER

In previous discussions, some assumptions of fuel equivalent factors were made for power and steam. You may ask: What is the basis for making these assumptions? How do you determine fuel equivalent values for power and steam in your plant? Let us consider the calculation of fuel equivalent for power first.

3.5.1   FE Factors for Power (FE_{power})

As mentioned in Chapter 2, FE_{power} is expressed as

$$FE_{power} = \frac{Q_{fuel}}{Q_{power}} = \frac{1}{\eta_{cycle}} \; (\text{Btu/Btu}), \tag{3.4}$$

where η_{cycle} is the cycle efficiency of power generation and thus $\eta_{cycle} = Q_{power}/Q_{fuel}$ with Q_{power} (in Btu/h) representing the amount of heat content associated with power with a conversion factor of 3414 Btu/kWh.

By using the conversion factor of $1\,\text{kW} = 3414\,\text{Btu/h}$, equation (3.4) can be converted to

$$FE_{power} = \frac{1}{\eta_{cycle}} \; (\text{Btu/Btu}) \times 3414 \; (\text{Btu/kWh}) = \frac{3414}{\eta_{cycle}} \; (\text{Btu/kWh}). \tag{3.5}$$

Rearranging equation (3.5) leads to

$$FE_{power} = \frac{3414}{\eta_{cycle}} = \frac{3414}{Q_{power}/Q_{fuel}} = \frac{Q_{fuel}}{Q_{power}/3414} = \frac{Q_{fuel}}{W} \; (\text{Btu/kWh}), \tag{3.6}$$

where $W = (Q_{power}/3414)$ and W (in kW) represents the amount of power. By converting the unit of FE_{power} from Btu/Btu in equation (3.4) to Btu/kWh in equation (3.6), the expression of FE_{power} in equation (3.6) becomes exactly the same as that of heat rate for power generation. Let us look at three cases for applying equation (3.6).

Case 1: Importing Power from Coal Power Plants   The average efficiency for today's coal-fired plants is 33% globally, while pulverized coal combustion can reach efficiency of 45% based on net low heating value (LHV) (IEA, 2012). Thus, the fuel equivalent factor FE_{power}^{ST} for purchased coal power are in the range of 7.58 MMBtu/MW (45% of power efficiency) and 10.34 MMBtu/MW (33%). For example, if assuming steam cycle efficiency is 37.56%, applying equation (3.5) yields

$$FE_{power} = \frac{3414}{\eta_{cycle}} = \frac{3414}{0.3756} = 9090 \; (\text{Btu/kWh}). \tag{3.7}$$

Note that 9090 Btu/kWh is the FE_{power} factor used in the previous assumption for power.

Case 2: On-Site Power Generation from Steam Turbines   For on-site power generation, usually heat rate is known and it should be used as FE_{power}. If unknown, a typical condensing steam turbine cycle efficiency of 30% could be used to yield

$$FE_{power} = \frac{3414}{\eta_{cycle}} = \frac{3414}{0.3} = 11{,}380 \; (\text{Btu/kWh}). \tag{3.8}$$

FE_{power} factors for back pressure steam turbines could be much higher than 11,380 Btu/kWh. What is the interpretation of a higher FE_{power} from on-site power generation than that of purchased power? The implication is that a commercial power plant can make power more efficiently than a process plant if cogeneration is not involved. Does it mean that the use of a motor is more energy efficient than using an on-site condensing turbine for process drivers? The answer is Yes. You may stretch out to think: The back pressure turbines could be even worse as process drivers. Is this true? The answer for this question relies on the steam balances. If the exhaust steam from the back pressure turbines is used for processes, the back pressure turbines have much higher cogeneration efficiency (power plus steam).

Case 3: On-Site Power Generation from Combined Gas and Steam Turbines - When power is generated by a gas turbine (GT), gas turbine exhaust is usually sent to the heat recovery steam generator for steam generation. Steam is then used for further power generation via steam turbines. A configuration such as this is known as a gas turbine–steam combined cycle.

The combined cycle efficiency can be expressed as

$$\eta_{CC} = \eta_{GT} + \eta_{ST} - \eta_{GT} \times \eta_{ST}. \tag{3.9}$$

By applying equation (3.5), the fuel equivalent factor for power generated from a combined cycle would be

$$FE_{power}^{CC} = \frac{3414}{\eta_{CC}}. \tag{3.10}$$

Suppose that a gas turbine cycle has an efficiency of 42%, which is a representative value for gas turbines, and the steam turbine has an efficiency of 30%. The combined cycle efficiency (η_{CC}) is 59.4% based on equation (3.9) and FE factor is 5747 Btu/kWh based on equation (3.10). In general, the combined cycle is much more efficient in power generation than the steam cycle alone.

3.5.2 FE Factors for Steam, Condensate, and Water

Steam headers are the central collection points where steam enters each header from different sources and distributes to different steam users. The total FE for each steam header is the summation of all FE's entering the steam header via different flow paths. The FE for each steam header is the total FE divided by the amount of steam generated from this header:

$$FE_{Header\,i} = \left.\frac{\sum FE\ consumed}{\sum steam\ generated}\right|_{Header\,i}, \quad i = (HP, MP, LP)\ MMBtu/klb. \tag{3.11}$$

A top–down approach is adopted for FE calculations. First, FE for HP steam is calculated and then by cascading down in the order of pressure levels, FEs for other steam headers are determined. Let us look at Example 3.1.

Example 3.1

Calculate the fuel equivalent values for the steam headers in Figure 3.3.

Solution. To determine the fuel equivalent for steam headers, different steam flow paths must be identified, which could have an influence on the fuel equivalent for the steam.

3.5.2.1   FE for HP Steam   There are two paths for making HP steam, namely, boiler 1 with 75% thermal efficiency and boiler 2 with 85% thermal efficiency. The FE factors for both HP generation sources can be calculated as

$$FE_{HP,boiler\ 1} = \frac{Q_{B1}}{M_{B1}} = \frac{179}{108} = 1.66\ \text{MMBtu/klb},$$

$$FE_{HP,boiler\ 2} = \frac{Q_{B1}}{M_{B1}} = \frac{158}{108} = 1.46\ \text{MMBtu/klb}.$$

The average FE for HP steam can be calculated as

$$FE_{HP} = \frac{Q_{B1} + Q_{B2}}{M_{B1} + M_{B2}} = \frac{179 + 158}{108 + 108} = 1.56\ \text{MMBtu/klb}.$$

For evaluating a base case scenario, the average FE factor for HP steam should be used. In the case when opportunities for steam saving or extra steam use are explored, the steam flow path based FE factors must be considered. For this example, when capturing the steam saving opportunity, steam generation should be reduced from boiler 1, the less efficient boiler. On the other hand, when extra HP steam is required from processes, it should be generated from boiler 2, the more efficient boiler.

In general, high-pressure steam is defined as steam produced from steam generators, mainly boilers. The fuel equivalent of high-pressure steam can be derived as:

$$FE_{HP} = \frac{Q_{\text{fuel}}}{M_{HP}}\bigg|_{\text{boiler}\ i} \quad \text{MMBtu/klb}, \tag{3.12}$$

where M_{HP} is the amount of HP steam generated from the boiler i.

In most cases, multiple boilers are used. In this case, equation (3.11) can be applied to derive the weighted average of fuel equivalent for combined HP steam going to the HP header:

$$FE_{HP} = \frac{\sum_{i}^{\text{Boilers}} M_{i,HP} FE_{i,HP}}{\sum_{i}^{\text{Boilers}} M_{i,HP}} \quad \text{MMBtu/klb}. \tag{3.13}$$

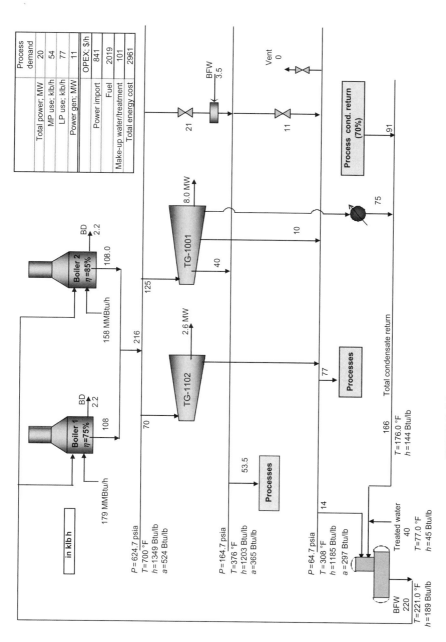

FIGURE 3.3. Steam system for Example problem 3.1.

		Process demand
Total power; MW		20
MP use; klb/h		54
LP use; klb/h		77
Power gen; MW		11
	OPEX: $/h	
Power import		841
Fuel		2019
Make-up water/treatment		101
Total energy cost		2961

27

3.5.2.2 FE for MP Steam

Three paths of MP generation are identified as follows:

- *Path 1*: 40 klb/h of MP extraction from TG-1001 with specific steam rate $m_{\text{HP-MP}}$ at 35.6 klb/MWh. The fuel equivalent for the MP steam exhaust can be calculated:

$$\text{FE}_{\text{MP-steam}} = \text{FE}_{\text{HP-steam}} - \frac{\text{FE}_{\text{power}}^{\text{import}}}{m_{\text{HP-MP}}}. \tag{3.14}$$

The reason why FE factor for power import is used in equation (3.14) is that power import is the marginal power source. In other words, if a steam turbine is replaced with a motor, purchased power will be used.

Assume fuel equivalent factor for purchased power as 9.09 MMBtu/MWh. Thus,

$$\text{FE}_{\text{MP-steam}}^{\text{P1}} = 1.56 - \frac{9.09}{35.6} = 1.30 \text{ MMBtu/klb MP}.$$

- *Path 2*: 21 klb/h of the letdown valve, $\text{FE}_{\text{MP-steam}}^{\text{P2}} = \text{FE}_{\text{HP}} = 1.56 \text{ MMBtu/klb}$ because a letdown is an adiabatic process and thus FE does not change through the letdown valve.
- *Path 3*: 3.5 klb/h of BFW addition for desuperheating, $\text{FE}_{\text{MP-steam}}^{\text{P3}} = \text{FE}_{\text{BFW}} = 177 \text{ Btu/lb}$. The FE factor for BFW is calculated based on equation (3.15), which assumes that low-pressure (LP) steam is used for BFW preheat. It must be pointed out that the FE of the pumping power is ignored as it is very low ($\sim$10 Btu/lb of BFW):

$$\text{FE}_{\text{BFW}} = \text{FE}_{\text{LP Steam}} \times \frac{h_{\text{BFW}} - h_{\text{Ambient water}}}{h_{\text{LP Steam}} - h_{\text{Ambient water}}}. \tag{3.15}$$

Thus, the average FE for the mixed MP steam can be calculated based on equation (3.11):

$$\begin{aligned}
\text{FE}_{\text{MP}}^{\text{av}} &= \frac{M_{\text{MP}}^{\text{P1}} \times \text{FE}_{\text{MP}}^{\text{P1}} + M_{\text{MP}}^{\text{P2}} \times \text{FE}_{\text{MP}}^{\text{P2}} + M_{\text{MP}}^{\text{P3}} \times \text{FE}_{\text{MP}}^{\text{P3}}}{M_{\text{MP}}^{\text{P1}} + M_{\text{MP}}^{\text{P2}} + M_{\text{MP}}^{\text{P3}}} \\
&= \frac{40 \times 1.30 + 21 \times 1.56 + 3.5 \times 0.177}{40 + 21 + 3.5} = 1.32 \text{ MMBtu/klb}.
\end{aligned}$$

3.5.2.3 FE for LP Steam

There are three paths for making LP steam:

- *Path 1*: 70 klb/h from the TG-1002 turbine with a specific steam rate of 26.8 lb/kWh. The fuel equivalent for the MP steam exhaust can be calculated:

$$\text{FE}_{\text{LP-steam}} = \text{FE}_{\text{HP-steam}} - \frac{\text{FE}_{\text{power}}^{\text{import}}}{m_{\text{HP-LP}}} = 1.56 - \frac{9.09}{26.8} = 1.22 \text{ MMBtu/klb}. \tag{3.16}$$

- *Path 2*: 10 /h from the TG-1001 LP extraction with a specific steam rate of 22.9 lb/kWh.

$$FE^{P2}_{LP\text{-}steam} = 1.56 - \frac{9.09}{22.9} = 1.16 \text{ MMBtu/klb LP.}$$

- *Path 3*: 11 klb/h of the letdown valve, $FE^{P3}_{LP\text{-}steam} = FE_{MP} = 1.32$ MMBtu/klb because a letdown is an adiabatic process. BFW desuperheating is not needed for the LP steam in this case because the superheated fraction in LP steam is very small.

Thus, the average FE for the mixed LP steam is calculated as:

$$FE^{av}_{LP} = \frac{70FE^{P1}_{LP} + 10FE^{P2}_{LP} + 11FE^{P3}_{LP}}{(70 + 10 + 11)} = \frac{(70 \times 1.22 + 10 \times 1.16 + 11 \times 1.32)}{91}$$

$$= 1.22 \text{ MMBtu/klb.}$$

What about the FE for vented LP steam? In this case, FE_{LP} should also be calculated based on the path from which this vented LP steam is generated. This is because a certain amount of fuel equivalent is consumed to make the LP steam no matter whether it is used or vented or not. For vented LP steam, the economic value is zero but FE is not.

3.5.2.4 FE for Condensate

Condensate temperature is typically around 200 °F. The condensate FE_{Cond} can be determined by the difference of condensate temperature and raw water temperature (ambient). FE_{Cond} is usually in the range of 100–150 Btu/lb of condensate. Although FE_{Cond} is small relative to steam, accumulated loss could be significant for a large amount of condensate loss. Also, condensate loss is costly due to the extra chemicals required to treat makeup water.

3.5.2.5 FE for BFW Water

The energy required for providing boiler makeup water includes the heat content and the pump power used to elevate its pressure. The BFW heat content is the major portion of FE_{BFW}, which is determined by the difference of BFW temperature (typically around 250 °F) and raw water temperature (ambient). The FE_{BFW} is in the range of 150–200 Btu/lb of BFW.

3.5.2.6 FE for Cooling Water

Energy for providing cooling water includes pump power and the fan power in running the cooling tower fans. The FE_{cw} is in the range of 60–130 Btu/lb of cooling water.

3.6 ENERGY PERFORMANCE INDEX (EPI) METHOD

Naturally, one would think that the first and second laws of thermodynamics should be used as methods for assessing energy efficiency for industrial processes. If we apply the first law of thermodynamics to Table 3.2, we have

$$\eta = \frac{\text{Useful energy}}{\text{Energy input}} = 1 - \frac{\text{Energy loss}}{\text{Energy input}} = 1 - \frac{741.5}{819.2} = 10\%. \qquad (3.17)$$

Clearly, this efficiency is not very insightful as energy input is provided at very high quality in terms of temperature, pressure, and composition, while the energy losses are at a much lower quality; however, they are compared on the same basis.

In contrast, an efficiency based on the second law of thermodynamics could make more sense as it takes energy quality into account. Exergy analysis is the method developed for industrial applications based on the second law of thermodynamics. Exergy could be done in reality, but the effort in data requirement could be prohibitive. More importantly, the second law of thermodynamics is a difficult concept to grasp for many process engineers, which is not common for applications in the process industry.

Instead, a much simpler yet effective method is presented and discussed here. This method is built on the concept of guideline energy performance, which is used as a benchmark against which actual energy performance is compared. The rational of using this concept as the basis for assessing process energy efficiency is revealed in the following.

Let a ratio of actual energy performance (AEP) and guideline energy performance (GEP) be defined as below. This ratio shall be labeled the energy performance index (EPI):

$$\text{EPI} = \frac{\text{Actual energy performance}}{\text{Guideline energy performance}} = \frac{\text{AEP}}{\text{GEP}}. \qquad (3.18)$$

By definition, EPI represents the energy efficiency for the process unit on the basis of GEP. In this way, any improvements in operation, design, equipment, and technology upgrade can be measured using EPI. Application of the EPI method is discussed below.

Generally speaking, an EPI gap of less than 5% between AEP and GEP belongs to an operational gap. In other words, better operating practices and control could close this gap. An EPI gap of larger than 10% may require small energy retrofit projects, which can feature a quick payback, for the gap to be closed. If the EPI gap is on the order of 20%, it may require significant energy and process retrofit projects to close the gap.

3.6.1   Benchmarking: Based on the Best-in-Operation Energy Performance

By applying the method for calculating specific energy use, you can obtain a plot of specific energy versus time based on the historical data. This plot can pinpoint the best-in-operation energy performance (OEP) that your process unit has achieved at a time when there was an institutionally dedicated effort for operation performance and with technical know-how available. You could confirm this by talking to engineers and operators who have worked in the plant during this period. As a result, you will be able to determine the OEP as the guideline performance. Assume the specific energy use on the basis of OEP is 438.7 kBtu/bbl feed, which can be obtained from the historian. With the actual energy use of 480.9 kBtu/bbl calculated as above, EPI for the process unit can be calculated:

$$\text{EPI} = \frac{\text{AEP}}{\text{OEP}} \times 100\% = \frac{480.9}{438.7} \times 100\% = 109.6\%. \tag{3.19}$$

Equation (3.19) indicates a 9.6% deviation of AEP from the OEP. Such a gap is significant, which should alert you to initiate investigations for root causes. The mere fact of determining EPI gives you an immediate indicator as to where your process unit stands in energy performance, so that you can quickly spot problematic areas.

3.6.2 Benchmarking: Based on Industrial Peers' Energy Performance

In the industry, there are peer survey groups organized based on industrial sectors and process technology. Organizers for the survey groups send questionnaires to survey members to gather sample data yearly and conduct performance calculations. Consequently, the peer performance results are shared among survey members. If your plant belongs to a certain survey group, you could obtain the best peer energy performance (PEP) via the representative in your organization. For large companies, there are CoP (community of practice) networks based on process technology. You should seek out the best PEP for your process unit via the CoP in your company.

Assume the specific energy for PEP is 430 kBtu/bbl for the example. Based on the actual energy use of 480.9 kBtu/bbl, we can calculate EPI for the process unit:

$$\text{EPI} = \frac{\text{AEP}}{\text{PEP}} \times 100\% = \frac{480.9}{430} \times 100\% = 111.8\%. \tag{3.20}$$

Usually, the survey group is divided into a tiered performance structure such as first, second, third, and fourth quartiles. Based on the energy performance index calculated above, you can find out which performance quartile your process unit belongs to. This indicates where your process unit stands among your peers.

3.6.3 Benchmarking: Based on the Best Technology Energy Performance

With technology advancements in catalyst, equipment, process design, and control, process energy efficiency could improve. It is not difficult to gather the performance data for state-of-the-art technology. In some cases, the data are published in public by government offices and you can find them via web search. If not available in public, you can contact technology companies—they are often eager to share this information with customers.

Assume the operation is improved for the example process and the energy use is reduced from 480.9.8 to 438.7 kBtu/bbl. The plant management is interested to know the scope of further energy improvement by applying better process technology and design. Assume the technology energy performance (TEP) is 380 kBtu/bbl. Thus, the

EPI for the process unit can be calculated by

$$\text{EPI} = \frac{\text{AEP}}{\text{TEP}} \times 100\% = \frac{438.7}{380} \times 100\% = 115\%. \tag{3.21}$$

Technology updates can make big step changes in both production and energy performance but usually with high capital costs and long implementation periods. Therefore, it should be applied very selectively.

At this point, we have a very good starting point. You know three essential facts: the energy intensity for your process unit, the performance target, and the gap against the target. Your mind may be racing with questions like: What has gone wrong with my process unit? How can the AEP be reduced to OEP for the process unit? These questions will be answered in Chapter 4.

3.7   CONCLUDING REMARKS

Three fundamental concepts are discussed in this chapter. The first one is the concept of converting all energy back to fuel equivalent. This concept makes all forms of energy on the same basis, that is, fuel fired or fuel equivalent. The second one is specific net energy, which describes the energy intensity for production. The third concept is guideline energy performance (GEP) as the best alternative for comparison with actual performance.

Applying these three concepts makes it a much simpler yet effective task to assess the energy performance of a process unit and require minimal data. The EPI method is developed based on these concepts and designed for practical applications.

The strategy for achieving the energy target can involve changes to operating practice, new control strategy, process modifications, or technology upgrade or combinations of the above. In general, closing the gap between the average and the best potential performance of an individual unit involves operational and maintenance improvements. Eliminating the gap between an existing unit and its peers in industry often involves retrofitting with modifications to both process and equipment design. Reaching the state-of-the-art performance usually involves technology upgrade.

You may have questions during data extraction. Which data periods should be used as the basis for energy benchmarking? What data are more representative than others? Although the general guideline is to collect data that represent the most common operation, specific guidelines are provided below.

3.7.1   Criteria for Data Extraction

- The most commonly used feed rate
- Use middle-of-the-run historian data
- Use 24 h rolling average based on hourly average data to smooth out fluctuation

- One year of data could be a good representation; get rid of bad data by all means

The reason for using middle-of-the-run historical data is that it represents an "average" operation performance. In contrast, both start-of-the-run and end-of-the-run represent two extreme operation modes would give biased indications of energy use.

Annual data can cover changes in season and operation modes while monthly data can zoom into focus on a particular operation mode.

3.7.2   Calculations Precision for Energy Benchmarking

At this stage, you need to start focusing on important data. Make quick estimates for small consumption users as they usually do not have meters. Do not chase decimal point of precision as the key is getting the order of magnitude right for small users. Some guidelines could be helpful to you:

- Ask instrumentation engineers to recheck critical meters to ensure they are functioning properly.
- Major consumptions need to be verified. Use design data for small consumptions if meters are not available. Corrections may be necessary to reflect the difference in temperature, pressure, and mass flow.
- Fill missing data by simulation or heat and mass balances.
- All forms of energy must be converted to fuel equivalent.
- Specific energy can be on feed volume or mass basis depending on the norm used in the industry. Specific energy can also be on a product basis, which is the ratio of total net energy usage to a desirable product rate on either volume or mass basis. For a process involving both reaction and separation, use feed as the basis for calculating specific energy use. If a process only involves separation and makes a single product, use product as the basis for calculating specific energy use.

NOMENCLATURE

AEP	actual energy performance
BFW	boiler feed water
EPI	energy performance index
FE	fuel equivalent; amount of fuel at the source to make a unit of energy utility (power, steam)
GEP	guideline energy performance
h	enthalpy
m, M	mass flow
OEP	best-in-operating energy performance

PEP   peer's energy performance
 Q heat content

Greek Letter

η_{cycle} power generation efficiency; the amount of fuel (energy input) require to make a unit of power (energy output)

Subscript

CC   combined gas and steam cycle
GT   gas turbine
 ST   steam turbine

REFERENCE

IEA (2012) Technology Roadmap: High-Efficiency, Low-Emissions Coal-Fired Power Generation, December 4.

4

KEY INDICATORS AND TARGETS

Knowing what key operating parameters to monitor and defining the targets and limits for these parameters is an important step for energy optimization. We also need to know the economic values of closing gaps between actual and targeted performances to create incentives for improvement. The system of key indicators is the cornerstone of a sustainable energy management system.

4.1 INTRODUCTION

If you ask operators and engineers how their plant is doing, they would tell you that the plant is under good control. Although true, the process performance could become much better economically and in terms of energy efficiency. The root cause is the lack of monitoring key indicators and no process optimization capability available to operators and engineers in the face of many variables and strong interactions that are typical in operation.

What really needs to happen is the indication of key parameters to monitor and the target values to achieve for these parameters, to gain better energy performance. Although process energy benchmarking in Chapter 3 gives a measure of process energy intensity for process units, the energy intensity does not provide indications of the root causes and the operating parameters to turn to improve energy performance. To determine how well a process unit is doing, a system of performance metrics

Energy and Process Optimization for the Process Industries, First Edition. Frank (Xin X.) Zhu.
© 2014 by the American Institute of Chemical Engineers, Inc. Published 2014 by John Wiley & Sons, Inc.

should be developed so that actual energy usage can be compared with a consumption target. Only then is it possible to conduct root cause analysis and take appropriate remedial actions. To accomplish this goal, the concept of key energy indicator (KEI) is introduced (Zhu and Martindale, 2007), which is the foundation of systematic performance assessment and optimization.

The rationale of introducing key energy indicators is to seek answers to this critical question: "How can engineers characterize energy use in a process unit with an emphasis on *major energy users* in terms of *their needs and the reasons and practical ways to minimize energy use for the needs*?" The application of the key energy indicators in reality follows a methodology based on three steps: defining key indicators, setting targets, and identifying actions to close gaps. Using this methodology, a process unit can be described by a small number of key indicators to measure energy performance, which can be developed based on process knowledge and experience. The application of key indicators will allow us to focus on important issues and prevent us from falling into a trap of details.

4.2   KEY INDICATORS REPRESENT OPERATION OPPORTUNITIES

The intention of defining key indicators is to describe the process and energy performance with a small number of operating parameters. A key indicator can be simply an operation parameter. Some examples of key indicators are reaction temperature, distillation temperature and pressure, column overhead (ovhd) reflux ratio, column overflash, spillback of a pump, heat exchanger U value, and so on. The parameter identified as a key indicator is important due to its significant effect on process and energy performance.

In defining key indicators, one needs to understand the strong interactions between process throughput, yields, and energy use. In the traditional view, energy use is regarded as a supporting role. Any amount of energy use requested from processes is supposed to be satisfied without question and challenge. This philosophy loses sight of synergetic opportunities available for optimizing energy use for more throughput and better yields. The following discussions will provide insights into these kinds of opportunities, which form the basis for defining key indicators for process units to capture specific opportunities.

4.2.1   Reaction and Separation Optimization

Optimizing energy use in reaction and separation systems could lead to significant energy savings because both reaction and product separation consume the majority of overall energy use. A lot of effort is commonly put into reducing energy losses incurred in heat exchangers, furnaces, steam leaks, insulations, and so on; however, little effort is spent on minimizing energy use for reactions and separations, which are the heart of the processes. Very often, energy demands in these systems are considered as "must meet" with the expectation of no challenges from engineers and operators. In reality, however, there is a large scope in minimizing energy use in these areas.

Reaction condition optimization considers reaction severity in terms of temperature and pressure profiles in accordance with catalyst performance in the entire run length. Optimizing reaction conditions, selecting better catalysts, and maintaining catalyst performance in operation have significant effects on both yields and energy efficiency. Consider reaction temperature as an example. In the catalyst cycle, the catalyst performance deteriorates, which affects the reaction conversion. To compensate, the reaction temperature may be increased. However, more severe reaction conditions require more heat from hot utilities such as fired heaters, while severe conditions also produce more desirable products as well as undesirable by-products. The question is how to determine the optimal reaction temperature, which is a function of reaction conversion, production rate, and energy use.

Separation optimization is to achieve product recovery and quality with minimum energy use. Consider overflash for a given separation. Overflash is the feed vaporized in excess of the products drawn above a flash zone. Overflash is typically expressed as a percentage of a distillation column feed. Higher overflash leads to more excess heat in the distillation column, which is rejected in overhead cooling leading to higher reflux rates. Overflash is typically provided by heat input from a hot utility on the feed stream to the distillation column. In a reboiled column, excess heat in the distillation column is provided by a reboiler. A higher overflash can achieve better separation, but at the expense of extra hot utility duty such as a reboiler or feed heater. On the other hand, a too low overflash will make poor separation and product quality will suffer. Yet, in another circumstance, a lower overflash can allow higher throughput with distillation corrections taking place further downstream. Achieving the optimal overflash or optimal vapor-to-liquid ratio (V/L) for a separation process is a critical operation optimization issue.

Minimizing recycle is another optimization example in separation. A common observation in a process plant is that unconverted streams or off-specification streams are recycled back to the front of a process. Recycle streams undo separation. Minimizing recycle presents a significant opportunity for process plants to smartly reduce energy consumption. However, minimizing recycle requires proper operation of separation columns across the whole plant from upstream to downstream.

4.2.2 Heat Exchanger Fouling Mitigation

Fouling mitigation represents a large opportunity for increased heat recovery in operation. Fouling occurs in many heat exchange services, and fouling reduces heat transfer duty significantly. Effective fouling mitigation can save a substantial amount of energy. The overall heat transfer coefficient, or the U value, is the single most important parameter for fouling monitoring for a stand-alone heat exchanger. However, for a complex exchanger network involving many exchangers, determining which exchangers should be selected for cleaning and the frequency of cleaning is not trivial. Selection of the most fouled heat exchanger for cleaning could lead to a suboptimal solution. The primary objective of heat exchanger fouling mitigation should be to minimize energy use. It is possible to select the exchanger for cleaning, which may not be the most fouled, but it could yield the greatest reduction in energy use than cleaning the most fouled one due to the configuration of current heat exchanger network. The methods for optimizing fouling mitigation are discussed in detail in Chapter 7.

4.2.3 Furnace Operation

Fired heaters provide a major part of heat source for reaction and separation. Reliability is the major concern for furnace operation with heat flux for large heaters and tube wall temperature (TWT) for small heaters as the most important reliability parameters. Increasing either heat flux or TWT could increase furnace efficiency. When operating heat flux is much lower than the maximum limit, this is an indication of the furnace being underutilized and thus presents an opportunity for increased feed rate. Increase feed rate is a win-win operation since more feed also results in reduced energy intensity.

Besides reliability, efficient furnace operation is a major part of an energy management system. Oxygen content (O_2%) in the flue gas and furnace stack temperature are the two key operating parameters. Correct measurement of O_2% in the flue gas is the first step, while good control of air intake is the necessary action to achieve low O_2%. Maintenance is essential for burners to function properly and to eliminate air leaks into combustion zones. However, too low O_2 content could promote uneven distribution of the combustion flame, damage furnace tubes, and cause reliability incidents. Typically, 3% of oxygen is the industrial average, although furnaces with state-of-the-art burners and control systems could achieve less than 3%. When reducing the air intake to minimize oxygen content in operation, it is imperative to minimize air leaks and have proper O_2% measurement.

Reducing stack temperature could yield greater reduction in combustion fuel and the limit for the stack temperature is usually set by the sulfur dew point. For furnaces with a convection section, the reduction of stack temperature could be achieved via adding an "economizer" service into the existing convection section. Economizer services are typically water heating services for boiler feed water preheat or saturated steam generation. For furnaces without a convection section, installing a convection section at the grade could be the answer. However, these options need to go through a thorough feasibility evaluation of foundation strength, and thermodynamic and hydraulic considerations. Preheating combustion air is another opportunity in reducing stack temperature.

4.2.4 Rotating Equipment Operation

Rotating equipment used in the process industry includes pumps, compressors, motors and turbines. For pumps and compressors with fixed speed motors, minimizing spill backs could result in significant power savings. Optimizing steam turbine operation and maintenance could save steam. The selection of process drivers, namely, motors versus steam turbines, could save energy costs.

4.2.5 Minimizing Steam Letdown Flows

Steam letdown valves are used to give steam supply flexibility and temperature control. However, letdown steam represents lost opportunity for power generation. Steam balance optimization could minimize the letdown steam flow and hence reduce the loss in power-generating potential.

4.2.6 Turndown Operation

Poor turndown operation implies that when feed rate reduces, energy use does not reduce accordingly. For example, air intake for furnaces should reduce accordingly when feed rate drops. Stripping steam for separation columns should be reduced based on steam-to-feed ratio instead of fixed steam flows. Some of the hand valves for steam turbines can be partially closed when feed rate drops significantly so that the governor could be maintained wide open to minimize steam rate. Proper turndown operation could generate significant energy savings for plants operating under large feed variations.

4.2.7 Housekeeping Aspects

Do not forget the basic energy management, which includes maintenance, steam leaks, steam trap management, condensate return, and insulation. These elements are very visible on site and the energy loss can be estimated. Typically, around 10% steam production can be reduced in steam boilers from better housekeeping management.

There are many other examples of combining energy optimization with process conditions and yields. These opportunities can be monitored by determining key indications with targets, which is a starting point in developing process- and energy-integrated optimization strategies.

4.3 DEFINE KEY INDICATORS

The previous discussions reveal the fact that major improvement opportunities can be captured by key operating parameters. The method for determining key indicators follows a thinking process: understand the process objectives → understand energy needs → develop the measures for the needs → define the key indicators representing the measures. As an example of how to define key indicators for a process, let us consider a single-stage hydrocracking process in the oil refining industry.

The single-stage flow scheme as shown in Figure 4.1 is a commonly used hydrocracking unit as it allows greater than 95% conversion of a wide variety of feedstocks to high-quality products such as gasoline, jet fuel (kerosene), and diesel. To reduce the charge heater duty, the reactor effluent stream is heat exchanged against both feed and recycle gas to recover process heat. Afterward,

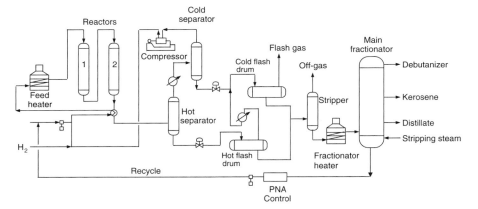

FIGURE 4.1. Typical single-stage hydrocracking unit.

reaction effluent is sent to the hot separator and then the high-pressure separator (or cold separator) where the gas containing a high concentration of hydrogen is recycled back to mix with raw feed for reaction. The conversion products as the liquid of the high-pressure separator go to the cold flash drum first and then join with the hot flash drum liquid going to the stripper where lighter components including H_2S are separated in the overhead, while the bottom liquid products go to the main fractionator to obtain kerosene and diesel. The fractionator overhead product goes to the debutanizer column where LPG and naphtha are made off the debutanizer column top and bottom, respectively. Majority of the main fractionation column bottom is recycled back to the reactor to achieve nearly complete conversion.

4.3.1 Simplifying the Problem

Clearly, hydrocracking is a complex process. To simplify the overall task of defining key indicators for the whole unit, we can divide it into three sections: reaction, product fractionation, and debutanizer or naphtha stabilizer. The naphtha stabilizer section is not shown in Figure 4.1. The goal is to define a set of key indicators for each section, which can be used for monitoring and optimization.

4.3.2 Developing Key Indicators in the Reaction Section

4.3.2.1 Understanding the Process In this example, the process objective is to achieve 95% conversion of feedstocks to high-quality products such as naphtha, diesel, and jet fuel. The reactor contains catalysts that allow maximum production of desirable products. Hydrocracking reactions are highly exothermic and thus requires cold hydrogen quench injection to control the reactor temperature. The energy efficiency for this unit will largely depend on how effectively the reaction effluent heat is recovered.

4.3.2.2 Understanding the Energy Needs   The following items are identified as major energy users in the reaction circuit and their distinct roles and significances are discussed next:

- The feed heater is used to increase the feed temperature and control the reactor inlet temperature. Although the heater efficiency depends on how it is designed and operated, the heater duty is determined by feed preheating requirement. The feed heater outlet temperature can also be a function of the reaction temperature profile—that is, an ascending temperature profile will lower the heater outlet temperature. For a given feed preheat, the heater duty is mainly a function of the heat of reaction and heat recovery. A process engineer can determine the ways to maintain process heat recovery, heater efficiency, and heat flux.

- The compressor for recycle gas is a large power user. The role of a recycle gas compressor is to provide the required amount of hydrogen to the reaction and to provide quench for reaction heat release. The recycle compression depends on the gas flow and its molecular weight. The gas flow depends much on the hydrogen purity in the recycle gas, while the molecular weight of the recycle gas depends on the amount of the light ends brought into the recycle gas from the high-pressure separator. The role of a process engineer is to optimize gas compression ratio.

- Stripping steam is injected to the stripper column for the purpose of removing H_2S and noncondensable light components from the bottom product. The role of a process engineer is to determine the dew point approach, which is the difference of the stripper overhead temperature and the water dew point. The dew point of water is a function of the amount of stripping steam and stripper overhead compositions. The dew point of water is important because the hydrogen sulfide in liquid water is corrosive to the stripper internals. There are two handles for the stripper operation. One is to maintain a certain feed temperature such that there is enough enthalpy in the feed to generate sufficient reflux to keep the overhead vapor above the water dew point. The second handle is to maintain a certain amount of stripping steam to remove H_2S and light ends from the bottom product. If the feed temperature is below a limit, the overhead vapor temperature could be below the water dew point and corrosive water could accumulate in the top tray. On the other hand, if too much stripping steam is injected, particularly in turndown operation, the water dew point could increase and reach a point where the water containing hydrogen sulfide in vapor condenses out.

4.3.2.3 Effective Measures for the Energy Needs   Based on the above understanding of the energy needs, we can go one step further to develop efficiency measures in providing these needs:

- *Heat of Reaction:* A higher heat of reaction occurs when dealing with feeds containing more aromatics, higher sulfur concentrations, or higher olefin concentrations. In these cases, the hydrocracking reaction becomes more severe resulting in shorter catalyst life, higher H_2 consumption, and compression work. Thus, specific heat of reaction is the parameter that connects reaction severity, catalyst life, H_2, and power consumption based on feed compositions and products.

- *Heater Reliability:* Heat flux (for large heaters) or tube wall temperature (for small heaters) is the key reliability parameter for a heater. When heater operation is higher than the limit of heat flux or TWT, a heater is under risk of reliability because the tube life is shortened for higher fluxes. On the other hand, operating a heater much lower than the limit makes the heater underutilized, which represents an opportunity for increased feed rate or higher process severity. Flame impingement is another reliability measure that is usually caused by too low O_2 content. Combustion flame becomes longer with too little O_2 and could reach the tube and pose a serious reliability risk.

- Heater efficiency can be affected by the excess O_2 content or extra air for combustion and a high stack temperature. Inappropriate O_2 content could be caused by lack of control, air leaks, and poor burner performance, while a high stack temperature corresponds to high heat loss in flue gas. A heater approach temperature, defined as the temperature difference between flue gas to the stack and heater feed inlet, could be caused by heater fouling in operation and by heater design.

- The feed is mainly heated by the reaction effluent in feed exchangers before the charge heater. The hot end approach temperature in feed exchangers is a good indication of heat recovery performance by the feed preheating system.

- *Reactor Effluent Air Cooler (REAC) Inlet Temperature:* After the reaction effluent transfers its heat to the feed and recycle gas, the reactor effluent goes to a REAC. Thus, the REAC inlet temperature on the reaction effluent side is a good indication of how effective the reaction effluent heat is recovered.

- *Hydrogen to Hydrocarbon Ratio (H_2/HC):* The recycle gas containing a high percent of hydrogen from the high-pressure separator is recycled back to the reactor. The recycle gas rate is determined by desirable hydrogen partial pressure for the reaction purpose, which affects compression, yields, and catalyst lifetime. Too high H_2/HC ratio could cause greater power usage due to increased recycle rate, while too low ratio could impact yield and shorten the catalyst life. Thus, hydrogen ratio is a parameter affecting energy and yields.

4.3.2.4 Developing Key Indicators for the Energy Needs   Through the above exercise of simplifying the problem and developing an understanding of the major energy needs and measures of efficiency in providing the needs, we can define the following key indicators for the reaction section:

- Specific heat of reaction
- H_2/HC ratio

- Combined feed exchanger hot end approach temperature
- REAC inlet temperature

Specific indicators for heaters could include the following:

- Heater O_2 content
- Heater stack temperature
- Heat flux
- Flame impingement

4.3.3 Developing Key Indicators for the Product Fractionation Section

4.3.3.1 Understand Process Characteristics H_2, H_2S, and NH_3 and light ends are removed from reaction effluents through a series of separation and flashes, resulting in the reaction products in a liquid form, which goes to the stripper, the feed heater, and then to the main product fractionator. The task of the product fractionation is to separate different products based on their product specifications such as distillation endpoint, ASTM D-86 T90% or T95% point, and so on. Side draws from the column go to the product strippers where kerosene and diesel products are made. The net draw from the column bottom is called unconverted oil (UCO), which is recycled back to the reaction section for nearly complete conversion. There are two pump-arounds, namely, kerosene and diesel pump-arounds, as a main feature of heat recovery from the main fractionation column.

It is essential to avoid flooding and dumping, which could severely affect fractionation and thus energy efficiency. Fractionation efficiency can be monitored by column internal V/L ratios. Desired V/L can be achieved jointly by optimizing feed heater outlet temperature, fractionation stripping steam, and overhead reflux rate together with pump-around heat duties, which are used to control excess heat in the column.

4.3.3.2 Understand the Energy Needs

- The main fractionator feed heater provides the driving force for the fractionator by vaporizing feed partially to generate sufficient vapor traffic within the column. The heater duty and stripping steam in the column bottom determine the level of product recovery from the column bottom, which can be monitored by the bottom 5% boiling point. For a given product recovery, the heater duty is a function of feed rate, temperature and compositions, feed preheat, and the enthalpy of vaporized products.
- *Fractionation Side-Stripper Steam:* Stripping steam is used to remove the lighter materials in diesel to control the flash point. This steam can be minimized, and in some cases even eliminated until the diesel flash point requirement is met.
- *Column Pump-Arounds:* Heat is recovered by the pump-arounds and trans- ferred to other process streams. The pump-around duty affects the downcomer

flow and temperature profile in the pump-around section. Thus, a pump-around not only affects the heat recovery but also the fractionation efficiency below the pump-around.

4.3.3.3 Effective Measure for the Energy Needs

- *Feed Heater Efficiency:* The discussions for heaters are similar to the reaction charge heater given previously.
- *Heater Outlet Temperature:* This temperature can affect the lift of diesel out of UCO. A too low heater outlet temperature would cause a slump of diesel into UCO, which degrades the value of diesel into fuel oil. Too high of a heater outlet temperature will cause unnecessarily high reflux rate at the expense of extra heater duty. The heater outlet temperature is mainly a function of the hydrocarbon partial pressure in the flash zone because the distillation cut point is a function of the diesel distillation specification and fractionation efficiency. An optimal heater outlet temperature could be determined by the fractionation overflash, which measures the internal reflux rate.
- Fractionation efficiency can be determined by the gap of diesel 95% cut point and UCO 5% cut point.

4.3.3.4 Developing Key Indicators for the Energy Needs

Based on the understanding of major energy needs and measures of efficiency in providing the needs, we can define the following key indicators for the fractionation section:

- Recycle oil combined feed ratio
- Fractionator over head pressure
- Fractionator stripping steam
- Fractionator column reflux ratio
- Fractionator column overflash
- Diesel stripping steam to diesel ratio
- The distillation gap of diesel 95% cut point and UCO 5% cut point
- Heater O_2 content
- Heater stack temperature
- Heat flux
- Flame impingement

4.3.4 Developing Key Indicators in the Naphtha Stabilizer Section

The main equipment in the naphtha stabilizer section is the debutanizer. Similar to the previous discussions, we can identify key indicators for the debutanizer column:

- *Reflux Ratio:* Affecting separation in the column and can control C_5 in the LPG (overhead product) and $C_4\%$ in naphtha (the bottom product).

- $C_5\%$ *in LPG:* Too much C_5 in LPG is a liquid loss as C_5 could be the blending stock for gasoline.
- $C_4\%$ *in Naphtha:* Too much C_4 could cause higher RVP (Reid vapor pressure) than specification, which is an indication of combustion instability in car engines.

4.3.5   Remarks for the Key Indicators Developed

- For a typical hydrocracking unit, there could be a couple of thousands of operating data measured. The key indicators identified above only accounts for a very small fraction of the overall data but capture the key performance, which contributes to the major portion of operating costs. If these indicators can be monitored and optimized, the unit can operate close to optimal performance.
- In most cases, parameters related to feed and product yields are measured and controlled using basic control systems and APC (advanced process control) systems. However, traditionally, the process indicators are not integrated with energy use. Furthermore, many energy parameters are not measured.
- By identifying the key process and energy indicators and optimizing them together, the optimization does not only reduce energy cost but may also allow increasing throughput when needed, improve product quality, and minimize product specification giveaway.

4.4   SET UP TARGETS FOR KEY INDICATORS

To improve from the current performance, targets must be established for the key indicators, and these targets provide a standard against which existing facilities are measured and equipment improvement is evaluated. The difference between a target and the current performance for each key indicator defines the performance gap. Each performance gap should be associated with a dollar value, which represents an opportunity to be captured. Each indicator is correlated with a number of parameters, including process and equipment conditions together with equipment limits. In this way, energy optimization is connected with process conditions and constraints.

How would one make the concept of key energy indicators working for a process unit or the company? Let us use the example of the debutanizer for the hydrocracking unit aforementioned.

Problem
The debutanizer column sketch is given in Figure 4.2. The reboiling at the bottom of the tower is to provide sufficient vapor flow on the trays for separating C_4 from C_5 and heavier components in the feed. C_4 and lighter components will be withdrawn at the overhead while the C_5 and heavier components leave at the bottom. A certain amount of C_4 in the bottom product is allowed based on the maximum RVP specification for gasoline. Reboiling duty is the main variable in controlling the C_4 amount in the bottom product. In other words, the reboiling duty must increase when C_4 amount in the bottom

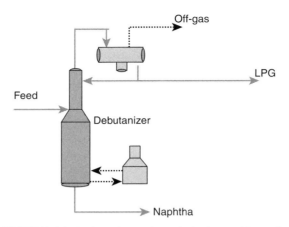

FIGURE 4.2. Debutanizer column in hydrocracking unit.

exceeds the specification. However, it is undesirable to reduce C_4 amount to lower than the specification as this would be product specification giveaway at the expense of additional utilities. The operating objective is to minimize the reboiling duty, while achieving the maximum RVP specification at all times.

Rationale

The task at hand is to develop a relationship between reboiling duty and $C_4\%$ in the debutanizer bottom product so that $C_4\%$ can be controlled by adjusting the reboiling duty. However, other operating parameters also affect the reboiling duty, which include feed conditions (rate and composition), feed preheating (feed temperature), and tower conditions (overhead temperature and pressure). If a correlation of reboiling duty against the above influencing parameters could be generated, reboiling duty can be adjusted according to any of the changes in the related parameters and thus avoid the need for trial and error.

Solution

There are a couple of ways to develop such correlations. The simplest way is the use of a data historian. This method can be applied if three conditions are met: (i) The related parameters are measured and data available in the historian. (ii) The measured data must reflect the operation at the time the butane content was measured. (iii) The historian data cover all possible operating scenarios. After all, online data are the true representation of real simulation!

Development of a correlation using the historian data can be conducted readily in a spreadsheet using regression techniques. After gathering the data from the historian, multiple variable regression can be applied to develop such a correlation. The overall correlation coefficient must be higher than 85% for sufficient regression fidelity.

The second option is to use the step test method usually for developing parametric relations for control systems. By making a small step change to the manipulatable (independent) variable (for example, feed rate), a response from the control variable (dependent variable; reboiling duty in this case) can be recorded after reaching the steady-state condition. This response can be called an energy response. Finally, the regression method is applied to derive the correlation of reboiling duty against all related variables.

However, in many cases, the conditions above for using a data historian are difficult to satisfy. It could be also labor intensive and inconvenient in operation to adopt the step test method. Thus, the most common method is to use the simulation method for developing relationship correlations. To do this, a simulation model for the tower can be developed readily based on the feed conditions (rate and composi-tions) and tower conditions (temperature, pressure, theoretical trays) with product specifications ($C_4\%$ in the bottom and $C_5\%$ in the overhead) established as set points in simulation. Operating parameters such as reflux rate and reboiling duty can be adjusted to meet product specifications. The simulation model is verified and revised against high-quality performance test data.

To evaluate the effect of individual parameters, simulation cases can be developed by prespecifying the values for independent variables of interest; the energy response (reboiling duty) will be recorded automatically. For example, to evaluate the effect of feed preheating, the UA value of the feed preheating exchanger is varied with prespecified values. During simulation runs, the feed temperature before the tower will change according to the UA values, which cause the reboiling duty to vary automatically in the simulation. A set of four curves for such one-to-one relationships can be obtained as shown in Figure 4.3a–d. Brief explanations are given for each figure.

Reducing reflux drum pressure will reduce reboiling duty with the trend as shown in Figure 4.3a. $C_4\%$ in the bottom is the specification that the gasoline product must meet. However, too low $C_4\%$ is not necessary because it does not generate commercial benefit for the cost of extra reboiling duty due to the steeper part of the curve (Figure 4.3b). This product specification giveaway operation must be avoided by all means. $C_5\%$ in overhead product (Figure 4.3c) is the indication of gasoline blending component lost in LPG, which should be avoided as well. Feed preheat also reduces reboiling duty but raises condensing duty (Figure 4.3d).

One must be aware of the capacity limit at the existing condenser when increasing feed preheating, which requires extra condensing. On the other hand, when the feed composition changes such as butane concentration, the reboiling duty is affected for the separation of C_4 from the C_5+ materials. There is no control for the feed composition for the debutanizer operation because the feed composition is the consequence of different raw feeds processed in the hydrocracking unit and the processing severity. For the other four parameters above, operators can make changes to reflux drum pressure, $C_4\%$ in the bottom product, $C_5\%$ in the overhead product, and feed preheating. Optimizing these parameters could give around 5% reduction in reboiling duty than operated based on experience only, which is very significant.

Reboiling and condensing duty can be described based on the relationship with individual parameters as shown in Figure 4.3a–d. If assuming a polynomial form of correlations with order of 3 is used, we have

$$R_i = a + bx_i + cx_i^2 + dx_i^3, \quad x_i = (C_4\% \text{ in bottom}, C_5\% \text{ in ovhd}, \text{preheat}, \text{drum } P),$$

$$(4.1)$$

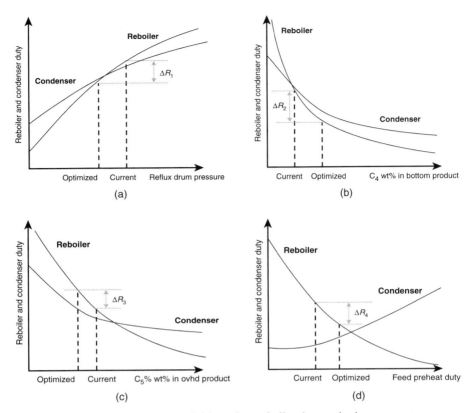

FIGURE 4.3. Correlations of debutanizer reboiler duty and other parameters.

where x_i is one of the four operating parameters, and equation (4.1) describes the relationship of reboiling duty (R_i) and x_i.

The incremental effect (ΔR_i) from individual parameter (Δx_i) can be determined as

$$\Delta R_i = R_{i,\text{new}} - R_{i,\text{base}} = a + b(x_{i,\text{new}} - x_{i,\text{base}}) + c(x_{i,\text{new}} - x_{i,\text{base}})^2 + d(x_{i,\text{new}} - x_{i,\text{base}})^3.$$

(4.2)

If there are no interactions among these four operating parameters, the total effect of changes in these parameters on reboiling duty would be the simple summation of individual effects:

$$\Delta R = R_{\text{new}} - R_{\text{base}} = \sum_{i=1}^{4} \Delta R_i.$$

(4.3)

However, in many cases, there could be strong interactions among operating parameters. In this case, two or more parameters could appear together in one term and the bilinear ($x_1\ x_2$) is the simplest form of interaction. To develop a relationship of parameters with interactions, several parameters need to vary at the same time in plant test or simulation, and the effect on reboiling duty can be seen as the statistically significant result of the interaction parameters. A set of data with changes to the operating parameters and the energy response can be obtained, and regression is subsequently applied to derive correlation involving interactions.

When dealing with correlations involving multiple variables, economic sensitivity analysis is essential to determine the most influential parameters. For example, feed preheat and $C_4\%$ in the bottom are very sensitive to reboiling duty more than other operating parameters for the debutanizer. Getting the most sensitive parameters right in operation can get the greatest economic and technical response.

The correlation developed can be implemented into the control system so that reboiling duty can be controlled automatically to achieve the minimum at all times. On the other hand, the correlation can be used as a supervisorial tool. Whenever a variation is expected, adjustments to operating parameters need to be made to optimize the reboiling duty. This reboiling duty is the minimum with all things considered and it is the target for the conditions at hand. This target and dollar value for closing the gap must be communicated with board operators in each shift so that actions will be taken for achieving targets, while dollar value saved could give operators a sense of pride as a recognition of their actions.

4.5 ECONOMIC EVALUATION FOR KEY INDICATORS

Operation variability is a major cause of operation inefficiency. In general, there are two kinds of variability, which can be observed in reality, namely, inconsistent operation and consistent but nonoptimal operation. Figure 4.4 presents the operating

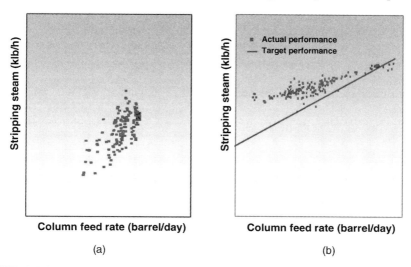

FIGURE 4.4. Two common operating patterns. (a) Inconsistent operation. (b) Consistent operation but nonoptimal.

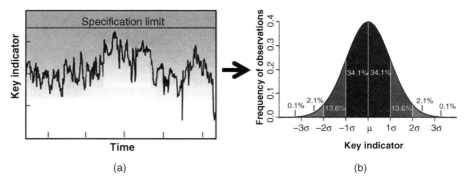

FIGURE 4.5. Operating data. (a) Historian. (b) Frequency distribution.

data of a stripping steam rate in the main fractionator in a hydrocracking unit. In Figure 4.4a, the stripping steam rates appear to be randomly scattered showing an example of inconsistent operation. This is usually caused by either poor control strategy or different operating policy used by operators for running the tower. In contrast, Figure 4.4b shows a consistent operation but nonoptimal. In this case, a consistent operating strategy was adopted, but it was far away from the target for adjusting reboiling duty against column feed rate. The target operation represents the minimum reboiling duty to achieve product specification.

The variability of any operating parameter occurs due to various reasons. The question is how to identify operation variability and the economic value of minimizing variability.

Variability assessment starts with a simple statistical analysis of operating data. For example, the operating data for $C_5\%$ in the debutanizer column overhead product under normal conditions can be extracted from the historian as shown in Figure 4.5a with specification limit provided. To understand the variability, data in Figure 4.5a are converted to a normal distribution curve, which represents frequency of observations as shown in Figure 4.5b. In many cases, the operating data mimic the normal distributions.

Two parameters describe the normal distribution, namely, mean or average (μ) and variance or variability (σ). σ defines the shape of normal distribution. The larger (smaller) the σ value, the fatter (thinner) the curve. μ and σ can be calculated via

$$\mu = \frac{\sum x}{N}, \tag{4.4}$$

$$\sigma = \frac{\sum (x - \mu)^2}{N}, \tag{4.5}$$

where x is the value of the key indicator obtained from the historian, while N is the number of sample data points for the key indicator.

Referring to the example discussed in White (2012) as shown in Figure 4.6a, there are two shortcomings in the operation performance. The first one is the large

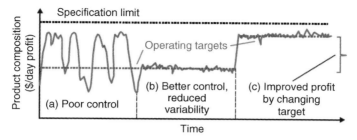

FIGURE 4.6. Operation performance. (a) Current operation. (b) Reduced variability. (c) Increased profit. (From White (2012), reprinted with permission by AIChE.)

variability, while the second is its being too conservative in reaching the specification limit. If the operation or control strategy improves, the variability could be minimized to achieve more consistent operation (Figure 4.6b) but still far away from the limit. In general, a limit usually means a physical limit such as product purity specification, maximum temperature or pressure, maximum valve opening, maximum vapor loading in a separation column, maximum space velocity in a reactor, and so on. The operation can be improved further (Figure 4.6c) by moving the average closer to the limit by adopting a better control strategy. Time series data in Figure 4.6 can be converted to normal distribution curves as shown in Figure 4.7.

If the economic value for the key indicator is known as C_i for the numerical value (x_i) of the key indicator of interest, the economic value (V_i) for x_i at frequency (f_i) of observations can be calculated via

$$V_i = C_i \cdot x_i \cdot f_i. \tag{4.6}$$

For example, $C_5\%$ in the debutanizer overhead product represents the high value component C_5 lost in LPG. The key indicator could be defined as the difference of actual $C_5\%$ in LPG and C_5 specification. If LPG produced from the column is

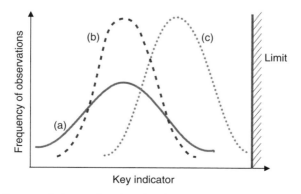

FIGURE 4.7. Converted time series data in Figure 4.6 into normal distribution curves. (a) Current. (b) Reduced variability. (c) Increased profit.

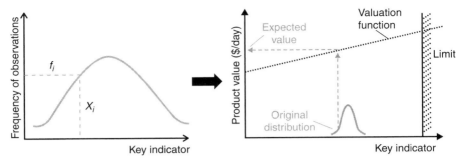

FIGURE 4.8. Converting the normal distribution curve to economic curve. (From White (2012), reprinted with permission by AIChE.)

1000 bpd with $C_5\%$ in LPG at 1% higher than the specification or $x_i = 1\%$ with frequency of occurrence as $f_i = 30\%$, and C_5 is valued at \$75/barrel, the economic value to avoid this occurrence is:

$$V = C_i \cdot x_i \cdot f_i = \$75/\text{bbl} \times 1\% \times 1000\,\text{bpd} \times 30\% = 225\,\$/\text{day}.$$

Similarly for other occurrences when $C_5\%$ in LPG could be lower or higher than 1%, the economic values can be calculated accordingly.

In this way, the normal distribution curve can be converted to economic curve as shown in Figure 4.8. The conversion is calculated to economic difference.

With the statistics-based economic evaluation method mentioned previously, improved operation can be quantified with economic values based on statistical distribution of operating data. The current operation with large variance (Figure 4.9a) is improved by more consistent operation and better control strategy to reduce variability (Figure 4.9b), while optimized operation (Figure 4.9c) utilizes the potential capability available in the process and equipment, and pushes the economic value even higher.

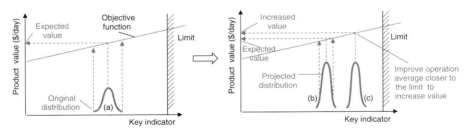

FIGURE 4.9. Economic curves generated based on normal distributions. (From White (2012), reprinted with permission by AIChE.)

4.6   APPLICATION 1: IMPLEMENTING KEY INDICATORS INTO AN "ENERGY DASHBOARD"

The concept of KEI and targets can be readily implemented into an energy dashboard, which can quickly show the performance gaps between current and targets on the computer screen. The level of a gap indicates the severity of deviations and forms the basis to assign a "traffic light" for each KEI—that is, a green light implying the current performance is acceptable as it is within the target range; a yellow light, a warning sign indicating that a gap occurs and requires attention; or a red light, an alarm sign urging for taking action at the earliest time possible. An example tool of monitoring key indicators is Honeywell's Energy Dashboard (Sheehan and Zhu, 2009). This tool could be tremendously valuable to operators and engineers as to what to watch, what to focus on, and which knobs to turn and when.

As a system, KEIs can be defined in a hierarchical structure, from overall site to each process unit down to major equipment and individual operating parameters. The sum of all incentives (opportunity gap between current and targets) represents the total opportunity for the entire process and the overall site. This hierarchical structure allows engineers to drill down from overall performance to specific parameters and thus identify specific actions.

- *Overall site view* shows the site-wide energy consumption and greenhouse gas (GHG) emission versus overall targets. At the same screen, the overall site view shows the energy consumptions and GHG emissions in each process unit. Traffic light color is assigned to indicate which processes are furthest away from the targets.
- *Process unit view* indicates the process performance, which can be measured by around 20 key energy indicators. These key energy indicators are developed from a combination of design, process simulation, and historical data. These predicated energy targets are automatically adjusted to reflect current operating conditions such as feed rate and compositions, operating mode, product yields, and so on. Color coding is assigned to each KEI, which could indicate the need for drill down in the next level of key indicators for identifying root causes and actions.
- *Equipment view* describes equipment performance via several key operating parameters with indications of current values versus corresponding targets. The operators may decide to perform a more detailed investigation for the root causes if the gap is large.
- *Deviation trends view* allows operators to review over the time periods when the KEI deviates significantly from the targets and to determine the major causes of the deviation. By building up a history of causes, operators are able to look back over time and see the most common causes of deviations. This can lead to recommendations about remediable actions for improving equipment performance and, hence, overall process performance.

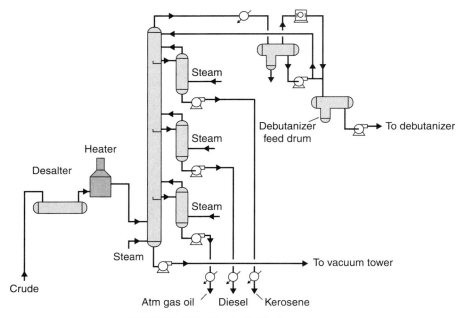

FIGURE 4.10. Crude distillation system.

For each key indicator, a target is established as the basis to compare with current performance. The difference between the target and the current performance for each key indicator defines the performance gap. Different gap levels indicate the severity and level of urgency for actions.

Gap analysis is then used to identify root causes—potential causes include inefficient process operation, insufficient maintenance, inadequate or lack of operating practices, procedures and control, inefficient energy system design, and outdated technology. Gap analyses are translated into specific corrective actions to achieve targets via either manual adjustments or by automatic control systems. Finally, the results are tracked to measure the improvements and benefits achieved.

A successful application of such an energy dashboard is shared here. A company was interested in reducing energy consumption in an existing crude and vacuum fractionation system (Figure 4.10).

The first step was working with unit engineers to define key indicators. One of the key indicators is the atmospheric distillation tower pressure. The distillation tower had been operated under much higher pressure than design pressure of 30 psig for an extended period of time. The consequence is that the tower now requires much higher energy to compensate for the negative effect of high pressure on distillation efficiency. The next step was developing targets for key indicators including the tower pressure based on the simulation and previous experience. From the DCS (distributed control system) data, a clear pattern was observed that the tower pressure was running consistently at 35 psig quite a few years ago. Then the tower pressure increased to 48 psig. What happened?

To communicate with operators for the key indicators and their targets, three workshops were conducted for operators of three shifts. Discussions with operations revealed that the APC was put out of operation due to frequent changes in feeds in the last few years because more slops were recycled back to the tower as a result of heavier feeds being processed. The APC was working under unstable conditions frequently. Therefore, operators put the tower from automatic to manual control. However, the shift lead operators had different experiences in operating the tower. One lead operator liked to ramp up the feed heater to obtain high vaporization because it could lift extra diesel out of the tower bottom. However, high feed vaporization led to high vapor load in the overhead and increased tower overhead pressure. In contrast, another shift lead operator liked to use more stripping steam in the tower as it could reduce partial pressure for the hydrocarbon at the flash zone, which could lift more diesel. However, extra stripping steam increased tower pressure. Both operating experiences caused high tower pressure. As a result the feed heaters became apparently restricted and could not accommodate typical feed rates because distillation under higher pressure requires much higher energy to achieve the same distillation efficiency. In the past, fractionation specialists from the company headquarters came to work on the unit and managed to reduce the pressure close to the design. However, a short while after they left, the tower pressure went up again resulted from the conflicting operating practices by the two lead operators.

Once the root causes for the high pressure were identified, a training effort was justified in view of significant value lost from the high-pressure operation. The effort involved training of operators on the fundamentals of the distillation column. The focus was not just on the key indicators and targets. Much time was spent on potential operating changes and their effects on both the operation and the targets. Operating limits were also reviewed in depth. The ultimate training objective was to get the operating staff on the same understanding of key operating parameters and to agree on the operating policy and procedure to optimize the tower operation. Once key energy indicators and their targets were well understood, operators were trained on the use of the dashboard for operation monitoring.

The implementation was conducted at the control room where manual adjustments were made by lead operators using the DCS system supervised by technical specialists. To achieve smooth manual implementation, a stepwise implementation plan was developed. The sequence of steps was designed to bring the pressure down gradually. A series of adjustments were made in the implementation including heater outlet temperature to control the flash zone temperature in a preferred range, setting key product draw rates, and maximizing the rate for a key pump-around stream to maintain good product level and stripping steam control in the bottom section of the column.

By monitoring the key indicators and making operating adjustments to achieve the targets, the tower pressure was reduced. Online data were gathered to enable a comparison of results before and after implementation. The net result was the ability to process 5000 bpd of additional crude due to unloading of the charge heaters. The value of the increased rate was estimated as $1.5 million per year. To sustain the benefits, operating procedures and targets were further developed and communicated to ensure continuous monitoring and optimization.

4.7   APPLICATION 2: IMPLEMENTING KEY INDICATORS TO CONTROLLERS

Many opportunities for energy improvement can be achieved from changing the plant conditions by adjusting the set point of key variables. In some cases, these opportunities for energy improvement may be possible by incorporating these key variables into an APC if the investment for such an APC is justified by the value to be captured.

Multivariable, predictive control, and optimization applications have been commonly applied in the process industry. The ability to take models derived from process data and simulations and configure the models in a highly flexible manner allows the engineers to design controllers that can be suitable for multiple purposes. The same controller can be used to maximize throughput, maximize yields, and/or minimize energy use just by changing the cost factors in the objective function. This APC environment is suitable for incorporating energy strategies into overall operating objectives. In fact, adding energy operating costs into the existing objective function and inserting related KEIs with corresponding correlations and operating limits is generally advisable. In this manner, minimizing energy cost will not be accomplished at the expense of the valuable product yields.

There are many energy-saving opportunities that can be incorporated into APC applications:

- Furnace pass balancing
- Distillation column controls combined with pressure minimization and flash zone temperature control to maintain yields of the most valuable products while minimizing energy use in reboiling or feed heater
- Reaction conversion control
- Feed preheating maximization
- Separation column reboiler duty control
- Recycle minimization
- Water dew point control for steam strippers

One example of a single variable control strategy is applied to a stripper in a hydrocracking unit. The main purpose of the stripper is to remove H_2S and noncondensable components from the bottom product. One of the key indicators identified is the water dew point at the top of the stripper column. As a matter of fact, the dew point is a function of column overhead vapor composition and the amount of water. There was no monitoring capability available for the dew point temperature. If the column top temperature is lower than the dew point, the hydrogen sulfide will dissolve in the condensed water and cause corrosion to the column overhead system.

To avoid this, operators usually run the column with a high dew point approach to make sure the top temperature is sufficiently higher than the dew point. By applying a new control device recently developed for dew point for this type of stripper columns, the actual dew point can be calculated accurately. With measured column top temperature, a tight dew point approach can be maintained, which results in 10% reduction of the stripping steam.

Another example is a large multivariable control strategy, which was applied to an ethylene complex (Sheehan and Zhu, 2009). This involves 17 multivariable controls that were linked together by an optimization strategy that included the use of a nonlinear cracking model to predict product yields. The result of the APC applications enabled the operators to increase the feed rate by 3% over the previous best rate by being able to operate the process up against multiple constraints simultaneously. In addition, the application is able to reduce energy consumption by 3.3% by reducing steam consumption in the fractionators and minimizing excess O_2 in the furnaces. This results in a payback of less than 5 months for the APC investment.

4.8 IT IS WORTH THE EFFORT

As demonstrated previously, the concept of key indicators and targets can play important roles for process- and energy-integrated optimization. Process optimization without taking energy use into account will lead to high energy costs, while energy optimization without fully addressing process needs will cause penalty in processing capacity, product quality, and yields. With appropriate work process fitting into the existing technical management system, the concept of key indicators and optimization can become the cornerstone of energy management.

Because of this significance, developing key indicators and targets should become a corporate concerted effort as management's support is critical. First of all, significant effort is required in developing technical targets for key indicators. Setting up targets needs modeling of major equipment. Once a base case of operation is defined and simulated, operation variations can be simulated and correlations can be developed for key indicators in relation with other operating parameters. Using the debutanizer as an example, the reboiling duty is the key indicator that can be affected by feed rate and compositions, reflux drum pressure, preheat, and $C_4\%$ in the bottom and $C_5\%$ in the overhead products. Regression analysis of simulation results may be required to develop the correlation of the reboiling duty and these operating parameters. Whenever operating conditions change, the correlation could be applied to determine the minimum reboiling duty under the new conditions.

The engineering effort is appreciable in developing a system of key indicators and technical targets at different operating conditions. It may take 1–2 man-years to develop such a system for processes and major equipment. In one example of a large refinery complex, 2 man-years were required to develop such a target system. Implementing this target system in day-to-day operation is essential and integrating the target system into existing technical management system is the key for success. The confidence will grow when people observe the operating cost savings of applying such a target system. The benefits will pay for the investment alongside the development. The lessons learned and experience gained will be disseminated to other process units.

NOMENCLATURE

C economic value for a key indicator
f frequency of observations
N the number of sample date points
R reboiling duty
V economical value
x measured value of key indicator

Greek Letters

μ mean value in normal distribution
σ variance in normal distribution

REFERENCES

Sheehan B, Zhu XX (2009) The first step in energy optimization, *Hydrocarbon Engineering Journal*, pp. 25–29, April.
White DC (2012) Optimize energy use in distillation, *Chemical Engineering Progress (CEP)*, pp. 35–41 AIChE, March.
Zhu XX, Martindale D (2007) Energy optimization to enhance both new projects and return on existing assets, 105th NPRA (National Petroleum Refining Association) Annual Meeting, March, Texas, USA.

PART 2

ENERGY SYSTEM ASSESSMENT METHODS

5

FIRED HEATER ASSESSMENT

Fired heaters operate under very severe conditions, which make reliability and safety the highest priority. High reliability leads to higher performance and longer life. At the same time, efficient operation of fired heaters can improve fuel efficiency. Proper design and operation can achieve both high reliability and energy efficiency.

5.1 INTRODUCTION

Fired heaters are used to provide high-temperature heating when high-pressure steam is unable to satisfy process heating demand in terms of temperature. The primary role of industrial fired heaters is to provide heat required for reaction and separation processes. In a fired heater as shown in Figure 5.1, the process fluid enters the tubes at the top of the convection section and flows down countercurrent to the flue gas flow. The fuel mixes with the combustion air in the burner. The hot combustion gases need residence time to transfer the heat to the tubes. The shock tubes are often the hottest tube in the fired heater. The shock tubes receive the full radiant heat transfer of about 12,000 Btu/h ft^2 or higher, plus the hot gases flowing over the tubes result in an additional convective heat transfer rate of about 5,000 Btu/h ft^2. Since the firebox operates in very high temperature, refractory lining is required to prevent heat loss to the atmosphere.

Due to the fact that fired heaters operate under severe conditions, fired heaters are designed with careful consideration of the high-temperature characteristics of

Energy and Process Optimization for the Process Industries, First Edition. Frank (Xin X.) Zhu.
© 2014 by the American Institute of Chemical Engineers, Inc. Published 2014 by John Wiley & Sons, Inc.

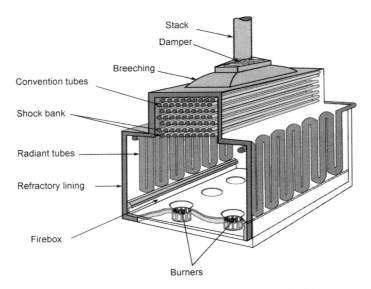

FIGURE 5.1. Schematic view of a typical process fired heater.

the alloy. With proper maintenance and operation, a fired heater can have a long operating life. However, the life of a fired heater can be greatly shortened due to creep, carbonation, fatigue, corrosions, and erosion due to lack of maintenance and reliability considerations. Fired heater failure can lead to significant production loss; in the worst case, it can cause damage to human life.

Therefore, maintaining a fired heater in reliable operation is the highest priority. With this priority in place, process plants can strive to maximize fired heater efficiency and hence reduce its running costs. This is because of a simple fact, fired heaters are the largest energy consumers in process plants and accounts for majority of total energy use.

5.2 FIRED HEATER DESIGN FOR HIGH RELIABILITY

The single most important measure of fired heater reliability is availability, and the goal is that a fired heater needs to be online almost 100% of time. What does it take for a fired heater to achieve this high reliability from the design point of view?

In this section, we will discuss the critical issues that a highly reliable fired heater must acquire and shed light on the fundamentals for these features so that they can be used as the benchmark for assessing a fired heater. The critical reliability issues are discussed below.

- Flux rate
- Burner to tube clearance
- Burner selection
- Fuel conditioning system

5.2.1 Flux Rate

Radiant heat flux is defined as heat intensity on a specific tube surface. Thus, heat flux represents the combustion intensity and is analogous to "how hard a fired heater is run." More specifically, keeping the firing rate within safe limits is equivalent to maintaining the peak heat flux at less than the design limit because high firebox temperatures could cause tubes, tube-sheet support, and refractory failures. What is the peak flux and why is it so important to keep it within the limit? These questions will be answered next.

Flux rate is influenced by combustion characteristics and heat distribution. While the combustion characteristics can be described by combustion intensity, the heat distribution is explained by heat flux. The heat flux is defined as the heat transferred to the process feed, while combustion intensity is the heat released from flame divided by the flame's external surface area. Clearly, combustion intensity is related to combustion flame, while heat flux is related to process. Another difference is that combustion intensity is inevitably an average value, while heat flux is either average or local values. Local heat flux requires more attention in design and operation.

Figure 5.2 shows a typical pattern of a heat flux profile. It can be observed that the flux distribution is not uniform. The heat intensity nearest flames is the highest (peak flux) and declines away from flames. The peak flux can exceed twice the average value and should be maintained below the flux limit all the time for safe operation. The nonuniformity of heat flux throughout the radiation section is described by the ratio of peak flux to average flux. A good heater design and operation should have a heat flux profile featuring a high average and low peak flux values.

Jenkins and Boothman (1996) reported an operating case with average flux at only 700 Btu/h ft^2, but the peak flux was over 20,000 Btu/h ft^2, nearly three times the average. The peak flux exceeded the safe limit of the process, and residual oil in the tubes in the peak flux area was cracked and coke was deposited in the inner tube surface. As a consequence, the coke acted as an insulating layer and caused the tubes

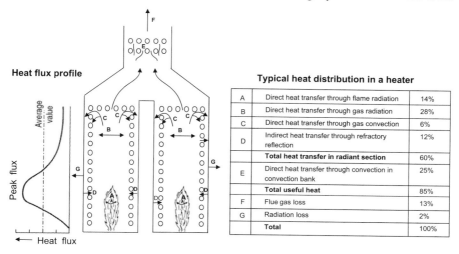

Heat flux profile

Typical heat distribution in a heater

A	Direct heat transfer through flame radiation	14%
B	Direct heat transfer through gas radiation	28%
C	Direct heat transfer through gas convection	6%
D	Indirect heat transfer through refractory reflection	12%
	Total heat transfer in radiant section	**60%**
E	Direct heat transfer through convection in convection bank	25%
	Total useful heat	**85%**
F	Flue gas loss	13%
G	Radiation loss	2%
	Total	**100%**

FIGURE 5.2. Flux profile and heat distribution in a heater.

to overheat, which was measured by tube wall temperature (TWT). This was dangerous as the tubes in the peak flux area could rupture. To mitigate the safety risk, the plant operators had to reduce the process feed rate and hence cut down the firing rate to keep the peak flux below the safe limit. Furthermore, the heater had to be shut down at regular intervals so that tubes could be cleaned to remove the carbon deposits. These mitigation actions cost a lot of money to the plant. The problem was solved fundamentally by improving air distribution between the burners together with changes to the burner gas nozzle and flame stabilizer. These changes resulted in a lower air pressure loss and improved fuel and air mixing. After these changes, the burners produced a more even flux profile: the average flux was increased from 700 to 10,000 Btu/h ft^2, while the peak flux reduced to 18,000 Btu/h ft^2. As a result, the unaccepted coking inside tubes was eliminated, which allowed feed rate to increase by 4%, and the heater to run continuously between scheduled outages.

The flux distribution around the tube is not uniform as well. As indicated in Figure 5.3, the radiating plane is the flame. The diagram on the left shows the flux profile for a single-fired heater. The front of the tube facing the fire picks up most of the heat. The diagram on the right shows the profile for a double-fired heater with flames on both sides of the tubes. The flux pattern is close to uniform.

The nonuniformity of heat flux around tube is described by the circumferential flux factor, which is the ratio of peak flux to average flux. Peak flux determines the maximum TWT. The peak flux is typically 1.5–1.8 times the average for a single-fired heater, while it is 1.2 times the average for a double-fired heater. That explains why the double-fired heater has longer run length as it has lower flux rate and hence lower TWT than the single-fired heater.

The tube thinning follows the same pattern of flux distribution. Figure 5.4 shows a fired heater tube with severe thinning creep caused by internal coking especially on

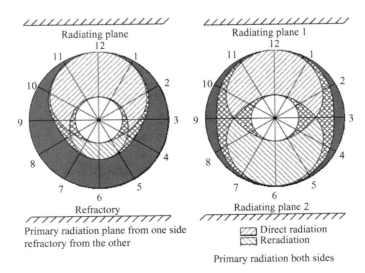

FIGURE 5.3. Flux distribution around fired heater tube.

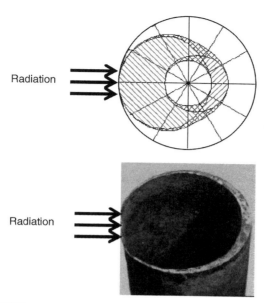

FIGURE 5.4. Tube thinning follows the flux distribution.

the fireside of the tube. The internal coking follows the same pattern with much greater coke thickness at the front face facing the flame. This is why inspectors concentrate their tube inspections on the fireside of the tube. As a reference, Table 5.1 gives the typical maximum heat flux.

Heat distribution throughout the fired heater is not even. The radiation section makes up 70–75% of total process heat transfer, while the convection section accounts for 25–30%, which can be observed in Figure 5.2. Different fuels have different heat distributions. For a gas-fired heater, one third of the heat transfer in the radiant section is flame radiation and two thirds is hot gas radiation. If the flame height is too high, there is not enough residence time for the hot gas cloud represented as "B" in Figure 5.2 to transfer heat to the tubes. This situation occurs when a long flame burner is placed in a short firebox. Oil firing is different. The oil flame has very high flame radiation, so approximately two thirds of the heat transfer in the radiant section is flame radiation and one third is hot gas radiation.

Oil and gas firing have different combustion characteristics. Oil firing is governed by flame radiation with the presence of visible flame light waves. In contrast, hot gas radiation produced by combustion is governed by gas firing. Oil has high emissivity close to one and thus able to drive the heat through the ash resistance.

TABLE 5.1. Maximum Flux Rate Used in an Operating Company; Btu/h ft^2

Vertical cylindrical with tube length 20–30 ft	12,000
Vertical cylindrical with tube length >30 ft	13,000
Cabin	14,000
Double fired U-tube	22,000

5.2.2 Burner to Tube Clearance

Burner to tube clearance is very important in heater design because flame radiation is directly proportional to the square of the distance to the tube. Small burner to tube clearance can result in flame impingement, hot spots, and tube failure. That is why most heater failures can be traced to flame impingement due to burners placed too close to the tubes. For example, consider a 5 ft burner to tube clearance versus 3 ft spacing, the smaller spacing case results in 2.8 $[=(5/3)^2]$ times the flame radiation as the larger spacing.

5.2.3 Burner Selection

There are four types of burners, namely, standard, premixed, staged air/fuel (low NO_x), and next generation (ultralow NO_x). Standard gas and premixed burners have luminous flames. The combustion reaction occurs within the visible flame boundaries. Ultralow NO_x burners have nonluminous flames, and much of the combustion reaction is not visible.

5.2.3.1 Flue gas Emissions NO_x emission is an important environmental issue for the process industry today. The NO_x is formed by nitrogen and oxygen reacting at the peak temperatures of the flames. A standard gas burner produces 100 ppm NO_x; premixed gas burner 80 ppm; staged gas burners 40 ppm; ultralow NO_x gas burners 30 ppm; and the latest generation ultralow NO_x gas burners produce 8–15 ppm NO_x. SO_x is controlled by the sulfur in the fuel. Many plants have sulfur limits that require burning low sulfur fuel oil. Carbon monoxide (CO) should be less than 20 ppm.

5.2.3.2 Objective of Burner Selection   The objective of burner selection is to determine burner type and configuration to obtain the desired heat flux profile to meet process heating demand. The combustion space and shape may be determined by physical, mechanical, or structural factors, but that space must be able to accommodate efficient aerodynamic mixing and combustion of the fuel, and generate the desired heat flux profile for the product. The heat release and hence heat flux generated from burner flames is not even. It is generally high in the region near the burner port, where fuel and air are plentiful, and reduces as the flame develops, owing to the depleting fuel content, and from losing heat to its surroundings. The burner designer can adjust this profile from burner type and configuration and flame envelope although it never achieves uniform flux distribution.

5.2.3.3 Flame Envelope   The flame envelope is defined as the visible combustion length and diameter. The flame length should be one third to one half of the firebox height. The hot combustion gases need residence time to transfer the heat to the tubes. Many burners have flame diameters that are between 1 to1.5 times the diameter of the burner tile. Since the tile diameters are often larger for ultralow NO_x and latest generation burners, the flame diameters at the base of the flame may be slightly larger. The flame diameter often expands, giving a wider flame at the top.

Ultralow NO_x and the latest generation burners have longer flame lengths than conventional burners. Longer flame lengths change the heat transfer profile in the firebox and can result in flame impingement on the tubes.

5.2.3.4 Physical Dimension of Firebox

Optimized designs have burner spacing that is designed to have gaps between the flame envelopes. Since the tile diameters are often larger for ultralow NO_x and latest generation burners, retrofits can result in closer turner-to-burner spacing and flame interaction. Flame interaction can produce longer flames and higher NO_x. Flame interaction can interrupt the flue gas convection currents in the firebox, reducing the amount of entrained flue gas in the flame envelope. This condition increases the NO_x levels. Ultralow NO_x and latest generation burners should be spaced far enough apart to allow even flue gas recirculation currents to the burners.

The burner centerline to burner centerline dimension is one of the most important dimensions in the firebox tube. Many tube failures are caused by flame and hot gas impingement. When ultralow NO_x and the latest generation burners are being retrofitted, the larger size of the flame envelope must be evaluated. Firebox convection currents can push the slow burning flames into the tubes.

Flame impingement on refractory often causes damage. When ultralow NO_x and the latest generation burners are being retrofitted, the larger burner diameter may result in the burners being spaced closer to the refractory. Unshielded refractory may require hot face protection.

Many heaters are designed for flame lengths that are one third to one half the firebox height. Ultralow NO_x and the latest generation burners typically have flame heights of 2–2.5 ft/MMBtu. Longer flame heights from ultralow NO_x and the latest generation burners may change the heat transfer profile in the firebox. The longer flames may result in flame or hot gas impingement on the roof and shock tubes. In this case, the solution is to change burners. Some older heaters have very short firebox heights and may not be suitable for retrofits to ultralow NO_x and the latest generation burners.

5.2.3.5 Process Related Parameters

Ultralow NO_x and the latest generation burners have longer flames that change the heat flux profile. This is especially important on thermal cracking heaters such as cokers and visbreakers in oil refineries. The longer flames may increase the bridge wall temperature (BWT) and change the duty split between the radiant section and convection section.

The location of the maximum tube metal temperature changes as the heat flux profile changes. Retrofitting ultralow NO_x and the latest generation burners in short fireboxes can result in high metal temperatures for roof and shock tubes.

Ultralow NO_x and the latest generation burners may have less turndown capability than conventional burners. High CO levels can occur when firebox temperatures are below 1240 °F. Flame instability and flameout can occur when firebox temperatures are below 1200 °F. Since ultralow NO_x and the latest generation burners are often designed at the limit of stability, a fuel composition change may cause a stability problem.

The proper design basis for the burner selection is extremely important. Sometimes the process requirements have changed significantly since the fired heater was designed. Important design basis items include: (i) emission requirements; (ii) process duty requirements; (iii) turndown requirements; (iv) fuel composition ranges; (v) fuel pressure; and (vi) startup considerations.

The guideline for burner selection is to select the most appropriate burner technology while meeting the NO_x emission limit. Reliability should be placed as higher priority than cost in burner selection because industrial applications show that 90% of fired heater problems come from poorly maintained and operated burners. Although it could be more expensive with the best burner technology, the money spent is worthwhile as burners cost only 5–10% of the fired heater overall cost, but it could avoid 90% of fired heater problems.

5.2.4   Fuel Conditioning System

Poor fuel conditioning could cause problems in burners and combustion. While many conventional burners have orifices 1/8 in. (3 mm) and larger, ultralow NO_x and the latest generation burners often have tip drillings of 1/16 in. (1.5 mm). These small orifices are extremely prone to plugging and require special protection. Most fuel systems are designed with carbon steel piping. Pipe scale forms from corrosion and plugs the burner tips. Although tip plugging is unacceptable for any burner, it is even more important not to have plugged tips on ultralow NO_x and the latest generation burners because plugged tips can result in stability problems and higher emissions.

Many companies have installed austenitic piping downstream of the fuel coalescer/filter to prevent scale plugging problems:

- Coalescers or fuel filters are required on all ultralow NO_x and latest generation burner installations to prevent tip plugging problems. The coalescers are often designed to remove liquid aerosol particles down to 0.3–0.6 μm. Some companies install pipe strainers upstream of the coalescer to prevent particulate fouling of the coalescing elements.

- Piping insulation and tracing are required on fuel piping downstream of the coalescer/fuel filter to prevent condensation in fuel piping. Some companies have used a fuel gas heater to superheat the fuel gas in place of pipe tracing. Unsaturated hydrocarbons can quickly plug the smaller burner tip holes on ultralow NO_x and the latest generation burners.

5.3   FIRED HEATER OPERATION FOR HIGH RELIABILITY

Fired heater capacity for critical processes is usually pushed hard for more production and thus the fired heaters are operated near or at the operation limits. It is essential to make sure the fired heater is running in a safe and reliable manner with the following key operating reliability parameters within acceptable limits:

- *Draft Control*: Avoid positive pressure to prevent safety hazards and provide sufficient primary air for burners.

- *High BWT*: BWT directly relates to flux rate and indicates how hard a heater is running.
- *TWT or Tube Metal Temperature (TMT)*: Identify root cause for high TWT operation.
- *Flame Impingement*: The most common reliability hazard for fired heaters.
- *Tube life*: High TWT operation shortens tube life significantly.
- *Excess Air or O_2 Content*: Optimal $O_2\%$ is the balance between reliability and efficiency.
- *Flame Pattern*: Visualize the flame shape, height, and color to identify abnormal combustion problems.

5.3.1 Draft

There are two types of draft: one is natural draft and the other is forced draft. For natural draft, draft depends on the density difference between hot flue gas and ambient air. Thus, stack height must be sufficient to provide adequate draft, while stack damper opening must be adjusted properly in operation at the same time. For forced draft, stack height can be short as fan is used for providing air. Thus, stack height is only set based on dispersion requirements. Similar to natural draft, the stack opening must be adjusted properly in operation. The key objective of draft control for both natural and forced draft is to avoid positive pressure inside the heater to prevent damage or safety hazards and provide sufficient combustion air (primary air) at the same time. A proper draft control is to maintain the draft at the range of 0.1–0.2 in. water column vacuum measured underneath the convection tubes or at the bridge wall (line Y; Figure 5.5a) . This can be achieved by the adjustment of both stack damper and air register. With this draft, sufficient air can be drawn in through the burners as primary air to obtain flame stability, while secondary air is provided by air register for O_2 control. However, too high or low draft must be avoided. A too high draft could occur when the damper is widely open and register fully closed. This could result in a too high vacuum in the stack and could increase cold air leakage into the heater (Figure 5.5b). Excessive draft could cause flame to lift off the burners touching the tubes and this could lead to serious damage to the heater. A too low draft corresponds to the case when the damper is almost closed and the air register is widely open. In this case, positive pressure at the bridge wall could be developed, which forces hot flue gases flowing outward through leaks in the convection section (Figure 5.5c). This could lead to serious structural damage. The draft profiles for these three cases are provided in Figure 5.5d.

Weather change could cause draft fluctuation. For example, when strong winds occur and cause draft fluctuation, the damper opening should be increased gradually to maintain flame stability. On the other hand, in windy weather, if the heater faces toward the wind with the highest static atmospheric pressure, this may result in a too high draft. In this case, the damper should be closed slightly.

5.3.2 Bridge Wall Temperature

A high BWT measurement indicates the heater operates at high radiant flux rates. BWT is the key reliability parameter for fired heaters as high BWT or TWT can cause

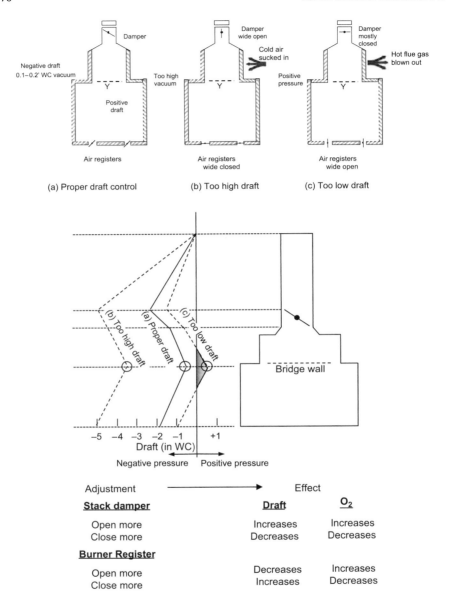

FIGURE 5.5. Correct and incorrect draft.

mechanical failures on tube sheet supports and refractory. A majority of heater failures is accompanied by high BWT. The general guideline for BWT is not to exceed the mechanical design limit that tube sheet supports and the BWT limit depends on design. High BWT could be caused by long flames, not enough flue gas residence time, and external fouling on the tubes.

5.3.3 Tube Wall Temperature

The skin temperature is the process temperature inside the tube, plus the temperature differences across the film and metal resistances. The film resistance is usually larger than the metal resistance. It is calculated by taking the peak flux and dividing by the heat transfer coefficient. The heat transfer coefficient is usually 200–500 Btu/(h ft^2 °F), which provides a typical film resistance of 45–80 °F. The metal resistance is much smaller than the film resistance. It is calculated by taking the peak flux and dividing by the thermal conductivity of the metal. The thermal conductivity is usually 12–16 Btu/(h ft °F), which results in a typical metal resistance of 15–20 °F (8–11 °C). The exception is for thick-walled tubes, which could have a metal resistance as high as 80 °F (44 °C).

Tube wall or skin temperature is an important reliability parameter and should be closely monitored, and guidelines for tube life can be developed. Guidelines should be effectively communicated to operators so that appropriate tube temperature can be determined that could meet the production requirement while minimizing the risk of tube damage. It is important for operators to know that over-firing is the main cause of tube damage. Process plants use skin thermocouples and infrared pyrometers to monitor TWTs.

It is very important to monitor the amount of scale on tubes to measure coking/fouling/corrosion rates. This can be achieved by a thermocouple and infrared pyrometer monitoring program. The scales on tube increase TWT or skin temperature. 0.01 in. scale on tube could raise tube surface temperature by 100 °F. The common way to get rid of scale is to sandblast the scale off the tubes while ceramic coating on tubes is a preventive measure; however, the later is expensive.

5.3.4 Flame Impingement

Flame impingement could be caused by low air as well as burner tip fouling, which could be avoided by adjusting excessive air and fuel pressure. Figure 5.6 shows a fired heater operating with severe flame impingement in which a long flame reaches tubes and the tube front receives almost six times as much heat as the back side of the tube. The best way to know if hard flame impingement is formed is to view the firebox using the glasses especially for that purpose. These glasses eliminate the glare and bright haze and make it possible to view real flame positions.

The following guidelines for better mixing could be used to determine the root causes for flame impingement:

- Primary air is used for achieving flame stability, while secondary air for O_2/NO_x control. Thus, primary air should be increased via damper opening to a limit beyond which the flame will lift off the burners. Excess air is provided by adjusting secondary air via register. Too much and too little secondary air gives poor combustion. This is because a minimum excess air is required for flame stability and too much excess air reduces flame temperature and hence efficiency drops and NO_x increases as a result.

FIGURE 5.6. An example of flame impingement.

- Close ignition ports, peep doors, and other holes around burners. Combustion air only mixes well with fuel gas when it flows through the air registers.
- At turndown operation, some of the burners may be blanketed off and do not forget to close the air registers for the idle burners. Burners work more satisfactorily close to the design capacity.
- Plugged burners require more excess air for combustion, but too much excess air could lift off flame. Sulfur deposit is the common cause of burner plugging, and a solution is to prevent oxygen from entering the fuel gas system as it could combine with hydrogen sulfide in the fuel gas to form NH_3Cl.

5.3.5 Tube Life

Realistic average tube life can be assessed based on creep measurement and metallurgic examination. The guidelines derived from assessment should be illustrated to operators for the serious damage that could occur by operating a fired heater over the TWT limit. In general, 18 °F increase over the TWT limit could halve the life of a heater. 30 °F over the TWT limit could shorten a heater's life substantially and cause rapid failure when a heater is in the creep range. It is important to know that it is the peak TWT that should not exceed the limit instead of the average TWT.

A fired heater is not operated uniformly over the entire run as it could run light in turndown operation and harder in full capacity and toward the end of run for reaction heaters. To estimate the effects of changing tube wall temperature, corrosion rates, and pressure, a metallurgic examination can be applied to estimate the remaining life of tubes. Knowing the tube life not only prevents premature tube failure, but also identifies the need for metal upgrade if the operating skin temperature increases over time.

5.3.6 Excess Air or O_2 Content

It must be stated that optimal $O_2\%$ is the balance between safety and efficiency. There are several signs visible when a firebox is short of combustion air: a hazy flame, regular thumping sound, and long flame touching the tubes.

One of reasons causing insufficient air is aggressive $O_2\%$ management regardless of burner conditions. Another root cause of insufficient air is the O_2 measurement based on the flue gas sample taken from the stack. This measurement is not accurately representative of the oxygen available in the firebox. Leaks in the convection section allow air to bypass the firebox and exit in the stack and contributes to the $O_2\%$ measured in the stack. When air registers are adjusted based on the oxygen level measured from stack, the firebox could be short of air. On the other hand, air leak is wasteful of hot flue gas for heating up cold air that is sucked into the convection section.

The cost-effective activities include seal welding of the casing, mudding up header boxes, and using high-temperature sealants. Leaks through roof penetration are also a major source of air leak, which should be inspected during turnaround. These activities are especially important for NO_x control.

5.3.7 Flame Pattern

Proper control of combustion air is the key to make complete combustion and stable flame and thus avoid flame impingement. Lower fuel pressure also helps to avoid flame impingement. When the amount of excess air is appropriate, flame is orange and flue gas from stack is light gray. With sufficient air, if flame is long with much smoke, burners may have problems.

Figure 5.7 shows a good combustion with orange color and a proper flame height about one third to one half of the firebox height. In contrast, Figure 5.8 displays a poor combustion with plugged gas tips on the first burner. There is a strong haze from the flame of the first burner indicating incomplete combustion. The burner tip plugging could be reduced by using a fuel gas coalescer and steam heater.

5.4 EFFICIENT FIRED HEATER OPERATION

Operators understand the importance of maintaining fired heaters in a safe and reliable condition. The response from operators to this priority could go to another extreme: run fired heaters with too much excess air. The result of much excess air is much reduced flame length and thus the risk of flame impingement is minimized.

FIGURE 5.7. Good flame color and height.

FIGURE 5.8. Poor flame pattern from the first burner.

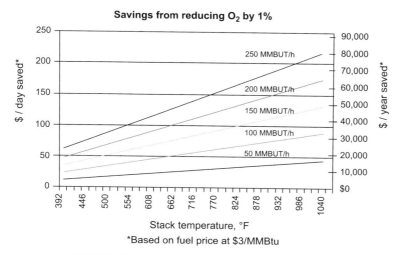

FIGURE 5.9. Dollar value for reducing $O_2\%$ by 1%.

However, the price for too much excess air is higher operating cost from burning extra fuel. Therefore, there is an optimization need for excess air.

Too much excess air is waste of fuel as cold air needs to be heated up from ambient to stack temperature. Figure 5.9 shows the fuel savings by dropping 1% of O_2 from reduced excess air. For example, for a heater with an operating duty of 200 MMBtu/h with stack temperature at 500 °F, a reduction of 1% oxygen saves 1 MMBtu/h of fuel, which is worth $72/day or $26,280/year for fuel priced at $3/MMBtu. Reducing $O_2\%$ from 7 to 3%, the savings could be worth around $100,000 per year. If fuel price is at $6/MMBtu, dropping 4% of O_2 could save $400,000 per year. 3% of O_2 is used as the basis for benefit calculation here as 3% is a typical limit for industrial fired heaters. However, do not start O_2 reduction before burners are in good working condition, and O_2 analyzers are installed and calibrated with correct readings.

Reducing stack temperature could improve heater efficiency more than $O_2\%$ optimization. Every 40 °F reduction in stack temperature is equivalent to 1% fuel efficiency improvement. For example, a small heater with duty of 50 MMBtu/h does not have a convection section and stack temperature is at 1250 °F. If the flue gas is routed to the convection section of a large heater in a close location, the stack temperature could be reduced to 500 °F. The capture of this waste heat could be worth $60/day and $220,000/year for fuel priced at $6/MMBtu. In general, reducing stack temperature is more of a design issue; for example, installing a steam generator and economizer in the convection section to recover waste heat. In contrast, O_2 level is an operation issue, which can be controlled by adjusting secondary air via the air register.

5.4.1 O_2 Analyzer

Fired heaters have either forced draft fans or induced draft fans to control air to the burners. This allows control of oxygen amount by direct measurement of air and fuel flow rates. Large and efficient process fired heaters with natural draft burners usually have induced draft fans. It is desirable to have control systems devised to maintain the desired amount of excess air. With O_2 analyzers, the control system adjusts damper openings automatically to control O_2 subject to a limitation on absolute draft level. Relatively small fired heaters can also justify O_2 analyzers for energy savings.

To obtain more uniform O_2 reading, every 30 ft should have one sample point, and sample points should be installed downstream from the convection section. The requirement is that there should be minimum air leakage into the convection section to avoid false O_2 readings. In general, sample points should not be located in the radiation section for the reason that flue gas from different burners are not well mixed. Otherwise, the O_2 reading would mainly reflect the operation of the burners close to the sample points. The exception to placing the oxygen analyzer downstream of the convection section is for fired heaters with high tube temperatures in the convection section. This is because it is desirable to monitor radiant section oxygen to avoid afterburning.

5.4.2 Why the Need to Optimize Excess Air

In an ideal combustion of fuel purely based on stoichiometric conversion, fuel is burnt to CO_2 and H_2O 100% with 0% excess air so that there is no oxygen left in the combustion flue gas. However, in reality, industrial fired heaters require excess air. To achieve complete combustion, a minimum of 10–15% excess air (2–3% O_2 in flue gas) is required for fuel gas. Otherwise, carbon monoxide and unburned hydrocarbon could appear in the flue gas leaving stack. Fuel oil usually requires 5–10% higher excess air than fuel gas. In other words, a minimum of 15–25% excess air (3–5% O_2) is required for fuel oil for complete combustion.

Older heaters with poor burner conditions could have O_2% higher than 5%. This is because many older heaters are not designed for low O_2 operations. The burner flame will be very poor. High excess air is required for these operations. For fuel oil used as fuel, black smoke is visible from the stack under incomplete combustion. For fuel gas and natural gas, smoke is not visible from the stack, but incomplete combustion can be measured by CO concentration in the flue gas.

Typically, 1% CO measure in the stack flue gas implies that 3–4% of fuel is wasted. Because O_2% is measured online, efficient and reliable operation of heaters should maintain O_2% as close (but not less than) to the limit as possible. It is important to make sure that O_2% is not a false indication as air coming from leaking could contribute to the O_2% measured. Too little excess air available for combustion could cause flame impingement to tubes and cause local hot spots, and coking on the tube and could eventually cause severe tube damage. Another consequence of too little excess air is afterburning in the convection section, which could result in elevated tube temperature, which is the root cause for premature tube failure and sagging of horizontal tubes. This is because the fired heater undergoes incomplete combustion and thus the combustibles or CO in the flue gas increases. The incomplete combustion makes lazy flames. These long flames can reach tubes in the radiation section and even convection section. In the worst case, flames could reach the exit of the stack.

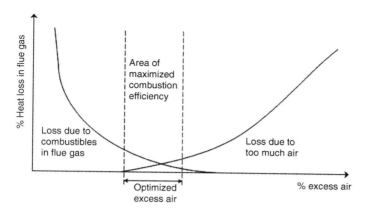

FIGURE 5.10. Optimizing excess air.

Therefore, what is the optimal excess air or $O_2\%$? The basis is to achieve complete combustion. For reliability considerations, optimal $O_2\%$ should be determined with a safety margin on top of the minimum excess air when burners are in good condition. The safety margin depends on specific technology, design, and conditions for each heater as well as measurement. Figure 5.10 is commonly used to explain qualitatively the existence of optimal excess air.

The more rigorous way compared to O_2 measurement is to measure CO in the flue gas. This can be accomplished by measuring combustibles in the flue gas. Combustibles here refer to the products of incomplete combustion including CO, hydrogen, and trace hydrocarbons, while CO accounts for the majority of combustibles. For consistency with O_2 measurement, the combustibles measurement should be taken in the same location as the O_2 analyzer. With reliable combustibles measurement available for ppm concentration, it allows the $O_2\%$ level to be reduced safely (safety margin) until the combustibles start to increase (Figure 5.11). This is the optimal $O_2\%$ for the heater.

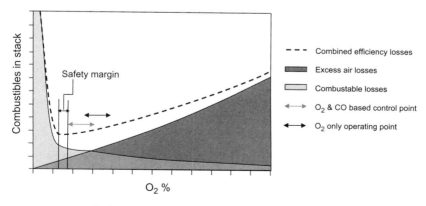

FIGURE 5.11. Determining optimal $O_2\%$ level.

5.4.3 Draft Effects

Efficient heater operation requires that excess air entering the convection section be minimized, which is indicated by a very small negative pressure at the convection section inlet. To achieve this, it should have a well balanced draft pressure profile between the firebox and stack. The hot gas pushes so that the pressure is always greatest at the firewall while the stack draft pulls. When this draft is correctly balanced, the pressure at the bridge wall should be around 0.1–0.2 WG (water gauge). Too much draft allows cold air leakage into the fired box resulting in wasted fuel.

5.4.4 Air Preheat Effects

Air preheating is a classic example of upgrading low-valued heat. This is done by providing heat to raise the combustion air temperature from ambient temperature using waste heat. Air preheat can be accomplished via low-pressure steam or flue gas. Typically, air preheat can increase fired heater efficiency up to 5%, which is more significant than reducing $O_2\%$.

5.4.5 Too Little Excess Air and Reliability

Too little excess air could result in flame impingement and afterburn in the convection section, which imposes reliability risks. With too little excess air, incomplete combustion occurs and reduces flame temperature, which might encourage operators to increase fuel flow to increase heater duty. Increased fuel with too little excess air enhances afterburn and could be dangerous.

5.4.6 Too Much Excess Air

This is inefficient operation and should be avoided. According to Kenney (1984), the common causes of too much excess air are:

- Improper draft control;
- Air leakage into the convection section;
- Improper calibration of O_2 analyzer;
- Faulty burner operation: (i) dirt burners; (ii) poor maintenance on air doors; and (iii) dual fuel burners needed.

5.4.7 Availability and Efficiency

Making fired heater in high availability is desirable for continuous production without interruption. As a consequence, a plant can achieve high profit and high energy efficiency at the same time. Experience from the industry indicates that high availability is the major contributor to improved energy efficiency.

5.4.8 Guidelines for Fired Heater Reliability and Efficient Operation

Draft and excess air control should be considered together in operation. This is because draft provides primary air, while the air register delivers secondary air for burners. As discussed before, both air supplies could affect reliability and efficiency. A systematic method for optimizing draft and excess air together is discussed in Figure 5.12.

I feel the need to provide additional comments on excess air as many plants have an O_2 reduction program. O_2 reduction (or minimum excess air) must be built upon the basis of proper draft control. Minimum excess air for the fired heater can be obtained when it is reduced to the point where combustibles begin to appear in the stack. For modern fired heaters, this occurs at 8% excess air equivalent to 1.8% of oxygen level in the flue gas. However, practical constraints prevent achieving this minimum excess air in operation, and these constraints include variations in fuel quality, feed rates, and other process variables. Thus, operation without flame impingement sets the limit for practical minimum excess air. The optimal flue

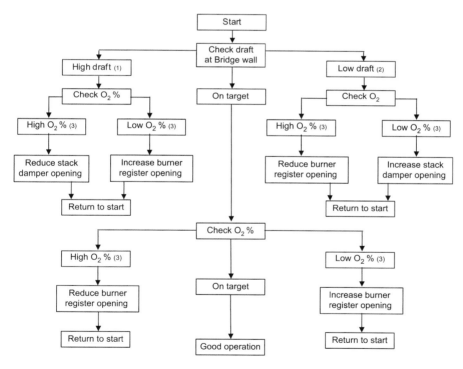

(1) High draft —fire box pressure more negative.

(2) Low draft —fire box pressure more positive.

(3) Low or high O_2% — O_2% is above or below target.

FIGURE 5.12. Integrated draft and O_2 control.

gas O_2 concentration depends on the heater duty, burner design, types of fuel, and burner performance.

To achieve the limit, the first step is monitoring $O_2\%$. O_2 measurement must reflect the true amount of excess air and air leaks must be eliminated. The following guidelines can be used for operation reference:

- O_2 analyzers should be installed below the convection section instead of the stack. If not, a correction factor must be developed for the readings with a portable analyzer. O_2 analyzers should be calibrated once per week.
- Efficiency based on stack temperature and corrected $O_2\%$ should be reported daily.
- Draft should be monitored and maintained as required for the specific fired heater design. Fired heaters without draft control should be periodically checked.
- Convective section air leakage should be measured once per shift and determined as the difference in convective inlet and outlet O_2. The source of leakage should be identified via inspection and eliminated. Ideally, all oxygen should enter the fired heater through burners.
- Coil flow paths should be balanced within ±5% accuracy once per shift to obtain equal outlet temperatures. On large fired heaters, this may be as often as every 2 hours, or continuously with control systems.
- In cases of turnarounds and large load changes, flue gas parameters (draft, $O_2\%$, etc.) should be checked and adjusted as necessary.
- Soot blowers in oil-fired heaters and boilers should be activated once per shift. The operator should observe which ones actually rotate and report those soot blowers that have failed. Where operability of soot blowers is less than 70%, an alternative plan for cleaning should be prepared and executed. This may include on-stream water-washing.
- The need for on-stream cleaning of outside tube surface should be evaluated. This may include water-washing of both convective and radiant sections.

5.5 FIRED HEATER REVAMP

In general, fired heaters are revamped for capacity expansion, process conversion changes, energy efficiency, and NO_x reduction. For capacity expansion revamps, the type of limitation for the revamp is usually the same as for the original design. In conversion revamps, one type of process technology is converted to another. Thus, in conversion revamps, a heater designed for one service may be used in a new service. Therefore, the type of heater limitation may be different for the new service.

Heaters encounter four major design limitations: heat flux, process pressure drop, TWT, and BWT. Heat flux limited heaters are usually characterized with high pressure ΔP (>20 psi) and most general service heaters fall into this flux-limited category. Typically, the flux limit for single-fired heaters is around 10,000 Btu/ft^2 h.

Small heaters (<10 MMBtu/h process duty) have lower flux rates. For revamps, the heat flux limit can go up to 12,000 Btu/ft² h.

Double-fired heaters are usually TWT-limited. However, flux limits are specified for revamps, which depend on specific services. The limits are provided by heater specialists.

Tube wall temperature limited heaters are characterized by low process ΔP (2–6 psi). Because of low ΔP, the heaters have low tube mass velocities, which result in low heat transfer coefficients, and thus high tube wall temperatures. TWT-limited heaters usually occur in high-temperature processes. For example, TWT of 800 °F is used for killed carbon steel heaters. The chrome limits are based on inhibiting tube oxidation and use a limit of 1075–1100 °F per recent data. For stainless steel, process temperature limits usually occur before reaching the TWT limit. There can be an exception for high-pressure heaters.

On the other hand, BWT-limited heaters are usually encountered when a heater with conventional burners is replaced with low NO_x, ultralow NO_x and the latest generation burners, or the revamp requires higher turndown ability. Flame instability and flameout can occur at low BWT. For fuel gas firing with ultralow NO_x and next generation burners, BWT should be greater than 1200 °F. For low NO_x burners, the BWT should be greater than 1000 °F. For oil firing burners, the BWT shall be greater than 1200 °F. When determining the BWT limit, the burner spacing and bridge wall temperatures at normal and turndown operations must be investigated on the stability of burners.

There is much more to discuss about heater revamp, which is beyond the scope of this chapter. As a general recommendation, heater revamp projects should be conducted by heater specialists who not only have good knowledge of the heaters, but also the process that the heaters serve.

NOMENCLATURE

BWT   bridge wall temperature
TMT   tube metal temperature
TWT   tube wall temperature

REFERENCES

Jenkins BG, Boothman M (1996) Combustion science: a contradiction in terms, *Petroleum Technology Quarterly*, pp. 71–76, Autumn.

Kenney WF (1984) *Energy Conservation in the Process Industries*, Academic Press, Inc.

6

HEAT EXCHANGER PERFORMANCE ASSESSMENT

Maintaining good performance of the heat exchanger is a major part of an energy efficiency program. However, how do we identify the root causes of inefficiency and what options do engineers have to resolve in the design and operation? These are the focus of discussions in this chapter.

6.1 INTRODUCTION

Often, exchangers do not perform as they should, and their performance deviates from optimum. Sometimes, they do not accomplish what they are capable of, and other times they are asked to perform what they are not capable of. The primary purpose of heat exchanger assessment is to identify the root causes if they are due to poor design, excessive fouling, or mechanical failure, and determine the required actions to improve the performance.

This chapter will provide the basic understanding of heat exchange assessment supported with examples considering whether the exchanger is designed correctly, evaluation of operating performance, evaluation of fouling, and the effects on heat transfer and pressure drop. On this basis, methods for improving exchanger performance are provided using examples associated with different application scenarios. The methods discussed in this chapter focus on shell-and-tube exchangers as they are the most commonly used in the process industry, although the assessment

Energy and Process Optimization for the Process Industries, First Edition. Frank (Xin X.) Zhu.
© 2014 by the American Institute of Chemical Engineers, Inc. Published 2014 by John Wiley & Sons, Inc.

methodology can be applied to other types of heat exchangers. Detailed assessment of heat exchange performance may be conducted using commercial software.

6.2 BASIC CONCEPTS AND CALCULATIONS

As is well known, the primary equation for heat exchange between two fluids is the Fourier equation expressed as

$$Q = UA\Delta T_M, \tag{6.1}$$

where Q is heat duty, MMBtu/h; A is heat transfer surface area, ft^2; U is overall heat transfer coefficient, Btu/(ft^2 °F h); and ΔT_M is effective mean temperature difference (EMTD), °F.

Let us define the U value first based on Figure 6.1 where h_i and h_o are film coefficients for fluids inside and outside of the tube, and they can be calculated from the physical form of the heat exchanger, physical properties of streams, and process conditions of streams. Thus, a clean overall heat transfer coefficient (U_C) can be determined based on:

$$\frac{1}{U_C} = \frac{1}{h_o} + \frac{1}{h_i}\left(\frac{A_o}{A_i}\right) + r_w, \tag{6.2}$$

where r_w is the conductive resistance of the tube wall. A_o and A_i are outside and inside tube surface area with subscripts "i" and "o" denoting inside and outside of the tube.

In reality, heat exchangers operate under fouled conditions with dirt, scale, and particulates deposited on the inside and outside of tubes. Allowance for the fouling

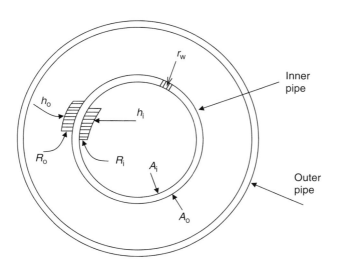

FIGURE 6.1. Location of h's and R's.

must be given in calculating the overall heat transfer coefficient. The graphical description of fouling resistances (R_o, R_i) and film coefficients (h_o, h_i) inside and outside of the tube is provided in Figure 6.1. Conceptually, R_i and h_i are equivalent to R_t and h_t (t for tube side) while R_o and h_o are for R_s and h_s (s for shell side).

The overall fouling resistance is then defined as

$$R_f = R_o + R_i \left(\frac{A_o}{A_i}\right). \tag{6.3}$$

By adding the overall fouling resistance to U_C, actual U_A is defined as

$$\frac{1}{U_A} = \frac{1}{U_C} + R_f. \tag{6.4}$$

More detailed discussions for U values are provided later in this chapter. Now let us turn our attention to ΔT_M or EMTD in equation 6.1. Several temperature differences can be used to calculate ΔT_M including inlet temperature difference, arithmetic temperature difference, and logarithmic mean temperature difference. Figure 6.2 is used for illustration.

Inlet temperature difference can be expressed as

$$\Delta T_1 = T_1 - t_2 \quad \text{(for countercurrent)}, \tag{6.5a}$$

$$\Delta T_1 = T_1 - t_1 \quad \text{(for cocurrent)}. \tag{6.5b}$$

This temperature difference could lead to a gross error in estimating true temperature difference over the entire pipe length.

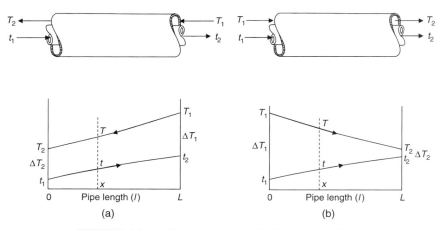

FIGURE 6.2. (a) Countercurrent and (b) cocurrent flows.

Arithmetic mean temperature difference is defined as

$$\Delta T_A = \frac{\Delta T_1 + \Delta T_2}{2} = \frac{(T_1 - t_2) + (T_2 - t_1)}{2} \quad \text{(for countercurrent)}, \qquad (6.6a)$$

$$\Delta T_A = \frac{\Delta T_1 + \Delta T_2}{2} = \frac{(T_1 - t_1) + (T_2 - t_2)}{2} \quad \text{(for cocurrent)}. \qquad (6.6b)$$

This temperature difference could give an erroneous estimate of true temperature difference when ΔT_1 (hot end approach) and ΔT_2 (cold end approach) differ significantly.

The logarithmic temperature difference (LMTD) is defined as

$$\Delta T_{LM} = \frac{\Delta T_1 - \Delta T_2}{\ln(\Delta T_1 / \Delta T_2)}. \qquad (6.7)$$

ΔT_{LM} represents a true temperature difference for a perfect countercurrent as well as cocurrent heat exchange. For illustration purposes, the countercurrent flow (Figure 6.2a) is used to show how ΔT_{LMTD} is derived.

First, applying the differential form of the Fourier equation to Figure 6.2a,

$$dQ = U(T - t)A'' dL, \qquad (6.8)$$

where A'' is the square feet of surface area per foot of pipe length (ft^2/ft length) and thus the differential surface area $dA = A'' dL$.

The heat balance for the differential surface area is

$$dQ = MC dT = mc dt, \qquad (6.9)$$

where M and C are the mass flow and specific heat capacity for the hot stream, while m and c are the mass flow and specific heat capacity for the cold stream.

The heat balance between two streams in the pipe length from $p = 0$ to $p = x$ is

$$MC(T - T_2) = mc(t - t_1). \qquad (6.10a)$$

The heat balance between two streams in the pipe length from $l = 0$ to $l = L$ is

$$Q = MC(T_1 - T_2) = mc(t_2 - t_1). \qquad (6.10b)$$

Solving equation (6.10a) for T,

$$T = T_2 + \frac{mc}{MC}(t - t_1). \qquad (6.11)$$

Let equations (6.8) and (6.9) be equal to each other and then substituting T with equation (6.11),

$$dQ = mc \, dt = U\left[\left(T_2 + \frac{mc}{MC}(t - t_1)\right) - t\right] dA. \qquad (6.12)$$

Rearranging equation (6.12) and colleting t and A gives

$$\int_0^L \frac{U dA}{mc} = \int_{t_1}^{t_2} \frac{dt}{T_2 - (mc/MC)t_1 + [(mc/MC) - 1]t}. \qquad (6.13)$$

Assume (U, M, m, C, c) as constants. Integrating both sides of equation (6.13) yields

$$\frac{UA}{mc} = \frac{1}{(mc/MC - 1)} \ln \frac{T_1 - t_2}{T_2 - t_1}. \qquad (6.14)$$

Let $\Delta T_1 = T_1 - t_2$ and $\Delta T_2 = T_2 - t_1$, combining equations (6.10b) and (6.14) gives

$$Q = UA \left[\frac{\Delta T_1 - \Delta T_2}{\ln(\Delta T_1/\Delta T_2)} \right]. \qquad (6.15)$$

The term in the bracket of equation (6.15) is named as ΔT_{LM} or LMTD, which is the same as that expressed in equation (6.7).

By comparing equations (6.1) and (6.15), we can obtain EMTD as

$$EMTD = LMTD. \qquad (6.16)$$

It must be emphasized that $EMTD = LMTD$ is only true for a perfect countercurrent (Figure 6.2a) or perfect cocurrent (Figure 6.2b) heat exchange.

At this point, the first question is: Why is the countercurrent pattern widely adopted in shell-and-tube exchangers? The answer is that the LMTD for countercurrent is always greater than the cocurrent LMTD. An example corresponding to Figure 6.2 is shown here.

Countercurrent			Cocurrent		
Hot fluid	Cold fluid		Hot fluid	Cold fluid	
$T_1 = 350°$	$t_2 = 230°$	$\Delta T_1 = 120°$	$T_1 = 350°$	$t_1 = 150°$	$\Delta T_1 = 200°$
$T_2 = 250°$	$t_1 = 150°$	$\Delta T_2 = 100°$	$T_2 = 250°$	$t_2 = 230°$	$\Delta T_2 = 20°$
		LMTD $= 109.7°$			LMTD $= 78.2°$

The second question is: What should be done if a heat exchange is not a perfect countercurrent? In fact, the flow pattern in most shell and tube exchangers is a mixture of cocurrent, countercurrent, and cross-flow. In these cases, $EMTD \leq$ LMTD. Thus, an LMTD correction factor F_t must be introduced:

$$EMTD = F_t \times LMTD, \qquad (6.17)$$

$F_t = 1$ for a true countercurrent heat exchange; otherwise, $F_t < 1$.

As a short summary, EMTD is obtained by calculating LMTD based on equation (6.7) first and then applying F_t to account for nonperfect countercurrent flow.

F_t can be obtained via equations or charts (Shah and Sekulić, 2003). For example, F_t for 1-2 heat exchangers can be numerically calculated by

$$F_t = \frac{\sqrt{R^2 + 1}\ln[(1 - P)/(1 - R \times P)]}{(R - 1)\ln\left\{\left[2 - P(R + 1 - \sqrt{R^2 + 1})\right]\Big/\left[2 - P(R + 1 + \sqrt{R^2 + 1})\right]\right\}},$$

(6.18)

where P is temperature efficiency and R is the ratio of heat flow, which are defined as

$$P = \frac{t_2 - t_1}{T_1 - t_1},$$

(6.19)

$$R = \frac{T_1 - T_2}{t_2 - t_1} = \frac{m \times c}{M \times C}.$$

(6.20)

F_t for 1-2 type (one shell pass and two tube passes) exchangers can also be found in the chart as shown in Figure 6.3. It can be observed from the figure that F_t values drop off rapidly below 0.8. Consequently, if a design indicates an F_t less than 0.8, it probably needs to be redesigned to get a better approximation of countercurrent flow and thus higher F_t value. Different F_t charts are available for each exchanger layout (TEMA 1-2, 1-4, etc.).

However, several assumptions are made in deriving equation (6.7), which are

(1) Constant overall heat transfer coefficient U.

(2) Constant specific heat capacity for both hot and cold streams. This assumption infers linear temperature change for both streams.

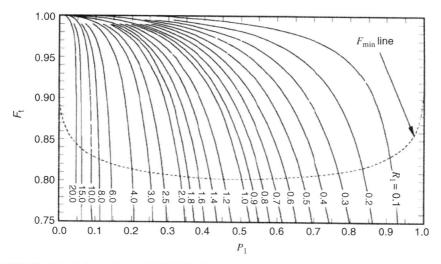

FIGURE 6.3. F_t factor for 1-2 TEMA E shell-and-tube exchangers. (From Shah and Sekulić (2003), reprinted with permission by John Wiley & Sons.)

(3) No partial phase change from either hot or cold stream. In other words, equation (6.7) is applicable to sensible heat exchange as well as vaporization or condensation.

The above assumptions may not be true for the whole length of a shell-and-tube heat exchanger, but these assumptions are true within each incremental section. Thus, in modern software for heat exchanger design, the whole length of the heat exchanger path is divided into N number of increments and the assumptions and calculations are performed for each increment. According to Bennett et al. (2007), modern heat exchanger design software calculate overall U values and EMTD based on increments via the following equations:

$$\frac{1}{U_{C,j}} = \frac{1}{h_{o,j}} + \frac{1}{h_{i,j}}\frac{A_{o,j}}{A_{i,j}} + r_{w,j}, \tag{6.21}$$

$$\frac{1}{U_{A,j}} = \frac{1}{U_{C,j}} + R_{f,j}, \tag{6.22}$$

$$A_{total} = \sum_{j=1}^{N} A_{o,j}, \tag{6.23}$$

$$Q_{total} = \sum_{j=1}^{N} Q_j, \tag{6.24}$$

$$LMTD_j = \frac{\Delta T_{1j} - \Delta T_{2j}}{\ln(\Delta T_{1j}/\Delta T_{2j})}, \tag{6.25}$$

$$\frac{1}{EMTD} = \frac{1}{Q_{total}} \sum_{j=1}^{N} \frac{Q_j}{LMTD_j}, \tag{6.26}$$

$$U_C = \frac{1}{A_{o,total}} \sum_{j=1}^{N} U_{C,j} A_{o,j}, \tag{6.27}$$

$$U_A = \frac{1}{A_{o,total}} \sum_{j=1}^{N} U_{A,j} A_{o,j}, \tag{6.28}$$

$$U_R = \frac{Q_{total}}{A_{total} \Delta T_M}. \tag{6.1}$$

U_R is called as required U value. The U values should follow the order: $U_C \geq U_A \geq U_R$. The main reasons for the inequality are the practical considerations of fouling, process variations, as well as inaccuracy in physical properties estimates and heat transfer calculations. The detailed discussions are provided here.

6.3   UNDERSTAND PERFORMANCE CRITERION—*U* VALUES

The critical question for operation assessment is: What is the "performance" indicator for heat exchanger. Because a heat exchanger is used to transfer heat, someone may naturally consider heat exchanger duty as the performance indicator. For verification, let us look at an example of a reaction effluent cooler. The operating data were obtained, which is shown against the design data (Table 6.1). Operating temperatures can be measured from instrumentation. Then, LMTD is calculated by equation (6.7) and F_t factor is obtained from F_t charts. UA is calculated via equation (6.1) and U is derived for a given surface area A of the cooler. The calculation results for operating data are shown in Table 6.2. The design data were obtained from the exchanger datasheet.

As can be observed in Table 6.2, the heat exchanger duty in operation is a little higher and the temperature changes are similar for both operation and design. What do you think of the performance of this exchanger in operation? If based on the exchanger

TABLE 6.1. Gathered Data for a Reaction Air Cooler

	Design	Operation
Q; MMBtu/h	16.3	18.0
M (effluent); lb/h	154,447	162,919
T_1; °F	296.6	341.6
T_2; °F	105.8	149.0
t_1; °F	89.6	81.0
t_2; °F	104.0	95.0
$\Delta T\ (T_1 - T_2)$; °F	190.8	192.6
$\Delta t\ (t_2 - t_1)$; °F	14.4	14.0

TABLE 6.2. Calculation Results for a Reaction Air Cooler

	Design	Operation
Q; MMBtu/h	16.3	18.0
M (effluent); lb/h	154,447	162,919
T_1; °F	296.6	341.6
T_2; °F	105.8	149.0
t_1; °F	89.6	81.0
t_2; °F	104.0	95.0
ΔT_1; °F	192.6	246.6
ΔT_2; °F	16.2	68.0
ΔT_{lm}; °F	71.3	138.7
F_t	0.85	0.97
UA; Btu/°F-h	269,896	133,759
A; ft^2	5,167	5,167
U; Btu/ft^2 °F-h	52	26
$U_{operation}/U_{design}$		0.50

(diagram at right of Table 6.2)
- T_1 (effluent in)
- T_2 (effluent out)
- Q
- ΔT_1 (hot end approach)
- ΔT_2 (cold end approach)
- t_2 (Air out)
- t_1 (Air in)

duty, we could conclude that it is performing better than design performance. However, when comparing overall heat transfer coefficient, surprisingly, the operation U value is only half of the design U value, although the heat duty in operation is 10% higher. If the operation U value could be maintained similar to the design U value, the heat duty could be increased much higher than 10%!

This example concludes that U value is a true performance indicator for heat exchanger under any process condition. *The higher the U value, the better perform-ance that a heat exchanger achieves.*

Clearly, a good understanding of U value is of paramount importance for appropriate assessment of heat exchanger performance as it is the most important characteristic of the heat exchanger representing its heat transfer capability. In view of the fact that many engineers are confused about the terminologies related to U, it is essential to get the basic understanding right before delving into the details of assessment methods.

6.3.1   Required U Value

The need for a heat exchanger is to satisfy process requirement in terms of heat duty (Q) and temperatures (LMTD or ΔT_{LM}). Thus, a heat exchanger is designed to have a certain surface area (A) to fulfill the process requirement. Based on the process temperature requirement, a certain amount of heat duty must be transferred. Under the basis of heat transfer duty and process temperatures, the required U value can be calculated from the Fourier equation (6.1):

$$U_{\mathrm{R}} = \frac{Q}{AF_{\mathrm{t}}\Delta T_{\mathrm{LM}}}. \qquad (6.1)$$

6.3.2   Clean U Value

Independent of the required U based on thermodynamics stated as in equation (6.1), U value can be calculated based on transport considerations without taking into account the fouling resistances. In other words, transport-based U is a function of film coefficients (h_{t} for tube side and h_{s} for shell side in Btu/h ft^2 °F) as expressed in equation (6.2):

$$\frac{1}{U_{\mathrm{C}}} = \frac{1}{h_{\mathrm{o}}} + \frac{1}{h_{\mathrm{i}}}\left(\frac{A_{\mathrm{o}}}{A_{\mathrm{i}}}\right) + r_{\mathrm{w}}. \qquad (6.2)$$

This U value is called as clean U value because fouling resistances ($R_{\mathrm{i}}, R_{\mathrm{o}}$) are not taken into account in equation (6.2). The film coefficients, h_{t} and h_{s}, can be calculated based on the fluids' physical properties and the geometry of the heat exchanger. For example, for U-tube exchangers with streams all liquid or all vapor (no boiling and condensing), the correlation (Dittus and Boelter, 1930) is used to estimate the tube side Nusselt number, Nu_t, and then tube side film coefficient, h_t:

$$Nu_t = 0.027 \left(\frac{C_p\mu}{k}\right)^{1/3} \left(\frac{\rho u d_i}{\mu}\right)^{0.8},$$

(6.29)

$$h_t = \frac{k}{d_i} Nu_t = 0.027 \frac{(C_p k^2)^{1/3}(\rho u)^{0.8}}{d_i^{0.2}\mu^{7/15}},$$

(6.30)

where C_p is fluid heat capacity, Btu/(lb °F); d_i is inner diameter of the tube, ft; k is fluid thermal conductivity, Btu/(h ft °F); u is fluid velocity, ft/h; ρ is fluid density, lb/ft^3; and μ is fluid viscosity, lb/(ft h).

Equation (6.30) states that physical properties of tube side stream (namely, conductivity k, specific heat capacity C_p), and mass velocity u have a positive effect on tube side film coefficient h_t. In contrast, viscosity μ and tube inside diameter d_i have a negative effect.

The Kern's correlation (Kern, 1950) is used to estimate the shell side Nusselt number, Nu_S, and then shell side film coefficient, h_S:

$$Nu_S = 0.36 \left(\frac{C_p\mu}{k}\right)^{1/3} \left(\frac{D_e\rho u}{\mu}\right)^{0.55} \left(\frac{\mu}{\mu_w}\right)^{0.14},$$

(6.31)

$$h_S = \frac{k}{D_e} Nu_S = 0.36 \frac{(C_p k^2)^{1/3}(\rho u)^{0.55}}{D_e^{0.45}\mu^{0.08}\mu_w^{0.14}},$$

(6.32)

where

$$D_e = \frac{4\left(p^2 - \frac{\pi d_o^2}{4}\right)}{\pi d_o} \text{ (for square tube pitch)},$$

(6.33)

where μ_w is water viscosity, lb/(ft h); D_e is shell side equivalent diameter, ft; d_o is outer diameter of the tube, ft; and p tube pitch, ft.

Equation (6.32) states that physical properties of shell side stream (namely, conductivity k, specific heat capacity C_p), and velocity u and tube outside diameter d_o have a positive effect on shell side film coefficient h_s. In contrast, viscosity μ and tube pitch p have a negative effect.

The above heat transfer equations provide the well-known observations: heat transfer coefficient on tube side is proportional to the 0.8 power of velocity, 0.67 power of thermo-conductivity, and −0.47 power of viscosity,

$$h_t \propto u^{0.8}$$

(6.34a)

$$h_t \propto k^{0.67}$$

(6.34b)

$$h_t \propto \mu^{-0.47}.$$

(6.34c)

That is the reason why cooling water has a very high heat transfer coefficient, followed by hydrocarbon, and then hydrocarbon gases because of the values of thermo-conductivities for these fluids. Hydrogen is an unusual gas due to its extremely high thermo-conductivity (greater than that of hydrocarbon liquids). Thus, its heat transfer coefficient is toward the upper limit of the range for the hydrocarbon liquid. The heat transfer coefficients for hydrocarbon liquids vary in a large range due to the large variations in viscosity, from less than 1 cP for ethylene to more than 1000 cP for bitumen. Heat transfer coefficients for hydrocarbon gases are proportional to pressure because higher pressure generates higher gas density resulting in higher gas velocity.

6.3.3   Actual U Value

In reality, heat exchangers operate under fouled conditions with dirt, scale, and particulates deposit on the inside and outside of tubes. The overall fouling resistance is defined in equation (6.3) as

$$R_f = R_o + R_i \left(\frac{A_o}{A_i} \right). \tag{6.3}$$

By adding the overall fouling resistance to U_C, actual U_A is defined in equation (6.4) as

$$\frac{1}{U_A} = \frac{1}{U_C} + R_f. \tag{6.4}$$

Clearly, U_C is the heat transfer capability that the exchanger can deliver when no fouling is included, while U_A takes into account the fouling resistances. U_A can be thought of as predicted or expected overall coefficient for actual heat transfer including the design fouling resistances.

   Fouling resistances for streams are based on the physical properties of the streams and the average fouling factors are documented in TEMA (2007). For illustration purposes, Table 6.3 shows the typical overall fouling resistances for hydrocarbon liquids based on the API gravity of the streams.

6.3.4   Overdesign (OD_A)

For a heat exchanger to satisfy process requirement under changing process conditions, U_A must be greater than U_R. Overdesign or design margin can be defined as

TABLE 6.3. Liquid Fouling Factors

API Gravity	R_f, ft^2 h °F/Btu
>40	0.002
<40	0.003
<15	0.004
<5	0.005

$$\%\text{OD}_A = \left(\frac{U_A}{U_R} - 1\right) \times 100. \qquad (6.35)$$

Overdesign is provided in the design stage beyond fouling factors to account for operation variations in fluid rates and properties as well as calculation inaccuracy for heat transfer and pressure drops. Some designers may use 5–10% overdesign for new heat exchangers if the designers have confidence in fluid properties and heat transfer calculation accuracy. Otherwise, 10–20% or higher overdesign might be used. In contrast, near-zero overdesign could be used for services with well-known fluid properties and accurate heat transfer calculations.

Statistically, heat exchangers are often designed with large overdesign intentionally because the designer wants to make sure they will satisfy process demand whatever occurs in operation. There are several uncertain factors that the designer has to consider in the design stage (Bennett et al., 2007).

First, the uncertainty in the accuracy of estimating fouling resistances to reflect the actual fouling. Furthermore, fouling resistances are static values, which are used in computation. In reality, fouling is a dynamic mechanism. The designer uses overdesign to account for this fouling dynamics based on his/her experience or company's best practices so that the exchanger can still satisfy the process demand under more severe fouling scenarios than estimated fouling resistances. The second factor is variations in process conditions. In particular, increasing feed rate is common as companies want to generate additional revenue using existing equipment. The designer provides overdesign to accommodate operating scenarios with increased feed rate. Third, the designer uses overdesign to account for the effects of inaccuracy in fluid properties and heat transfer calculations. These uncertainties become the basis for the designer to provide overdesign.

However, excessive overdesign can cause fouling and other problems with the exchanger. When too much overdesign in surface area is added, velocity reduces, which makes it easier for fouling deposits to accumulate. In some cases, a temperature-controlled bypass line may be required for critical services to avoid too much heat transfer than process requirements in the start of the run. Bypass operation could enhance fouling as fluid velocity reduces.

6.3.5 Controlling Resistance

If the total resistance (film resistance plus fouling resistance) of one side is much larger than the other, this side is referred to as the controlling side of resistance. For example in a heat exchanger with gas and liquid on either side, gas side is the controlling side as gas has a very low film coefficient. In design and operation, special attention is devoted to this controlling resistance as any incremental decrease to this controlling resistance will greatly increase the overall *U* value. On the other hand, incremental change to the noncontrolling side film coefficient has very little effect on the overall *U* value.

One way to minimize the adverse effect of controlling resistance is to use extended surface area to offset the effect. Another way is to increase the velocity on the controlling side. Furthermore, the most heavily fouling stream should be placed on the tube side for ease of cleaning. Use of fouling mitigation methods such as fluid treatment, antifouling additive, and regular cleaning to prolong the "clean" operation can help maintain high U value, which will be discussed in Chapter 7.

6.4   UNDERSTANDING PRESSURE DROP

In technical discussions on heat exchangers, pressure drop will naturally become an important topic. Process engineers usually prefer to keep pressure drop as low as possible to maintain sufficient suction pressure downstream of the heat exchanger and reduce pump power consumption and avoid process issues. For example, high pressure drops could cause feed flashing before fired heaters downstream. In contrast, reliability and design engineers would like to keep pressure drop as high as possible to reduce fouling and improve film coefficients. This helps to avoid operation issues and minimize overdesign.

Heat exchanger pressure drop is mainly a function of velocity, that is, tube velocity for tube side pressure drop and bundle velocity for shell side pressure drop.

6.4.1   Tube Side Pressure Drop

Pressure drop for tube side can be expressed as

$$\Delta P_t = \frac{1}{2}\rho(u_t)^2 \frac{4L}{d_i}f_t, \quad [f_t = f(\text{Re})], \tag{6.36}$$

where u_t is tube velocity, ft/h and f_t is tube side friction factor, (ft^2 °F h)/Btu.

From equation (6.36), we can observe that the major parameters affecting the tube side pressure drop include tube diameter and length, fluid density, viscosity, and velocity:

$$\Delta P_t \propto u^2 \tag{6.37a}$$

$$\Delta P_t \propto \rho \tag{6.37b}$$

$$\Delta P_t \propto L \tag{6.37c}$$

$$\Delta P_t \propto f_t \tag{6.37d}$$

$$\Delta P_t \propto d_i^{-1}. \tag{6.37e}$$

6.4.2   Shell Side Pressure Drop

The shell side flow path is more complex than that for tube; hence, the calculation of shell side pressure drop is more difficult. More accurate calculation of shell side pressure drop could be obtained by the Bell-Delaware method (Bell, 1973). For the purpose of providing an explanation of shell pressure drop conceptually, Kern's correlation (Kern, 1950) is used here. Based on bundle velocity, Kern's correlation for shell side pressure drop (equation (6.38)) mirrors equation (6.36) for tube side pressure drop:

$$\Delta P_s = \frac{1}{2} u_s^2 \frac{4 D_s (N_B + 1)}{\rho D_e} f_s, \qquad [f_s = f(\mathrm{Re})], \qquad (6.38)$$

where u_s is shell side cross-flow velocity, ft/h; D_s is shell diameter, ft; D_e is equivalent shell diameter, ft; N_B is number of baffles; and f_s is shell side friction factor, (ft^2 °F h)/Btu. f_s is a function of Reynolds number and f_s charts are available in Hewitt et al. (1994).

To transform the friction factor to a shell side pressure drop, the number of the fluid crossing the tube bundle should be given. Because the fluid crosses between baffles, the number of "crosses" will be one more than the number of baffles, N_B. If the number of baffles is unknown, it can be determined using the baffle spacing P_B and tube length L:

$$N_B + 1 = \frac{L}{P_B}. \qquad (6.39)$$

Equation (6.38) is then reduced to

$$\Delta P_s = \frac{1}{2} u^2 \frac{4 D_s L}{\rho D_e P_B} f_s. \qquad (6.40)$$

Clearly, equation (6.40) indicates major parameters affecting shell side pressure drop, which include baffle spacing, tube length, shell diameter, velocity, and viscosity. Some of the important observations are

$$\Delta P_s \propto u^2 \qquad (6.41a)$$

$$\Delta P_s \propto L \qquad (6.41b)$$

$$\Delta P_s \propto N_B \qquad (6.41c)$$

$$\Delta P_s \propto D_S^{-1}. \qquad (6.41d)$$

6.4.3 Effects of Velocity on Heat Transfer, Pressure Drop, and Fouling

Examination of equations (6.30) and (6.32) for heat transfer, and equations (6.36) and (6.40) for pressure drop, indicates that for a given heat exchanger and fluids, the fluid velocity is the most important parameter affecting heat transfer and pressure drop on both tube and shell sides. Thus, with increasing velocity, both pressure drop and heat transfer coefficient increase. The rate of pressure drop increase is faster than that of heat transfer coefficient. Since pressure drop is supplied by pumping (for liquid) or compression (for gas), higher pressure drop is at the expense of extra power cost, while increased heat coefficient results in smaller surface area.

Learning from the above equations can lead to the conclusion that a short (long) and wide (narrow) heat exchanger could have a low (high) pressure drop but a low (high) heat transfer coefficient for both tube and shell sides. Clearly, higher pressure drop (ΔP value) forces the fluids to flow faster through the heat exchanger leading to higher overall heat transfer coefficient (U value). However, this is at the cost of high pump power. On the other hand, for a large surface area, the U and ΔP do not need to be so high, but this is at expense of a larger heat exchanger. Therefore, there is an optimal velocity for each side in a heat exchanger, which can be obtained from the trade-off between the capital cost of a heat exchanger in terms of size and the operating cost in terms of power.

One common case is that actual pressure drop could be less than allowable pressure drop on either tube or shell side. This opportunity may be used to enhance the U value via increasing the fluid velocity. Velocity increase can be achieved by increasing flow passes on either tube side or shell side depending on which side is the controlling side on U value. Due to the fact that tube side pressure drop rises steeply with increase in tube passes, it often happens that pressure drop is much lower than the allowable value for a given number of tubes and two tube passes, but it exceeds the allowable value with four passes. In this case, the tube diameter and length could be varied to increase pressure drop with the result of a higher tube side velocity obtained.

Another common scenario is that hydraulics can impose constraints when a heat recovery opportunity is implemented. In this case, the fluid velocity could be reduced via parallel arrangement of new and existing heat exchangers by splitting a total flow into two flows. Assuming that the flow split is equal, the fluid velocity for each branch flow is reduced by half while pressure drops on both sides are reduced by four times.

Fouling has to be addressed in heat exchanger design and operation. When the heat exchanger is fouling, the fouling deposits build up additional resistance to heat transfer. At the same time, fouling deposits reduce cross-sectional flow area and increase pressure drop. Plugging could also reduce cross-sectional flow area and it could be treated the same as fouling in its effect on pressure drop. Fouling in liquids reduces heat transfer coefficient more rapidly than increase in pumping power. In contrast, fouling in gases reduces heat transfer in the range of 5–10%, but it increases pressure drop and fluid pumping power more steeply.

Increasing fluid velocity also reduces fouling tendency. Bennett et al. (2007) provided design guidelines for heavy fouling services with fluid velocity for shell-and-tube exchangers: tube side velocity ≥ 2 m/s (6.5 ft/s) and shell side B-stream (the main cross-flow stream through the bundle) ≥ 0.6 m/s (2 ft/s).

6.5   HEAT EXCHANGER RATING ASSESSMENT

When an evaluation is performed to assess the suitability of an existing heat exchanger for given process conditions or for new conditions, this exercise is called heat exchanger rating. Applications of rating can be for operational performance, for changes in process conditions, or in process design. There are three fundamental points in determining if a heat exchanger performs well for given operating conditions or for a new service:

(1) What actual coefficient U_A value can be "performed" by the two fluids as the result of their flow rates, individual film coefficients h_t and h_s, and fouling resistance?

(2) From the heat balance: $Q = M \cdot C \cdot (T_1 - T_2) = m \cdot c \cdot (t_2 - t_1)$, for given area A, and actual temperatures, required U value (U_R) can be calculated based on Fourier's equation (6.1).

(3) The operating pressure drops for the two streams passing through the existing heat exchanger.

The criteria can be established for the suitability of an existing exchanger for given or new services as two necessary and sufficient conditions:

(a) U_A must exceed U_R to give desired overdesign (%OD) so that the heat exchanger can meet changing process conditions for a reasonable period of service continuously.

(b) Operating pressure drops on both sides must be less than allowable pressure drops.

When these two conditions are fulfilled, an existing exchanger is suitable for the process conditions for which it was rated. When the process conditions undergo significant changes, a rating should be performed to make sure the exchanger can perform the task satisfactorily under the new conditions.

6.5.1 Assess the Suitability of an Existing Exchanger for Changing Conditions

When it is considered to use an existing exchanger for changing conditions or new services, rating assessment must be conducted well in advance for the suitability of existing exchangers for such services.

Example 6.1   Rating of an Existing Naphtha–Heavier Naphtha Exchanger to Operate Under Small Changes in Flowrates

124,600 lb/h (versus 122,500 in design) of a 56.3°API heavy naphtha leaves the naphtha splitter tower at 276 °F and is cooled to 174 °F by 193,000 lb/h (versus 188,000 in design) of 69°API naphtha feed at 116 °F and heated to 170 °F. There is 6.3% vapor in the naphtha at 170 °F. Pressure drops of 10 and 5 psi are permissible on tube and shell sides, respectively. Can this exchanger operate satisfactorily under new conditions?

The exchanger is a TEMA type AES (see appendix for TEMA Types) with a 21 in. shell inside diameter (ID) having 268 tubes with 3/4 in. tube outside diameter (OD), 14 BWG thickness and 20 ft long, which are laid out on a 1 in. triangle pitch. There are four tube passes and one shell pass with baffles spaced $11\frac{1}{4}$ in. apart and baffle cut 32% of shell diameter. The hot heavy naphtha is on the tube side.

Solution:

(1) *Heat Balance*:

For naphtha feed, $Q = m \times c \times \Delta t + q \times m \times \text{vapor}\%$
$$= 193{,}000 \times [0.53 \times (170 - 116) + 141.6 \times 6.3\%] =$$
7.23 MMBtu/h

For heavy naphtha, $Q = M \times C \times \Delta T = 124{,}600 \times 0.57 \times (276 - 174)/10^6$
$$= 7.23\,\text{MMBtu/h}$$

(2) ΔT_{lm} *and* F_t:

Tube Side		Shell Side	Differences	
Hot Stream		Cold Stream		
276	Higher temperature	170	106	ΔT_1
174	Lower temperature	116	58	ΔT_2
102	Differences	54	48	$\Delta T_1 - \Delta T_2$
ΔT		Δt		

$\Delta T_{lm} = (\Delta T_1 - \Delta T_2)/\ln(\Delta T_1/\Delta T_2) = 48/\ln(106/58) = 79.6$
$R = \Delta T/\Delta t = 102/54 = 1.89$
$P = \Delta t/(T_1 - t_1) = 54/(276 - 116) = 0.34$
$F_t = 0.83$
$F_t \Delta T_{lm} = 66.1°F$

(3) *Rating Summary:*

U_C	166.1
U_A	124.7
U_R	115.1
Overdesign	8%
R_f calculated	0.003
R_f required	0.002
ΔP_s calculated	3.8
ΔP_s allowable	5
ΔP_t calculated	9.9
ΔP_t allowable	10

The allowable fouling factor of 0.002 is assumed based on Table 6.3. U_A is calculated by taking fouling into account. The heat exchanger has 8% of overdesign over normal fouling conditions. Pressure drops on both sides of the exchanger are less than allowable pressure drops. Thus, this exchanger meets the two criteria. Therefore, it can operate satisfactorily to fulfill the new flow conditions.

From time to time, a process plant wishes to increase feed rate and/or make different product yields due to economic drivers. In feasibility evaluation, it is essential to assess the suitability of existing heat exchangers for new process conditions and find the most economical ways to handle significant changes.

Example 6.2   Rating of an Existing Naphtha–Diesel Exchanger to Handle Large Increase in Flowrate

A refinery plant plans to increase diesel production by 20% via revamping the hydrocracking unit. This is because diesel is a highly desirable commodity in today's energy market. Currently, the naphtha–diesel exchanger is located downstream of the naphtha–heavy naphtha exchanger, which is discussed in Example 6.1. Via the naphtha–diesel exchanger under the scenario of increased diesel production, 193,000 lb/h naphtha feed to the tower will increase vaporization up to 29.8% from 6.3% by 121,500 lb/h diesel product at 351 °F cooled to 260 °F. A 5 psi pressure drop is permissible on both sides. Can the current exchanger operate satisfactorily under new conditions?

The exchanger is a TEMA type AES with 16 in. shell ID having 130 tubes with 3/4 in. tube OD, 14 BWG thickness and 16 ft long, which are laid out on a 1 in. triangle pitch. There are two tube passes and one shell pass with single segmental baffles spaced 14 in. apart and baffle cut 40% of shell diameter. The hot diesel is on the tube side.

Solution:

(1) *Heat Duty*:

Naphtha feed, $\quad Q = m \times q \times \Delta\text{vapor}\% = 193,000 \times 141.7 \times (29.8\% - 6.3\%)$
$\qquad\qquad\qquad = 6.4\,\text{MMBtu/h}$

Diesel product, $\quad Q = M \times C \times \Delta T = 121,500 \times 0.587 \times (351 - 260)/10^6$
$\qquad\qquad\qquad = 6.4\,\text{MMBtu/h}$

(2) ΔT_{lm} *and* F_{t}:

Tube Side		Shell Side	Differences	
Hot Stream		Cold Stream		
351	Higher temperature	171	180	ΔT_1
260	Lower temperature	170	90	ΔT_2
91	Differences	1	90	$\Delta T_1 - \Delta T_2$
ΔT		Δt		

$$\Delta T_{lm} = (\Delta T_1 - \Delta T_2)/\ln(\Delta T_1/\Delta T_2) = 129.8$$
$$R = \Delta T/\Delta t = 91$$
$$P = \Delta t/(T_1 - t_1) = 1/(351 - 170) = 0.01$$
$$F_t = 0.99$$
$$F_t \Delta T_{lm} = 128.5°F$$

(3) *Rating Summary*:

U_C	135.4
U_A	106.6
U_R	123.5
Overdesign	-14%
ΔP_s calculated	10.7
ΔP_s allowable	5
ΔP_t calculated	5.6
ΔP_t allowable	5

The above rating calculations show that the existing exchanger alone cannot handle 20% increase in diesel flowrate. This is because the surface area is not sufficient, and the shell side pressure drop in particular is too large to be allowed. In addition, $U_A < U_R$. Thus, it violates the criteria for the suitability of an existing exchanger to fulfill changing process conditions. The following discussions will show how to assess practical solutions by use of spare heat exchangers.

6.5.2   Determine Arrangement of Heat Exchangers in Series or Parallel

In some plants where a large number of exchangers are used, certain size standards are usually established in-house for 1-2 type of exchangers so that future services can be satisfied by arranging standard exchangers in series or in parallel. Use of standard exchangers could come at a price because of the impossibility of utilizing the standard equipment in the most efficient manner. However, it does offer a great advantage of reducing spare parts, tubes, and tools for replacement. When tube bundles are retubed, the standard exchangers can provide services as new ones to meet process conditions.

There are two basic arrangements of exchangers, namely, series and parallel arrangements. When use of a single 1-2 exchanger cannot satisfy new process conditions or lead to a severe temperature cross signaled by a low F_t factor, it may be necessary to use two 1-2 exchangers in series. On other hand, when hydraulic limitation could be an issue for a 1-2 exchanger, placing multiple 1-2 exchangers in parallel could resolve the issue.

Example 6.3   (Continuing from Example 6.2)

The rating assessment for the existing naphtha–diesel exchanger in Example 6.2 showed that the single 1-2 exchanger is not sufficient to meet 20% increase in

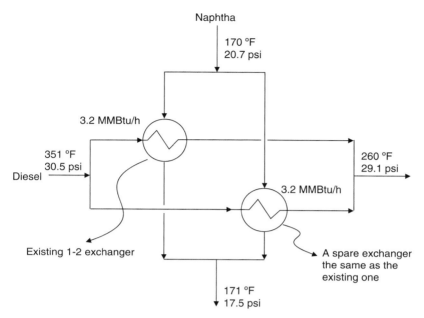

FIGURE 6.4. A parallel arrangement of two 1-2 exchangers.

diesel flowrate. A spare 1-2 exchanger was considered to add to the existing naphtha–diesel exchanger. What is the proper arrangement of this spare 1-2 exchanger in relation to the existing naphtha–diesel exchanger to handle the large increase in diesel flowrate?

Under increased diesel production, 193,000 lb/h naphtha feed (placed on the shell side of the exchanger) to the tower will increase vaporization up to 29.8% from 6.3% by 121,500 lb/h diesel product at 351 °F cooled to 260 °F. A pressure drop of 5 psi is permissible on both sides.

Based on the rating assessment in Example 6.2, it is observed that pressure drop on the shell side is too large to be allowed. Thus, a parallel arrangement is considered in this assessment as shown in Figure 6.4. The two exchangers are counted as one exchanger unit for the rating calculations below.

Solution:

(1) *Heat Duty*:

$$\text{Naphtha feed, } Q = m \times q \times \Delta\text{vapor}\% = 193,000 \times 141.7 \times (29.8\% - 6.3\%)$$
$$= 6.4 \text{ MMBtu/h}$$
$$\text{Diesel, } Q = M \times C \times \Delta T = 121,500 \times 0.587 \times (351 - 260)/10^6$$
$$= 6.4 \text{ MMBtu/h}$$

(2) ΔT_{lm} and F_t:

Tube Side		Shell Side	Differences	
Hot Stream		Cold Stream		
351	Higher temperature	171	180	ΔT_1
260	Lower temperature	170	90	ΔT_2
91	Differences	1	90	$\Delta T_1 - \Delta T_2$
ΔT		Δt		

$$\Delta T_{lm} = (\Delta T_1 - \Delta T_2)/\ln(\Delta T_1/\Delta T_2) = 129.8$$
$$R = \Delta T/\Delta t = 91$$
$$P = \Delta t/(T_1 - t_1) = 0.01$$
$$F_t = 0.99$$
$$F_t \Delta T_{lm} = 128.5°F$$

(3) *Summary*:

U_C	78.7
U_A	68.0
U_R	61.8
Overdesign	10%
R_f calculated	0.003
R_f required	0.002
ΔP_s calculated	3.2
ΔP_s allowable	5
ΔP_t calculated	1.4
ΔP_t allowable	5

Two 1-2 heat exchangers in parallel are adequate to satisfy process heat transfer requirements with 10% overdesign. Pressure drops on both sides of the exchanger are less than allowable pressure drops.

Example 6.4   Use of Spare Exchangers in Series to an Existing Acetone–Acetic Acid Exchanger

Acetone at 250 °F is to be sent to storage at 100 °F and at a rate of 60,000 lb/h. The heat will be received by 185,000 lb/h of 100% acetic acid coming from storage at 90 °F and heated to 150 °F. Pressure drops of 10.0 psi are available for both fluids, and an overall fouling factor of 0.004 should be provided.

Available for the service are several 1-2 exchangers having $21\frac{1}{4}$ in. shell ID, having 270 tubes with 3/4 in. tube OD, 14 BWG, 16 ft long and laid out on a 1 in. square pitch. The bundles are arranged for two tube passes with segmental baffle spaced 5 in. apart. Determine the suitability of these 1-2 exchangers for the specific service.

Solution:

(1) *Exchanger Data*:

Shell Side	Tube Side
ID = 21 1/4 in.	Number and length = 270, 16 ft
Baffle spacing = 5 in.	OD/BWG/pitch = 3/4 in./14 BWG/1 in. square
Shell passes = 1	Tube passes = 2

(2) *Heat Balances*:

Acetone, $Q = 60{,}000 \times 0.57(250 - 100) = 5{,}130{,}000$ Btu/h
Acetic acid, $Q = 168{,}000 \times 0.51(15 - 90) = 5{,}130{,}000$ Btu/h

(3) F_t *Factor*:

Tube Side		Shell Side	Differences	
Hot Stream		Cold Stream		
250	Higher temperature	150	100	ΔT_1
100	Lower temperature	90	10	ΔT_2
150	Differences	60	90	$\Delta T_1 - \Delta T_2$
ΔT		Δt		

$\Delta T_{lm} = 39.1°F$
$R = 150/60 = 2.5$
$P = 60/(250 - 90) = 0.375$

Thus,
One 1-2 exchanger, F_t is not on F_t charts
Two 1-2 exchangers, $F_t = 0.57$ (too small)
Three 1-2 exchangers, $F_t = 0.86$ (OK); $F_t \times \Delta T_{lm} = 33.6°F$.
To permit the heat transfer with the temperatures given by the process, three 1-2 exchangers are required.

(4) U_c: Calculated from heat exchanger rating software:

$h_t = 194$ Btu/h ft^2°F and $h_s = 242$ Btu/h ft^2°F
$U_c = h_t \times h_s/(h_t + h_s) = 194 \times 242/(194 + 242) = 107.7$ Btu/h ft^2°F

(5) U_R: $U_R = Q/(AF_t \text{ LMTD}) = 5{,}130{,}000/(2{,}540 \times 34.4) = 58.8$ Btu/h ft^2 °F

(6) R_f: $R_f = (U_c - U_R)/U_c U_R = (107.5 - 58.8)/(107.5 \times 58.8) = 0.0077$ h ft^2 °F/Btu

(7) ΔP_s and ΔP_t: Calculated from heat exchanger rating software

$\Delta P_s = 10.4\,\text{psi}$ (allowable $\Delta P_s = 10.0\,\text{psi}$) and $\Delta P_t = 5.2\,\text{psi}$ (allowable $\Delta P_t = 10.0\,\text{psi}$)

(8) *Rating Summary*:

U_C	107.5
U_A	75.2
U_R	58.8
OD%	28%
R_f calculated	0.0077
R_f required	0.004
ΔP_s calculated	10.4
ΔP_s allowable	10
ΔP_t calculated	5.2
ΔP_t allowable	10

Conclusion: Three 1-2 exchangers are more than adequate for heat transfer even though the pressure drop on the shell side is slightly higher than allowable which is tolerable. Fewer exchangers cannot fulfill the process requirement.

6.5.3 Assess Heat Exchanger Fouling

Heat exchanger fouling occurs in operation and its performance deteriorates over time. In some cases, it requires cleaning several times before the process is shut down for turnaround maintenance. It is an economic decision for selecting fouling mitigation methods and when to apply them. Heat exchanger rating can determine the level of fouling and if the heat exchanger in question requires attention.

Example 6.5 Calculation of Heat Transfer Performance for an Existing Vacuum Residue–Crude Exchanger

710,000 lb/h of a 31.1°API crude oil going through the tube side of the exchanger is heated from 359 to 375 °F by 213,500 lb/h of 11.1 °API vacuum residue entering at 503 °F and cooled to 449 °F. How is this heat exchanger performing?

The exchanger is a TEMA Type AES with 48 in. ID shell having 964 tubes with 1 in. OD, 12 BWG thickness and 24 ft long, which are laid out on a $1\frac{1}{4}$ in. square pitch. There are four tube passes and one shell pass with baffles spaced 9.5 in. apart and baffle cut 15%. Fouling factors of 0.003 and 0.01 are provided for crude and vacuum residue, respectively.

(1) *Heat Duty*:

Crude oil, $Q = 730{,}000 \times 0.60 \times (375 - 359) = 7.0\,\text{MMBtu/h}$

Vacuum residue, $Q = 213{,}500 \times 0.61 \times (503 - 449) = 7.0\,\text{MMBtu/h}$

(2) ΔT_{lm} *and* F_t:

Tube Side		Shell Side	Differences	
Hot Stream		Cold Stream		
503	Higher temperature	375	128	ΔT_1
449	Lower temperature	359	90	ΔT_2
54	Differences	16	38	$\Delta T_1 - \Delta T_2$
ΔT		Δt		

$\Delta T_{lm} = (\Delta T_1 - \Delta T_2)/\ln(\Delta T_1/\Delta T_2) = 107.9$
$R = \Delta T/\Delta t = 3.38$
$S = \Delta t/(T_1 - t_1) = 0.11$
$F_t = 0.98$
$F_t \Delta T_{lm} = 105.8°F$

(3) *Initial Assessment*:

U_C	49.6
U_A	30.2
U_R	11.1
OD%	171%
R_f calculated	0.070
R_f required	0.013

The required U value to achieve 7.0 MMBtu of heat transfer is only 11.1 in comparison with the actual U value of 30.2 based on fouling factors of 0.01 for vacuum residue and 0.003 for crude. In other words, the heat exchanger only accomplishes one third of the heat transfer capability offered by the heat exchanger, which warrants a more detailed investigation.

(4) *More Detailed Assessment*: As a follow-up, engineers conducted the performance comparison between operation and design and the results are given in the table below. It can be observed that the flowrates in operation are higher than those in design. The higher flowrates should have corresponded to a higher U value in operation. However, in this case, the U value in design is 41% higher than in operation.

	Design	Operation
Q; MMBtu/h	8.9	7.0
W(resid); lb/h	167,250	213,500
T_1; °F	505	503
T_2; °F	416	449
w(crude); lb/h	665,000	710,000
t_1; °F	360	359
t_2; °F	382	375
ΔT_1; °F	123	128

(*continued*)

(*Continued*)

	Design	Operation
ΔT_2; °F	56	90
ΔT_{lm}; °C	85	108
F_t	0.93	0.98
UA	112,658	66,074
A	5,936	5,936
U	18.98	11.13
$U_{\text{operation}}/U_{\text{design}}$		0.59

Field inspection was performed and pressure drop was measured. It was found that the pressure drop on the crude (tube side) was around 60 psi versus 6.8 psi under normal fouling conditions. It was concluded that the heat exchanger suffers severe fouling with loss of more than half the heat transfer capability. In addition, the much higher pressure drop caused crude feed flashing before the charge heater, which could be the potential safety issue for the heater. Thus, it was decided to clean the exchanger immediately online by means of bypass arrangement. After cleaning, a dedicated investigation was conducted to identify the root causes of this fouling.

The rating assessment indicated that the tube side crude velocity is a bit too low, which is 5.3 ft/s. It should be 7 ft/s for this hot crude heating service because precipitation fouling becomes more active under high temperatures. The change was made to the number of tube passes from four to six. As a result, the tube velocity was increased to 7.9 ft/s, but at the expense of higher pressure drop on the tube side. The tube side pressure drop was increased to 21 psi from 6.8 psi with four tube passes. This change reduced tube fouling and prolonged the operation of the exchanger between cleanings.

The lessons learned from this investigation indicate that fluid flowrate affects fouling behavior significantly. Flowrates much lower than design result in lower velocity, which can promote accumulation of fouling deposits. High temperature is another major cause for promoting fouling. The heat exchangers in the high-temperature region are more prone to be fouled due to inherent thermal coking tendency. Threshold conditions in terms of velocity and temperature should be identified beyond which fouling occurs in a faster pace.

6.6   IMPROVING HEAT EXCHANGER PERFORMANCE

The objective of heat exchanger operation management is to maintain good performance to fulfill process requirements for desirable periods of time. There are three major reasons why exchanger operation could deviate from design: poor design, excessive fouling, and mechanical failure. In any event, heat exchangers can deliver trouble-free services while meeting process requirements if the heat exchanger is designed well thermally and mechanically, stored carefully before use, installed correctly, operating within its design limits, and cleaned periodically depending on fouling formation. In contrast, it can be stressful if heat exchangers do not perform as expected in meeting process requirements. In the worst case, mechanical and performance failure of heat exchangers could cause undesirable unit shutdowns.

The methods for monitoring and troubleshooting are discussed briefly below with the focus on thermal and hydraulic performance. Fijas (1989) provides good discussions on mechanical problems often encountered with heat exchangers. With exchanger performance, the priority issues are good knowledge of fouling resistances to avoid poor design, continuous monitoring of U value to maintain good performance, pressure drop survey for troubleshooting, managing of two-phase flow, fouling mitigation, and heat transfer enhancements. The general methodology for improving heat exchanger performance is: monitor performance trends, identify opportunity, and develop and implement solutions.

6.6.1 How to Identify Deteriorating Performance

6.6.1.1 Fouling Resistances Inappropriate estimate of fouling resistances results in either too much or too little overdesign. Although the TEMA fouling resistances were originally only considered to be rough guidelines for heat exchanger design, they are often treated as accurate values. This may cause considerable errors because the transient character of the fouling process is neglected. Conditions in initially over-designed heat exchangers often promote fouling deposition, thus making fouling formation a self-fulfilling prophecy. Thus, one needs to be critical of the fouling resistances listed in the public domain and make proper adjustment based on historical fouling data. For existing services, obtain historical fouling trends and assess the characteristics of the system to determine the root causes for fouling accordingly.

6.6.1.2 U Value Monitoring Due to the fact that heat exchanger performance varies with flow rates, compositions, and fouling conditions, heat exchanger assessment must be conducted on a regular basis so that a performance trend can be obtained and problems can be detected at an early stage. A single rating of an exchanger is good for getting a baseline data on its performance, but it must be done on a regular basis to define trends. From a single point of rating, you can calculate a single U value, pressure drops of shell side and tube side, and calculate a single value of heat duty. However, single point assessment cannot provide insights into fouling evolution over time and sudden changes in U value due to process variations. A U value trend can help you for troubleshooting.

The most important point of a U value trend is its capability of showing the fouling behavior. The purpose of U-trend monitoring is to identify any abnormal fouling behavior. In general, fouling accumulation in heat exchanger depends on the type of fouling, the service (fluid compositions, temperature, and pressure), and the exchanger design and so on. Under normal operation, a U trend should display gradual changes in U value. However, operation changes could affect fouling, which include feed rate variations, fluid composition change, bypass operation, hydraulic head change, and so on. If an operation change suddenly distorts the normal fouling behavior (e.g., U value reduces sharply), this change must be investigated and appropriate actions must be taken.

When an exchanger is new, a detailed performance evaluation is warranted and it should be repeated after 6 months or so. One should trend data in-between and afterward. Process temperatures, pressures, and flows around the heat exchanger are measured in daily average. It is recommended to use distributed control trend logs for data collection. It is important to keep the data and the calculations for reference in the future. The potential actions for fouling mitigation are discussed in Chapter 7.

6.6.1.3 Pressure Drop Monitoring   The importance of pressure drop calculations cannot be overemphasized as it can help with analyzing performance problems and troubleshooting of heat exchanger malfunction. Calculated pressure drop for single phase flow can be reasonably close to measured pressure drop if there is no fouling. For two-phase flow, calculated pressure drop can also be reasonably close to measured pressure drop if pressure drop zones are used and flow patterns are considered. With these two assumptions as the basis, pressure drop calculations can be used as a tool for troubleshooting.

If measured pressure drops are significantly lower than calculated drops, this might indicate fluid bypassing, which could occur either on tube side or shell side. On the other hand, if measured pressure drop is too high, this is often caused by severe fouling, freezing, or slug flow for two-phase flows.

6.6.1.4 Avoid Poor Design   The chemical and petroleum industries have been plagued for decades with poorly operating and occasionally inoperable heat exchangers. One of the common causes is usually traced to poor design, which should be avoided in the design stage by all means. Careful considerations of major design choices must be made to obtain an "optimal" heat exchanger design. The design issues include fouling considerations, tube side design (tube counts, tube passes, tube length, tube pitch, tube layout), shell side design (shell diameter, shell types including TEMA Types E/F/G/H/J/K/X, shell flow distribution), and baffle design (baffle types, segmental baffle including single/double/triple, baffle spacing, and baffle cut).

The essential design task is to optimize velocity in both tube and shell sides by the best use of allowable pressure drop available. For example, when the number of tube passes is increased from one to two passes, the velocity could be twice that of one-pass velocity as the travel distance is doubled. Then the tube side heat transfer coefficient will increase according to the 0.8 power of velocity. At the same time, the tube side pressure drop will increase according to the square of velocity and the travel distance. Therefore, pressure drop will rise to the cubic of the increase in tube passes for a given tube count and tube side flowrate. When the pressure drop is higher than the allowable one, reduction of tube length could reduce pressure drop. Other design choices are the tube outside diameter, tube pitch, tube counts, and layout, which are the important design choices besides the number of tube passes and tube length.

Another example is design choices available to reduce shell side pressure drop. The number of baffles (N_B) is proportional to the baffle spacing. Baffle spacing and baffle cut have a profound effect on shell side pressure drop. In many cases, the shell side pressured drop is still too high with single segmental baffles in a single-pass shell even after increasing the baffle spacing and baffle cut to the highest values recommended. These cases may be accompanied with very high shell side flowrate. The next design choice is to consider double segmental baffles. When double segmental baffles at relatively high, baffle spacing cannot satisfy shell side allowable pressure drop; a divided-flow shell (TEMA J) with single segmental baffles could be considered. Since pressure drop is proportional the square of velocity (u^2) and to the length of travel (L), a divided-flow shell could have one eighth the pressure drop in an identical single-pass exchanger. This discussion can go on as there are other design choices that are available to deal with high shell side pressure drops, for which Mukherjee (1998) provides detailed explanations.

APPENDIX: TEMA TYPES OF HEAT EXCHANGERS

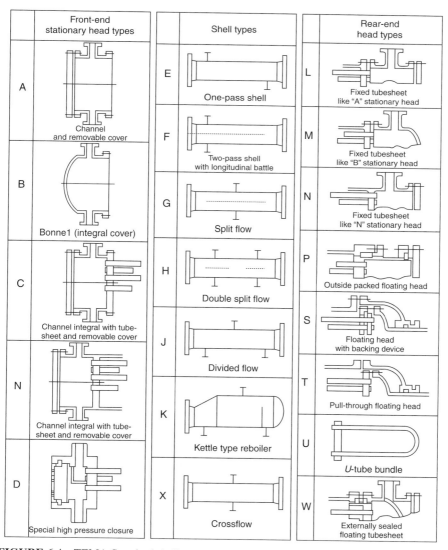

FIGURE 6.A. TEMA Standard shell types and front- and rear-end head types. (From Shah and Sekulić (2003), reprinted with permission by John Wiley & Sons.)

NOMENCLATURE

A	surface area, ft^2
BWG	Birmingham Wire Gauge, tube wall thickness
C, c	fluid heat capacity of hot and cold streams, Btu/(lb °F)
C_p, c_p	fluid heat capacity, Btu/(lb °F)
d_i	inner diameter of the tube, ft
d_o	outer diameter of the tube, ft
D_e	equivalent shell diameter, ft
D_s	shell diameter, ft
f_s	shell side friction factor, $(\text{ft}^2$ °F h)/Btu
f_t	tube side friction factor, $(\text{ft}^2$ °F h)/Btu
F_t	LMTD correction factor, fraction
h_t	tube side film coefficient, Btu/(ft^2 °F h)
h_s	shell side film coefficient, Btu/(ft^2 °F h)
k	fluid thermal conductivity, Btu/(h ft °F)
l, L	tube length, ft
LMTD	logarithmic time mean temperature difference, °F
M, m	mass flow rate for hot and cold streams, lb/h
N_B	number of baffles
p	tube pitch, ft
ΔP	pressure drop, psia
Q	heat duty, MMBtu/h
R	temperature ratio: $(T_1 - T_2)/(t_2 - t_1)$; dimensionless
Re	Reynolds number, dimensionless
R_f	overall fouling resistance $= R_t + R_s$, $(\text{ft}^2$ °F h)/Btu
R_t, R_s	tube and shell side fouling resistance, $(\text{ft}^2$ °F h)/Btu
P	temperature ratio: $(t_2 - t_1)/(T_1 - t_1)$, dimensionless
r_w	resistance of the inner tube referred to the tube outside diameter, $(\text{ft}^2$ °F h)/Btu
T_1, t_1	supply temperature of hot and cold streams, °F
T_2, t_2	target temperature of hot and cold streams, °F
ΔT_1	hot end temperature approach, °F
ΔT_2	cold end temperature approach, °F
ΔT_{lm}	logarithmic mean temperature difference (LMTD), °F
u	shell side cross-flow velocity or tube velocity, ft/h
U	overall heat transfer coefficient, Btu/(ft^2 °F h)
V_S	superficial gas velocity, ft/s

Greek Letters

ρ	fluid density, lb/ft^3
μ	fluid viscosity, lb/(ft h)

μ_w water viscosity, lb/(ft h)
τ_0 sheer stress, lb/ft^2

Subscript and Superscript

A   actual
C   clean
D   design
e   equivalent
f   friction
G   gas
i   inside of tube or shell
L   liquid
lm   logarithmic mean
o   outside of tube or shell
R   required
s   shell side
t   tube side

REFERENCES

Bell KL (1973) Thermal design of heat transfer equipment, in Perry RH, Chilton CH (eds), *Chemical Engineers Handbook*, 5th edition, pp. 10, McGraw-Hill.

Bennett CA, Kistler RS, Lestina TG, King DC (2007) Improving heat exchanger designs, *Chemical Engineering Progress*, pp. 40–45, April.

Dittus FW, Boelter LMK (1930) *Publications on Engineering*, University of California Berkley, Vol. 2, p. 443.

Fijas DF (1989) Getting top performance from heat exchangers, *Chemical Engineering*, pp. 141–145, December.

Hewitt GF, Shires GL, Bott TR (1994) *Process Heat Transfer*, CRC Press, pp. 275–285.

Kern DQ (1950) *Process Heat Transfer*, McGraw-Hill, p. 148.

Mukherjee R (1998) Effectively design shell-and-tube heat exchangers, *Chemical Engineering Progress*, AIChE, February issue.

Shah RK, Sekulić P (2003) *Fundamentals of Heat Exchanger Design*, John Wiley & Sons.

TEMA (2007) *Standards of TEMA*, 9th edition, Tubular Exchangers Manufacturer Association, New York.

7

HEAT EXCHANGER FOULING ASSESSMENT

Heat exchanger fouling costs a lot of money for process plants. It is essential to determine the root causes and find appropriate methods for fouling mitigation from fouling prevention to exchanger cleaning cycle optimization.

7.1 INTRODUCTION

Fouling is the accumulation of undesirable materials as deposits on heat exchanger surfaces. Fouling deposits come in many different forms, which could have a hard crystalline structure, a polymerized coating, or a weakly bonded particulate matrix. Chemicals may dissolve the deposit layer or be totally ineffective. No matter what material is contained in heat exchanger fouling deposits, it leads to similar consequences, which are reduction in thermal performance and an increase in pressure drop.

Fouling is very costly in terms of energy use. For example, fouling in oil refinery plants consumes an extra 0.2 quad of energy annually, which is enough to heat and power 2 million homes in the United States for a year. Furthermore, heat exchanger fouling can also cause decreased process throughput and extra maintenance cost in operation. It costs more than $2 billion a year in U.S. refineries alone.

Fouling should be considered during the design of a new heat exchanger and operation of an existing exchanger. There are complex factors causing heat

Energy and Process Optimization for the Process Industries, First Edition. Frank (Xin X.) Zhu.
© 2014 by the American Institute of Chemical Engineers, Inc. Published 2014 by John Wiley & Sons, Inc.

exchanger fouling such as physical and chemical properties of process streams, operating conditions, heat exchanger design, and operation. Since complex factors affect the choice of methods to reduce and prevent fouling, identification of root causes could derive more effective solutions. In this chapter, methods for fouling identification and mitigation will be discussed.

7.2 FOULING MECHANISMS

Since there are a great variety of fouling phenomena, it is useful to group them into six types of fouling mechanisms for better understanding (Bott, 1990; Melo et al., 1988):

(1) *Crystallization Fouling:* Precipitation and deposition of dissolved salts, which are supersaturated at the heat transfer surface. Supersaturation may be caused by

- Evaporation of solvent.
- Cooling below the solubility limit for normal solubility (increasing solubility with decreasing temperature, such as wax deposits, gas hydrates, and sulfur condensation). The precipitation fouling occurs on the cold surface (i.e., by cooling the solution).
- Heating above the solubility limit for inverse solubility (increasing solubility with increasing temperature, such as calcium and magnesium salts). The precipitation of salt occurs with heating the solution.
- Mixing of streams with incompatable compositions.
- Variation of pH, which affects the solubility of CO_2 in water.

(2) *Particulate Fouling:* Accumulation of particles from heat exchanger working fluids (liquids and/or gaseous suspensions) on the heat transfer surface. Most often, this type of fouling involves deposition of corrosion products dispersed in fluids, clay and mineral particles in river water, suspended solids in cooling water, soot particles of incomplete combustion, magnetic particles in economizers, deposition of salts in desalination systems, deposition of dust particles in air coolers, particulates partially present in fire-side (gas-side) fouling of boiler, and so on. If particulate fouling is gravitational settling of relatively large particles onto horizontal surfaces, this phenomenon is also called sedimentation fouling.

(3) *Chemical Reaction Fouling:* Deposit formation (fouling precursors) at the heat transfer surface by unwanted chemical reaction (such as polymerization, coking) within the process fluid, but heat transfer surface material itself is not involved in the chemical reaction. Thermal instability of chemical components such as asphaltenes and proteins can become fouling precursors. Usually, this type of fouling starts to form at local hot spots in a heat exchanger. It can occur over a wide temperature range from ambient to over 1000 °C (1832 °F) but is more pronounced at higher temperatures.

(4) *Corrosion Fouling:* The heat transfer surface itself reacts with chemical species present in the process fluid. Its trace materials are carried by the fluid in the exchanger, and it produces corrosion products that deposit on the surface. The thermal resistance of corrosion layers is low due to high thermal conductivity of oxides.

(5) *Biological Fouling:* Deposition and growth of macro-organisms and micro-organisms on the heat transfer surface. It usually happens in water streams.

(6) *Freezing Fouling:* It is also called solidification fouling, which occurs due to freezing of a liquid or some of its constituents to form deposition of solids on a subcooled heat transfer surface. For example, formation of ice on a heat transfer surface during chilled water production or cooling of moist air, deposits formed in phenol coolers, and deposits formed during cooling of mixtures of substances such as paraffin are some examples of solidification fouling. This fouling mechanism occurs at low temperatures, usually ambient and below.

There is no single unified theory to model the fouling process because combined fouling occurs in many applications and no single solution exists for fouling control. Appropriate theories and methods must be selected to tackle fouling issues for each application.

7.3 FOULING MITIGATION

Although fouling involves very complex mechanisms, it is the objective from the fouling mitigation point of view that a few fouling control variables are determined to control a fouling process. These variables are fluid velocity, fluid and heat transfer surface temperatures, physical and chemical properties of the fluid, and geometry of the fluid flow passage. Other important variables are concentration of foulant or precursor, impurities, heat transfer surface roughness, surface chemistry, fluid pressure, fluid chemistry (pH level, oxygen concentration, etc.), and so on (Shah and Sekulić, 2003).

The characteristics of different fouling mechanisms determine appropriate mitigation methods to employ. Thus, it is important to understand what type of fouling occurred and identify what methods to apply. In the end, the key to success for fouling mitigation is to determine the most effective method(s) to deal with the root causes of the fouling problem.

Removing fouling materials at the source is ideal. For example, filtration can remove particles effectively before heat exchangers, while allowing sufficient settling for feed in storage tanks could alter the propensity for fouling. The most effective technique for preventing water-side fouling is still the conventional water treatment. Not blending of crudes with chemical incompatibility could prevent fouling in the crude preheat train.

In general, there are four kinds of methods, namely, chemical, mechanical, operational, and process conditions based methods. These methods are briefly reviewed as follows:

(1) Chemical methods that can be applied online include:
- Chemical additives can hold the fouling deposit in suspension, by reducing their tendency to stick, and by weakening their bond to the surface. However, sometimes, chemical additives can contribute to fouling. The effectiveness of this method may depend on the match between chemical additives, chemical components of process streams, and conditions under which an exchanger operates. For example, a certain additive may work well with asphaltene but not wax deposits. In same cases, chemical additives can also poison catalysts or adsorbents.
- Solvent cleaning agents are added to the feed line and circulated through product lines, one product at a time. This effectively removes fouling formed in process lines and exchangers. This method can be performed for ongoing operations in comparison with other cleaning methods using aqueous chemical solutions, which have to be carried out during turnaround or maintenance.
- Biofouling is usually easy to control with biocides, but compatibility with the exchanger materials must be checked.
- Chlorination aided by flow-induced removal of disintegrated biofilm is the most common mitigation technique.

(2) Mechanical methods that can be applied online include:
- Installing mechanical enhancement devices such as tube inserts. In general, these methods require additional pressure drop. When they are applied in revamp, additional pressure drops may require a new pump, which could be expensive. Some of the example devices include the following:
  - Spirelf is a reliable vibration device that requires bare tube velocities above 0.8 m/s although the higher velocity is better if additional pressure drop is acceptable. System integrity is expected to last for one turnaround cycle and should be replaced at each turnaround. This device has been proven through several applications.
  - Turbotal is a rotation device that can be used between 0.5 and 1.5 m/s bare tube velocities, with about half of the pressure drop required by Spirelf. The life span is about 1.5 to 2 years as many springs will break after 3 years. Thus, it must be replaced in between turnarounds.
  - Fixotal is the only device that can handle mixed phase flow. Beneficial for condensers; however, pressure drop may not be acceptable for many applications.
- Installing mechanical cleaning devices. Some of the devices include:
  - Sponge rubber balls with abrasives on their outer surfaces circulated through tubes. This method is widely applied to surface condensers in power plants.

    ○ Brushes propelled through tubes and captured at the ends in cages. A system must be in place for reversing the flow in order for brushes to go back through tubes.

- Advanced heat transfer technologies. These technologies have two features for fouling mitigation: one is to increase turbulent flow while the other eliminates dead flow regions. Some of the technologies include:
  - ○ Helical baffle bundles have a helical flow pattern on the shell side and thus effectively remove dead flow zones. This helical flow pattern provides an effective way of fouling mitigation on the shell side.
  - ○ Spiral exchangers use spiral-shaped fluid channels for both sides and feature self-cleaning capability.

- For corrosion fouling, corrosion-resistant material is the best remedy. For example, use proper aluminum alloy to prevent mercury corrosion in a plate–fin exchanger.

- Offline cleaning. In some applications, flows are diverted in a bypass pipe or exchanger, and then the fouled exchanger is cleaned offline. The following methods are frequently used:
  - ○ Hydro-blasting: using high-pressure water jet for salt-related fouling.
  - ○ Hot light cycle oil for asphaltene related fouling.
  - ○ Chemical aqueous solutions.
  - ○ Baking compact heat exchanger modules in an oven (to burn the deposits) and then rinsing.

(3) Operational methods:
- Avoid bypass operation as bypass reduces velocity.
- Increase tube passes to increase flow velocity when it is too low.
- Sulfer dew point temperature control for column overhead to avoid sulfur condensation.
- Removal of particulates and sludge in the feed via storage settling and filtration.
- Better desalting operation for removing salts and acid ions.
- Proper feed blending to avoid mixing of incompatible components.

(4) Changing operating conditions to deal with gaseous fouling (Shah and Sekulić, 2003):
- Crystallization fouling can be prevented if the surface temperature is kept above the freezing of vapors from the gaseous stream; the solidification can be minimized by keeping a high velocity of freezable species, having some impurities in the gas stream, and decreasing the foulant concentration.
- Minimizing reaction effluent fouling by
  - ○ Maintaining the right temperature range in the exhaust gas within the exchanger.

- o Increasing or decreasing the velocity of the gaseous stream, depending on the application.
- o Reducing the oxygen concentration in the gaseous stream.
- Eliminating corrosion fouling by sulfur dew point control. The outlet temperature of the exhaust gas stream from the exchanger should be maintained in a very narrow range:
  - o Above the acid dew point of 150 °C (302 °F) for sulfuric or hydrochloric acid condensation. Since sulfur is present in all fossil fuels and some natural gas, the dew point of sulfur must be avoided in the exchanger, which is dependent on the sulfur content in the fuel.
  - o Below 200 °C (400 °F) for attack by sulfur, chlorine, and hydrogen in the exhaust gas stream.
- The pH value has a considerable role in the corrosion fouling rate and the corrosion rate is at minimum for a pH of 11 to 12 for steel surfaces.
- Low oxygen concentrations in the flue gases promote the fire-side corrosion of mild steel tubes in coal-fired boilers.

7.4 FOULING MITIGATION FOR CRUDE PREHEAT TRAIN

Fouling is a very serious issue for crude preheat train in oil refining plants and it is characterized by the following root causes:

- *Crude Properties:* Crude with high paraffin content is prone to cause wax deposition, which precipitates in the form of sticky scaling. This kind of fouling could occur at relatively low temperatures and develop very rapidly to a severe level in weeks. Predominantly, wax fouling occurs when different crudes are mixed together, which have incompatible compositions. On the other hand, asphaltenic components in the crude could generate fouling due to polymerization of asphaltene at high temperatures. This kind fouling builds up gradually to reach a severe level typically in months.
- *Low Velocity:* Low velocity can promote accumulation of fouling deposits. The desirable velocity for heat exchangers for the crude preheat train is between 6 and 7 ft/s.
- *High Temperature:* High temperature is another major cause for promoting fouling. The heat exchangers in the high temperature region are more prone to be fouled due to inherent thermal coking tendency. A threshold condition in terms of velocity and temperature can be identified beyond which fouling occurs in a fast pace.
- *Salt Removal:* If salt cannot be removed sufficiently from crude via desalting, severe corrosion fouling could be the consequence.
- *Exchanger Geometry:* Different exchanger designs could result in different rates of fouling accumulation. For fouling service, compact exchangers are

not recommended but helical baffle and spiral exchangers could be considered for the slowdown of fouling build up. When shell-and-tube exchangers are used, the more easily fouled stream should be placed in the tube side for easy cleaning. Horizontal heat exchangers are easier to clean than vertical exchangers. It is easier to clean square or rotated square tube layouts mechanically on the shell side than to clean other types of tube layouts.

A case study is used here to provide insights for how to find effective mitigation methods. The crude preheat train in a refining plant encounted a low desalting temperature at 115 °C on average versus design desalting temperature at 135 °C. It was found that the low desalting temperature was caused by exchanger fouling in three exchangers before the desalter. The energy-saving benefit of fouling mitigation for these three exchangers was estimated as 20 MMBtu/h (process absorb duty). To deal with the low desalting temperature, a lot of emulsion breakers is added to the crude, which is not only expensive but also negatively affects fractionation efficiency downstream. The million dollar question was: What causes fouling in these exchangers?

To find the root causes, lab tests were conducted and heat exchanger rating calculations were performed. The lab tests for two different feed conditions showed that the fouling materials are of two types, wax type (i.e., heavy paraffin) while olefins or oxygenates are below detection limit, and salt type. Wax precipitates due to the chemical incompatibility of crude blended. It was also found that large amounts of slops containing heavy paraffin were recycled and mixed with raw crude, which makes wax fouling even worse. In the case of fouling caused by salt, scaling salts in the raw crude precipitate when pH concentration and temperature are increased. As the crude is heated, scaling salts drop out and fouling occurs.

At the same time, rating calculations of three fouling exchangers revealed new insights to the plant engineers. These three exchangers were found having 50% overdesign which results in low fluid velocity around 1 m/s—only half of the recommended velocity for these services. Low velocity may contribute to fast fouling accumulation.

The above findings lead to the following recommendations. First of all, the plant should reduce heavy slop recycle by all means. Second, a better crude blend management should be adopted to avoid blending of incompatible crudes. To do this, a penalty should be assigned when selecting heavy wax crude. In the event that heavy wax crude will be processed due to the unbeatable price, two online methods for fouling mitigation could be applied. One is waxy deposits spalling and the other is water injection.

To remove the waxy deposits online, block in and bypass the tube side of the fouled exchanger, while continuing shellside flow, which heats the tubes. After a while, waxy deposits are melted off and harder deposits are spalled off due to thermal shock. After 20 min, restore normal flow. According to Lieberman (1997), this method works well in several cases.

For salt related fouling, it was recommended to have 1 to 2% wash water injection upstream of the desalter. Some trial and error may be required to get the injection strategy right and mix valve setting correct. The most appropriate wash water must be used, which should be neutral to slightly acidic, and low ammonia. The condensate could be the best. The water generated from the sour water stripper is also very good as it has low ammonia and is coustic free.

For the exchangers with too low velocity, it was recommended to increase tube passes. The message from this example is that the fouling mechanisms could be different even for the same process unit, and thus different mitigation methods should be employed.

7.5 FOULING RESISTANCE CALCULATIONS

The above discussions focus on fouling mechanisms and mitigation methods; however, the following questions are left to be answered before searching for fouling mitigation solutions: (i) how much fouling exists and (ii) how much heat transfer capability is lost due to fouling.

To answer the first question, one needs to determine the fouling resistance or fouling factor commonly called. There is one practical way of determining a realistic fouling factor—performance test run. The test run for this purpose is to measure temperature, pressure, and compositions in and out of the heat exchanger after the heat exchanger is operated for a certain period of time when fouling growth becomes slow and stable. Then the fouling factor is determined from the rating calculation, which is discussed below. The fouling factor for nominal operating conditions is referred to as nominal fouling factors. Fouling could become more severe beyond the nominal fouling condition. Fouling factors for severe fouling conditions can also be determined using the rating method below.

The overall heat transfer coefficient U_C for a clean heat exchanger can be described as

$$\frac{1}{U_C} = \frac{1}{h_t} + \frac{1}{h_s} + \frac{L}{k_m},\qquad(7.1)$$

where h_t and h_s are the film coefficients on tube and shell sides, respectively, k_m is the thermal conductivity, and L the tube length.

When a heat exchanger becomes fouled, fouling resistances inside and outside of tubes prohibit the heat exchanger from performing at the level as clean heat transfer coefficient. In practice, these two fouling resistances are lumped together into a total fouling resistance, R_f. The dirty heat transfer coefficient (U_D) can be expressed as

$$\frac{1}{U_D} = \frac{1}{U_C} + R_f.\qquad(7.2)$$

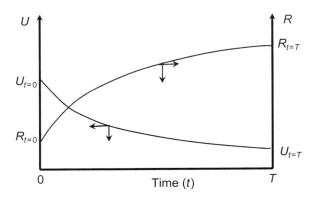

FIGURE 7.1. U and R of a fouling exchanger over time.

The fouling deposits build up very quickly in the first few months of service until a terminal velocity is reached. After this point, fouling continues but develops at a slower pace for most fouling cases. Thus, the overall fouling resistance behaves in the manner as depicted in Figure 7.1. Similarly, pressure drop increases as fouling deposits build up. The fouling deposits make the tube flow diameter smaller over time and pressure drop continues to increase as tube flow diameter reduces.

The fouling factor has to be determined from actual heat exchanger performance based on online measurements taken from a process unit test run. Heat exchanger clean performance is obtained from process flowsheet simulation software (e.g., Hysys by Aspen Tech or Unisim by Honeywell), while dirty performance from exchanger rating software (e.g., HTRI by Heat Transfer Research Institute).

First, heat exchanger heat balance calculations are conducted in a flowsheet simulation software, which has adequate thermal data and can describe process streams according to their physical properties and operating conditions. By providing measured temperatures, the simulation can determine the heat transfer duty from $Q = m \cdot C_p \cdot \Delta T$. At the same time, the simulation calculates transfer capability by lumping overall heat transfer coefficient and surface area together as $U \cdot A = Q/\Delta T_{\mathrm{LM}}$, where ΔT_{LM} is defined in equation (6.7) in Chapter 6.

Second, heat exchanger performance calculations are performed in a rating software. The thermal and physical property data for process streams are transferred from the flowsheet simulation to rating software. The dimensions and geometry of the heat exchanger are entered based on the manufacturing datasheet.

The rating software calculates two U values, namely, required (U_R) and actual (U_A). For given surface area (A), duty (Q), and temperatures, U_R value is obtained according to equation (6.1): $U_R = Q/(A \cdot \Delta T_{\mathrm{LM}})$. At the same time, the software calculates U_A value based on the fouling factors for both tube and shell sides using equation (6.4) in Chapter 6. While U_A denotes the capability of the exchanger based on its dimension and geometry, U_R indicates the requirement from the process heat transfer. The U_A must be larger than the U_R to counter fouling, which is accomplished by overdesign.

7.5.1 Determine Pressure Drop

As the tube wall thickness increases with fouling deposits, pressure drop measurement must be conducted and used as the basis for pressure drop rating calculations. In doing so, the tube wall thickness including fouling deposits are assumed and iterated until the calculated pressure drops from the rating software converges with the measured ones.

Typical fouled exchanger pressure drops are 1.3–2 times that of clean exchangers (Barletta, 1998). For extreme cases, fouled exchanger pressure drops are much higher than that of clean exchangers.

It is recommended that hydraulic calculations should be conducted in an exchanger rating software (e.g., HTRI) as the rating software is more rigorous in pressure drop calculations than flowsheet simulation software.

7.6 A COST-BASED MODEL FOR CLEAN CYCLE OPTIMIZATION[1]

An important part of mitigating a fouling problem is to determine the time interval for cleaning of heat exchangers. The cleaning interval is referred to as the cleaning cycle. Once cleaning cycle is decided, exchangers are subsequently cleaned by using the most appropriate cleaning method(s).

In general, there are three types of cleaning cycle strategies: fixed time based, reliability based, and cost based. With the fixed time-based strategy, exchangers are cleaned based on a regular maintenance schedule during process shutdowns. The reliability-based cleaning strategy is to restore exchanger performance for critical services according to a certain maintenance schedule. The cost-based cleaning strategy considers the costs associated with cleaning, financial penalty, and additional fuel/power consumption due to fouling. In situations in which the cost of operation and maintenance of critical services is an important factor, clean cycle can be optimized by developing cost as a function of reliability and then searching for a minimum cost-based solution (Zubair et al., 1997). This cost-optimized solution will also result in a desirable level of heat exchanger reliability of critical services.

What heat exchangers provide critical services? Critical services are the services that could comprise the processing objectives from poor performance. Consider an example of a distillation column with pump-arounds for heat removal from the column (Figure 7.2); these pump-around heat exchangers are critical services. If these pump-around exchangers fail to meet the design duty, the top section of the column could be flooded. To prevent this from happening, the feed furnace duty must be reduced via lower furnace outlet temperature to suppress the column overhead vapor loading. However, the reduced furnace outlet temperature will result in a loss of distillate yield from the bottom section of the column. Therefore, these pump-around exchangers must be cleaned when a critical level of fouling is reached. On the other hand, the kerosene product-feed exchanger is not a critical service. If the exchanger fouls, the cooling water exchanger has enough surface area to cool the kerosene product to the tank temperature.

[1]Reprinted by permission of BP plc for the publication.

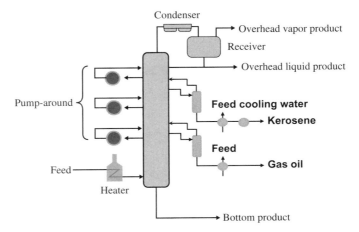

FIGURE 7.2. Critical services of heat exchangers.

The cost-based strategy with considerations of critical services is more appropriate for the process industry. Although several cost-based methods are available, the one developed by Smeenk and Kers (2001) is selected for discussion here as the method provides insights for cost–benefit trade-off in determining the clean cycle.

A heat exchanger network is a combination of several heat exchangers interconnected. Figure 7.3 shows a typical example of the heat exchanger network where feed is preheated by several process streams before being heated by a furnace to a desirable temperature required for fractionation. The terminal temperature of the exchanger network (T^N) is the same as the furnace inlet temperature (FIT). Clearly, the network terminal temperature represents collective heat recovery by individual heat exchangers in the network.

Thus, the overall performance for a heat exchanger network can be measured by the network terminal temperature (T^N). T^N declines over time due to exchanger fouling resulting in increased furnace duty.

To define a nominal base case to focus on fouling alone, typical feed rate, standard physical properties, and temperatures for all streams entering the network are used to model the heat exchanger network. The terminal temperature calculated for the nominal case is called nominal terminal temperature. To calculate the nominal terminal temperature at any point of time, the heat transfer coefficients for heat exchangers are calculated using spot online readings of temperature and flow rates based on equation (6.1) ($U_i A_i = Q_i / \Delta T_{\text{lm},i}$). The UA values calculated are then used

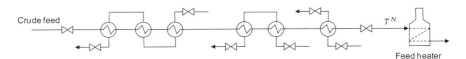

FIGURE 7.3. Heat exchanger network with a single train.

in the heat exchanger network based on the nominal case to determine the network terminal temperature.

Let us assume the plant starts operation with a clean set of exchangers. After t_1 days of operation, the exchangers appear to be much fouled and the plant engineers may debate if they should clean the exchangers to restore the network terminal temperature. Assume the process unit is shut down as t_2 days are required for cleaning. On one side of the dilemma, fouling increases the furnace duty and thus operating cost. On the other hand, cleaning will take t_2 days at lost margin and K dollars as cleaning costs. Therefore, does it make sense to do the cleaning now or should it be postponed? The quest is to search for the optimal cleaning cycle in terms of time interval t_1 with cleaning time t_2 known in advance.

It has been observed by Smeenk and Kers (2001) that the network terminal temperature declines linearly with time due to fouling. If denoting T^N for the nominal network terminal temperature at any time and T_0^N for the nominal network terminal temperature at the clean condition, respectively, the linear decline of T^N with time can be expressed as

$$T^N(t) = T_0^N - \lambda \cdot t, \tag{7.3}$$

where t stands for time (e.g., days) and λ is the decay of T^N due to fouling in °C/day. The SI units are used by Smeenk and Kers (2001) in deriving the equations below.

Similarly, the fact that furnace duty Q increases linearly as T^N decreases over time can be modeled via

$$Q(t) = Q_0^N - \beta \cdot T^N(t), \tag{7.4}$$

where β is the rate of heater duty reduction as T^N declines over time and Q_0^N is the heater duty at initial time of data points for generating the correlation in equation (7.4).

Thus, the furnace fuel cost can be expressed as

$$E(t) = c_f Q(t) = c_f [Q_0^N - \beta \cdot T^N(t)] = A - B \cdot T^N(t), \tag{7.5}$$

where c_f is the fuel price in \$/GJ, $A = c_f Q_0^N$, and $B = c_f \cdot \beta$.

The objective function deciding the clean cycle is based on the average net margin (M_n) over the total time length of operation period including cleaning cycle (t_1) and cleaning time (t_2). That is,

$$M_n = \frac{M_G - E - C}{t_1 + t_2}, \tag{7.6}$$

where M_G is the gross margin in dollars, and E and C are the costs of energy and exchanger cleaning, respectively.

Equation (7.6) can be expressed in an integral form as

$$M_n = \frac{\int_0^{t_1}\{m_1 - E(t)\}dt - \sum_i C_{L,i}}{t_1 + t_2} = \frac{\int_0^{t_1}\{m_1 - [A - B \cdot T^N(t)]\}dt - \sum_i C_{L,i}}{t_1 + t_2}, \quad (7.7)$$

where m_1 is the average gross margin in dollars per day. $C_{L,i}$ is the constant cleaning cost for exchanger i.

Let $K = \sum_i C_{L,i}$ and $T^N(t)$ be a linear function of T_0^N via equation (7.3), the integral of equation (7.7) yields

$$M_n = \frac{(m_1 - A + B \cdot T_0^N)t_1 - (1/2)\lambda B t_1^2 - K}{t_1 + t_2}. \quad (7.8)$$

Therefore, the optimal clean cycle t_1 can be determined from the derivative of M_n over t_1 from equation (7.8):

$$\frac{dM_n}{dt_1} = 0, \quad (7.9)$$

which leads to

$$(\tfrac{1}{2}\lambda B)t_1^2 + (\lambda B t_2)t_1 + [-K - (m_1 - A + BN_o)t_2] = 0. \quad (7.10)$$

If let $a = (1/2)\lambda B$, $b = \lambda B t_2$, and $c = -K - (m_1 - A + BN_o)t_2$, equation (7.10) becomes

$$at_1^2 + bt_1 + c = 0. \quad (7.11)$$

Solving equation (7.11) for t_1 gives

$$\hat{t}_1 = \frac{-b \pm \sqrt{b^2 - 4ac}}{2a}. \quad (7.12)$$

There are two solutions to equation (7.12) with one being negative, which is to be ignored. The positive one gives the optimal clean cycle denoted as $\hat{t}_1$.

The average energy cost (k\$/day) can be found from

$$\bar{E} = \frac{\int_{t_1}[A - BT^N(t)]dt}{t_1 + t_2}\bigg|_{\hat{t}_1} = \frac{(A - B \cdot T_0^N)\hat{t}_1 + (1/2)\lambda B\hat{t}_1^2}{\hat{t}_1 + t_2}. \quad (7.13)$$

The network terminal temperature at $\hat{t}_1$ just before cleaning is

$$T^N(\hat{t}_1) = T_0^N - \lambda \cdot \hat{t}_1. \quad (7.14)$$

Sample calculations from the method: The model is implemented in the spreadsheet (Table 7.1) in which various cases can be studied for the purpose of illustration. Note the numbers in bold are manual input, while the rest are calculated based on the model.

TABLE 7.1. Sample Spreadsheet for Optimal Cleaning Cycle Based on the Simple Model

		Case 1	Case 2	
λ	Decay of T^N	**0.2**	**0.15**	°C/day
T_0^N	Clean terminal temperature	280	280	°C
t_2	Down time for cleaning	2	2	days
K	Cost of cleaning (not margin loss)	100	100	k$
c_f	Cost of energy	3.5	3.5	$/GJ
	Specific gross margin	1	1	$/bbl
	Unit throughput	205,306	205,306	bbl/day
	Unit throughput during cleaning	0	0	bbl/day
m_1	Gross margin during normal operation	205	205	k$/day
m_2	Gross margin during cleaning	0	0	k$/day
A	Energy cost constant	118.6	118.6	k$/day
B	Energy cost constant	03	0.3	k$/day/°C
a	Constant	0.0	0.0	
b	Constant	0.1	0.1	
c	Constant	−434.3	−434.3	
$\hat{t}_1$	Optimal clean cycle	121	140	days
$T^N(\hat{t}_1)$	Optimal network terminal temperature	256	259	°C
Q	Furnace duty at $\hat{t}_1$	149	146	MW
$\bar{E}$	Average energy costs	40.9	40.6	k$/day
$\bar{m}$	Average gross margin	202.0	202.4	k$/day
	Yearly cleaning costs	302	261	k$/year
$\bar{M}$	Yearly net margin	58.5	58.8	M$/year

In this example two hypothetical cases are compared. The value of λ in the first case of 0.2 [°C/day] is on the high side, but it is not unrealistic as values as high as this one have been observed in processing very heavy feed in the crude unit in a refining plant. The spreadsheet finds an optimal cleaning cycle at 121 days. Given the maximum furnace duty of 160 MW, one finds that the cleaning after 121 days is justified purely on energy costs!

In the second case, the value of λ is reduced to 0.15 [°C/day] because heavy fouling exchangers were replaced with more advanced heat exchangers. At a lower fouling rate λ, three effects can be observed:

- The optimum run length found from the model is longer (141 vs. 121 days).
- T^N at $\hat{t}_1$ is higher than that of case 1 (259 °F vs. 256 °F).
- Lower λ (slower rate in fouling) yields both energy and cleaning cost savings.

7.7 REVISED MODEL FOR CLEAN CYCLE OPTIMIZATION[2]

The model described in the previous section assumes that the plant will shut down during cleaning of the exchanger train. This is usually not the case in reality where

[2]Reprinted by permission of BP plc for the publication.

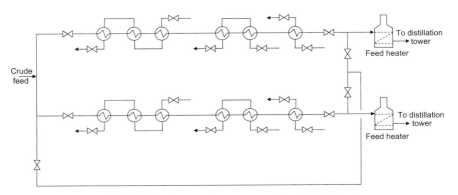

FIGURE 7.4. Heat exchanger network with dual train.

only some of the heat exchangers are taken out for cleaning, while the process unit is kept running at reduced throughput. Figure 7.4 is considered where there are two parallel trains of heat exchangers. It is assumed that during cleaning, one train is kept in operation while the other train of exchangers go through cleaning in rotation. To minimize throughput loss the bypass lines are opened slightly to maintain the terminal pressure above the bubble point pressure to avoid feed flashing before furnace. Minimum amount of vapor ($<3\%$) before furnace is allowed in order for control of furnace flow paths. The bypass is opened until the furnace hits the maximum duty.

In the case that one train is cleaned while the other train continues to operate, the net margin in period t_2 is no longer zero and therefore an extra term is introduced to equation (7.7) to yield

$$M_n = \frac{\int_{t_1}[m_1 - A + B \cdot T^N(t)] \cdot dt + \int_{t_2}[(m_2 - E_2(t)) \cdot dt - K}{t_1 + t_2}, \qquad (7.15)$$

where E_2 is the energy cost during the cleaning, which is a constant because we know that the furnace will be running at maximum duty during cleaning time t_2. The gross margin m_2 during the cleaning will be lower than m_1 since the throughput will be lower. How far the throughput has to be reduced will depend on the actual situation, and will have to be estimated from historical data. For simplicity, m_2 is considered as a constant since t_2 is much shorter in time compared to t_1.

The integral of equation (7.15) yields

$$M_n = \frac{(m_1 - A + B \cdot T_0^N)t_1 - (1/2)\lambda B t_1^2 + (m_2 - E_2)t_2 - K}{t_1 + t_2}. \qquad (7.16)$$

A pseudocost of cleaning, K'', is introduced as

$$K'' = K - (m_2 - E_2)t_2. \qquad (7.17)$$

This K'' can be negative when the cleaning cost is compensated by the unit margin made from the unit running on one preheat train.

Similarly, the optimal clean cycle can be derived from $\partial M_n/\partial t_1 = 0$ to give

$$\hat{t}_1 = \frac{-b \pm \sqrt{b^2 - 4ac}}{2a}, \tag{7.18}$$

where $a = (1/2)\lambda B$, $b = \lambda B t_2$, and $c = -K'' - (m - A + BN_o)t_2$. The positive solution gives the optimal clean cycle denoted as $\hat{t}_1$.

The average energy cost at the optimal cleaning time is calculated as

$$\bar{E} = \frac{\int_{t_1}(A - B \cdot T^N(t)) \cdot dt + \int_{t_2} E_2(t)dt}{t_1 + t_2}\bigg|_{\hat{t}_1} = \frac{(A - BT_0^N)\hat{t}_1 + (1/2)\lambda B\hat{t}_1^2 + E_2 t_2}{\hat{t}_1 + t_2}. \tag{7.19}$$

Sample calculation from revised model: The change can be made readily to the spreadsheet, which is shown in Table 7.2.

TABLE 7.2. Sample Spreadsheet for Optimal Cleaning Cycle Based on the Revised Model

		Case 1	Case 2	
λ	Decay of T^N	0.2	0.15	°C/day
T_0^N	Clean terminal temperature	280	280	°C
t_2	Down time for cleaning	2	2	days
K	Cost of cleaning (not margin loss)	100	100	k$
c_f	Cost of energy	3.5	3.5	$/GJ
	Specific gross margin	1	1	$/bbl
	Unit throughput	205,306	205,306	bbl/day
	Unit throughput during cleaning	150,960	150,960	bbl/day
$E(0)$	Energy cost during clean	38	38	k$/day
m_1	Gross margin during normal operation	205	205	k$/day
m_2	Gross margin during cleaning	151	151	k$/day
E_2	Energy cost at max duty (160 MW)	48.4	48.4	k$/day
K''	Pseudo cleaning costs	−105	−105	k$
A	Energy cost constant	118.6	118.6	k$/day
B	Energy cost constant	0.3	0.3	k$/day/°C
a		0.0	0.0	
b		0.1	0.1	
c		−229.2	−229.2	
$\hat{t}_1$	Optimal clean cycle	87	101	days
$T^N(\hat{t}_1)$	Optimal network terminal temperature	263	265	°C
Q	Furnace duty at $\hat{t}_1$	143	141	MW
$\bar{E}$	Average energy costs	40.8	40.5	k$/day
$\bar{m}$	Average gross margin	204.1	204.3	k$/day
	Yearly cleaning costs	418	361	k$/year
$\bar{M}$	Yearly net margin	59.2	59.4	M$/year

The input values are kept the same as in Table 7.1 to allow a fair comparison between the two models. The difference of the yearly net margin for these two models indicates that it makes sense to clean one train while keeping the other train running instead of shutting down the unit. Furthermore, the optimal cleaning cycle for the revised model is much shorter ($t_1 = 87$ days for $\lambda = 0.2$). Interestingly, this cleaning cycle is way lower than the current practice for crude unit preheat train, but well in line with practice in the early 1990s. Again the model justifies cleaning solely justified on energy loss.

As a remark, the above model can be adjusted to reflect different situations: (i) where only the most fouled exchangers are selected to clean; (ii) where time variant margin is estimated instead of average margin; and (iii) other significant costs are introduced.

7.8   A PRACTICAL METHOD FOR CLEAN CYCLE OPTIMIZATION

Although the above modeling discussions provide fundamental and insightful understanding for clean cycle optimization, the method is built based on the assumption that the fouling decay λ is a constant, which is available or can be determined. However, the reality does not echo this assumption. The methodology proposed here is more rigorous and practical for the purpose of fouling monitoring and cleaning cycle optimization.

A view of data transfer and function design is shown in Figure 7.5. This method is simulation based and Excel is used as an interface for data transfer with simulation while simulation is run in the back. Data are collected in Excel and pushed to simulation, while simulation sends converged data back to Excel where optimization is conducted. Results are displayed in Excel for root cause analysis, trending, and reporting.

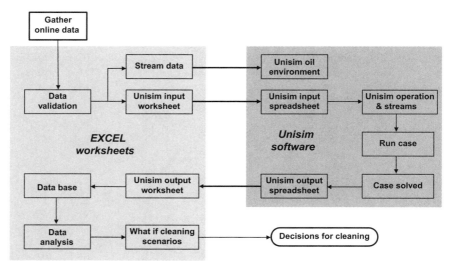

FIGURE 7.5. Data transfer and function design.

Some of the basic concepts, such as standardized streams, nominal terminal temperature, and cost-based cleaning are implemented in this methodology. The methodology features (i) data reconciliation; (ii) standard or normalized simulation of process unit and heat exchanger network; (iii) root cause analysis; (iv) optimization model for determining which exchanger to clean and when; and (v) U trend monitoring and forecasting. These features are discussed next in detail.

7.8.1 Data Reconciliation

It is common that field data quality could be a problem and not all exchanger temperatures are measured online. To resolve these issues, data reconciliation is developed. First, field data are reviewed by engineers and some critical temperatures that are not measured are estimated by engineers. Then the least squares error minimization is applied to make estimation of the rest of the temperatures. It is an interactive procedure and the estimated temperatures are subject to further data reconciliation.

7.8.2 A Standard Simulation Case

This standardized simulation covers the process and the heat exchanger network with all streams modeled based on standard physical properties. The standard simulation can determine the nominal network terminal temperature. The U values for the standard case are calculated from the online measurement and estimated temperatures as discussed above.

7.8.3 Root Cause Analysis Capability

When a heat exchanger is severely fouled, one needs to know the root causes. For example, it could be caused by low velocity or type of crudes or else. Identification of root causes lead to determination of the most appropriate fouling mitigation and cleaning method.

7.8.4 Optimization Model

The optimization model focuses on those exchangers prone to foul and select the exchangers to clean with the biggest effect on the network terminal temperature. A list of exchangers requiring clean is displayed with ranking as to which exchanger to be cleaned when at what benefits. The results provide the guidance for determining the cleaning schedule as to which one to clean and when.

7.8.5 *U* Value Trend and Fouling Forecasting

U value trends for major exchangers are provided for monitoring. Fouling forecasting predicts the fouling trend in the near future if some exchangers are left for operation as well as the trend of fouling built up after some exchangers are selected to clean but not immediately. The former case is to decide not to clean some exchangers, which can get fouled very quickly, and thus the benefit of cleaning does not justify the cost of cleaning. The latter case happens when it might be too early to clean the exchanger based on cost–benefit analysis, but the user can make a schedule to clean the exchanger in the future.

7.9   PUTTING ALL TOGETHER—A PRACTICAL EXAMPLE OF FOULING MITIGATION

The ultimate goal for fouling mitigation is to select the most cost-effective method to provide the best exchanger performance. To achieve this overall goal, a systematic method is discussed (Figure 7.6), which can identify the root causes

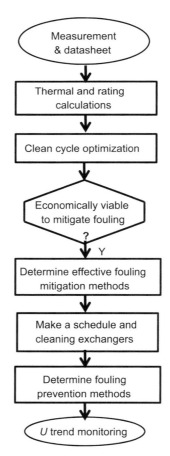

FIGURE 7.6. A complete methodology for fouling analysis and mitigation.

of fouling, determine the economic justification, and eventually determine the effective methods.

The loss in heat transfer largely depends on how fast fouling is developed. The U value trend should be able to indicate fouling development over time, while heat exchanger rating assessment can give the actual fouling as compared to nominal fouling factors. It is important to conduct a root cause assessment for the fouling problem in question, carry out chemical analysis of fouling materials if necessary, and consider appropriate fouling mitigation solutions, which include preventive methods via use of advanced heat transfer technology and online chemical or mechanical cleaning.

Fouling problems are process specific and thus unique solutions must be found. Let us use the crude distillation process in a petroleum refining plant to walk through the steps of the methodology from identifying root causes, rating assessment, to determining fouling mitigation methods.

In the example process, the crude unit has a charge rate of 200,000 bpd and crude density is 35.6 API. There is a crude preheat train consisting of 18 heat exchangers with 10 before and 8 after the desalting system. Crude oil from storage tanks is fed to the heat exchangers of the crude preheat train, initially at ambient temperature. The crude oil is then heated through the preheat train to 480 °F (250 °C) at entry to the feed furnace.

Out of the furnace, the crude is fed to a distillation column where valuable products, such as gases, kerosene, gasoline, and diesel are distilled and collected. After the process unit is operated for less than six months, severe fouling occurs in the heat exchangers and leads to a decline in furnace inlet temperature by as much as 65 °F, which results in burning of a significant amount of extra fuel in the furnace to make up for the temperature necessary for effective distillation.

7.9.1 Know the Price for Fouling Mitigation

The average heat recovery from the preheat train is 378 MMBtu/h versus the maximum heat recovery of 511 MMBtu/h if every exchanger is in relatively clean condition. This implies that 133 MMBtu/h is lost due to fouling and percentage of heat recovery is 74%. The plant engineers determined that if 50% of this lost heat could be captured by better fouling control, about extra 70 MMBtu/h of process heat can be recovered and thus 82 MMBtu/h of fuel could be saved in the feed heater. In monetary terms, it could save approximately $4.0 MM/year with natural gas price at $6/MMBtu. These savings do not take into account additional benefits of larger throughput achievable as the furnace capacity is limited for the unit. Thus, the plant management decided to approve a fouling control project to capture the benefit.

7.9.2 Rating Assessment

The rating assessment was conducted for heat exchangers in the preheat train and the assessment results indicated that fouling occurred, to a varying degree, to most of the heat exchangers, but the severe fouling happened only to the heat exchangers under high temperature located near the terminal of the preheat train. These heat exchangers mainly involve services provided by reduced crude, heavy vacuum gas oil (HVGO), and vacuum residue.

The severe fouling in these exchangers caused a higher pressure drop, which vaporized the crude more than 5% before the feed heaters with flow path control valves wide open. High crude flashing made it impossible to control feed flows with eight paths. This could be a safety hazard. The worst case could happen if the furnace was allowed to continue to operate in this unsafe mode. This is because stratified flow pattern forms in tubes, and tube walls could be directly exposed to combustion flames and become too hot. This could eventually lead to tube rupture. The finding of this reliability risk made the fouling mitigation project even more urgent.

7.9.3 Root Cause Analysis

Root cause analysis was conducted, which revealed the following issues:

- *Crude Properties:* During offline cleaning, the bundle from one of the very fouled heat exchangers was pulled out for inspection and chemical analysis of fouling materials was performed (Table 7.3). It was found that asphaltenic deposits were dominant (Figure 7.7), which is caused by polymerization of asphaltene. This explains the reason why the most severe fouling occurs under high temperatures. In comparison, the crude corrosive properties (e.g., reactive sulfur and naphthenic acids) were found to be much less responsible for depositional fouling.
- *Low Velocity:* Low velocity was another primary cause for promoting fouling, which occurs in some of the heat exchangers. The velocities in these exchangers were found to be 3 to 4 ft/s versus 6 ft/s recommended for these services.
- *Poor Desalter Operation:* It was found that salts were not removed sufficiently from the desalter, which makes fouling worse.
- *Exchanger Geometry:* The heat exchangers located in the terminal of the preheat train suffered the most severe fouling. Use of helical baffle bundles could help mitigate fouling much better than the shell-and-tube bundles currently used.

TABLE 7.3. Fouling Deposit Compositions from Lab Test

Materials	Weight %
Iron sulfides/oxides	15
Asphaltic material (soluble in benzene)	20
Asphaltic material (insoluble in benzene)	50
Inorganic salts (CO_3, SO_4, Cl)	15

FIGURE 7.7. Asphaltenic deposits found in the heat exchangers.

7.9.4 Explore Fouling Mitigation Options

Based on the fouling mechanisms and compositions found and fouling mitigation methods available, a fouling mitigation strategy was developed:

(a) *Offline Cleaning:* Proper cleaning of crude-HVGO exchanger (heavy gas oil) to restore the performance of this exchanger.
(b) *Operation Change:* Improve desalting operation to minimize salt related fouling.
(c) *Design Change:* Replace existing vertical baffle bundle with helical baffle for crude–residue exchanger. This design change can prolong the operation for this exchanger.
(d) *Flow Path Change in Maintenance:* Increase flow velocity from 3 to 5 ft/s for two exchangers.

Let us go through each of these fouling mitigation solutions in detail.

7.9.4.1 Offline Cleaning of Crude–HVGO Exchanger   Since this exchanger suffered severe fouling, it was difficult to clean using water-jetting alone. To have effective cleaning, a combination of different cleaning methods was employed for two shells of this exchanger:

- *Cleaning Method 1:* Use hot light cycle oil to clean two bundles for E1007-A/B.
- *Cleaning Method 2:* Hydro-blasting E1007-B using water jet.
- *Cleaning Method 3:* Combination of water-jetting and use of abrasive balls for cleaning E1007-A.

The performance trend in *UA* is plotted in Figure 7.8, which shows significant improvements from different cleaning methods (1–3). As a result, the bare metal performance for E1007 was established (Figure 7.9).

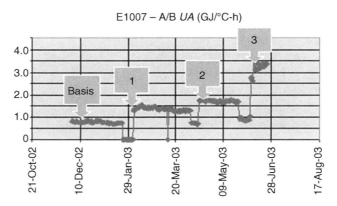

FIGURE 7.8. *UA* improvements for different cleaning methods.

7.9.4.2 Improving Desalting Performance   Crude requires desalting to minimize fouling and corrosion caused by salt deposition on heat exchanger surfaces and acids formed by decomposition of chloride salts. This crude unit has a one-stage desalting design, which can reduce the salt content in crude by 95%. The desalting and operating objectives are to remove salt, solids, and metal while minimizing the loss of oil in water phase due to oil entrainment. The key operating parameters such as desalting temperature, water injection, and pH control were optimized, which could achieve the target desalting performance as shown in Table 7.4.

7.9.4.3 Design Change: Change Bundle Type   E1011 and E1012, the crude–residue heat exchangers arranged in parallel, are the last two exchangers in the preheat train before fired heaters. The crude was placed on the shell side in design with the purpose of reducing pressure drop for the crude. The asphaltene deposit quickly built up from the dead flow zones (Figure 7.10) and then filled up the flow channels on the shell side where crude was placed.

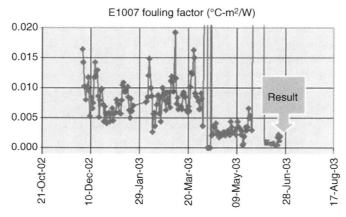

FIGURE 7.9. Bare metal performance established after cleaning.

TABLE 7.4. Benchmark Desalting Performance for the Single Stage Desalting

	Unit	Benchmark
Salt in desalted crude	ppm	0.6–6
Salt removal	%	90–98
Solids removal	%	50–80
Solids in desalted crude	%	0.1–1.0
Na removal	%	95+
Oil in effluent water	%	0–2
Iron removal	% inorganic iron	50–75
Slops reprocessing	% of crude charge	0.1–1.0

The design change was implemented to replace existing tube bundles with single helical baffle bundles (Figure 7.11). The helical flow eliminated dead flow zones as well as enhanced heat transfer on the shell side due to the countercurrent flow. Not only was heat transfer duty augmented greatly, but pressure drop was reduced at the same time.

7.9.4.4 Flow Path Change in Maintenance After implementing bundle replacement for the two crude–residue exchangers, plant engineers focused on the other two crude–residue exchangers, namely, E1009 and E1010 located upstream of E1011 and E1012. High pressure drop from measurement signaled that both E1009 and E1010 were also experiencing severe fouling. The rating assessment indicated that the velocity of the crude was too low only at 3.9 ft/s by design versus the required velocity of 6 to 7 ft/s. The solution to this problem was to increase the tube passes from four to

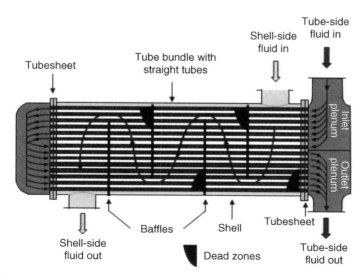

FIGURE 7.10. Dead flow zones in straight baffle shell-and-tube exchanger (one shell pass and two tube passes).

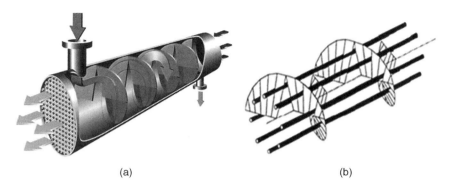

(a)   (b)

FIGURE 7.11. Helix shell-and-tube exchanger. (a) Outside view. (b) Inside view. (From Master et al. (2003), reprinted with permission from Engineering Conferences International.)

six and thus velocity increased to 5.8 ft/s at the expense of higher pressure drop than four tube passes. The extra pressure drop came from the replacement of the existing bundles with helical baffle bundles for E1011 and E1012.

7.9.4.5 Measuring and Sustaining Benefits The overall energy-saving benefit from the above improvements was about 80 MMBtu/h worth about $3 million per year on a sustainable basis. Importantly, the unsafe operation with feed flashing before heaters was resolved. The implementations required cleaning of one heat exchanger, replacement of four tube bundles for two exchangers, change of tube passes for two exchangers, as well as improvement for desalting operation. As a measure to sustain the benefits, a monitoring system and clean cycles were developed for these heat exchangers.

NOMENCLATURE

A_i surface area of heat exchanger, ft^2
A constant energy cost, $/day
B constant energy cost, $/day/°C
c_f energy cost, $/GJ
C heat exchanger cleaning cost, $
C_p fluid heat capacity, BTU/(lb °F)
E energy cost, $/year
E_2 energy cost during cleaning of heat exchanger, $/cleaning period
h_t heat transfer coefficient of tube side, BTU/(h ft^2 °F)
h_s heat transfer coefficient of shell side, BTU/(h ft^2 °F)
k fluid thermal conductivity, BTU/(h ft °F)
K lumped heat exchanger cleaning cost, $
K'' pseudoheat exchanger cleaning cost, $
M,m mass flow rate for hot and cold streams, lb/h
M_G gross margin, $/year

M_n average net margin, \$/year
m_1 gross margin during normal operation, \$/day
m_2 gross margin during heat exchanger cleaning, \$/day
Q heat duty, MMBtu/h or GJ
R_f heat transfer resistance, ft^2 °F h/Btu
T temperature, °F
T^N terminal temperature of heat exchanger network, °C
t_1 operation time before exchanger cleaning, days
t_2 exchanger cleaning time, days
U overall heat transfer coefficient, Btu/(ft^2 °F h)

Greek Letters

α constant, GJ/day
β constant, GJ/day/°C
λ decay of temperature, °C/day

Subscripts and Superscripts

0   starting time
A   actual
C   clean
D   design
f   fouling
lm   logarithmic mean
R   required
s   shell side
t   tube side

REFERENCES

Barletta AF (1998) Revamping crude units, *Hydrocarbon Processing*, February pp. 51–57.

Bott TR (1990) *Fouling Notebook*, Institution of Chemical Engineers, Rugby, UK.

Lieberman NP (1997) *A Working Guide to Process Equipment*, McGraw-Hill, New York.

Master B, Chunangad KS, Pushpanathan V (2003) Fouling mitigation using Helixchanger heat exchangers, presented at ECI Conference, Heat Exchanger Fouling and Cleaning: Fundamentals and Applications.

Melo LF, Bott TR, Bernardo CA (eds) (1988) *Advances in Fouling Science and Technology*, Kluwer Academic Publishers.

Shah RK, Sekulić DP (2003) *Fundamentals of Heat Exchanger Design*, John Wiley & Sons.

Smeenk G, Kers T (2001) In time cleaning of the crude preheat train, BP Nerefco Refinery.

Zubair SM, Sheikh AK, Budair MA, Badar MA (1997) A maintenance strategy for heat-transfer equipment subject to fouling: a probabilistic approach, *ASME Journal of Heat Transfer*, **119**, 575–580.

8

ENERGY LOSS ASSESSMENT

What are the major energy losses in a process? This is the first question that people should ask before embarking on a significant effort to improve process energy efficiency. The answer to this question could lead to identification of major improvement opportunities and help to define the need for large energy improvement effort.

8.1   INTRODUCTION

In a process, energy losses consist of both thermal and mechanical losses. Thermal losses are typically originated from column overhead condensers, product run down coolers, furnace stack, steam leaks, poor insulation of heat exchangers/piping and vassals and so on. Mechanical losses could also be significant, which usually occurs in rotating equipment, pressure letdown valves, control valves, pump spill back, heat exchangers, pipelines, and so on. Some of the thermal and mechanical losses are recoverable with a decent payback of investment but many others do not.

An energy loss audit, as discussed here, seeks to identify key recoverable losses. The audit is relatively quick and is designed to determine improvement potential. If the energy loss audit identifies large energy losses, more detailed energy assessment efforts will be undertaken later if so required. Detailed assessment methods will be discussed in the later chapters.

Energy and Process Optimization for the Process Industries, First Edition. Frank (Xin X.) Zhu.
© 2014 by the American Institute of Chemical Engineers, Inc. Published 2014 by John Wiley & Sons, Inc.

8.2 ENERGY LOSS AUDIT

The major loss assessment components are discussed below.

8.2.1 Heat Loss Assessment for Process Coolers

Heat loss assessment is conducted mainly based on temperature and pressure measurement. It is essential that a reference temperature be defined as the first step. The reason for defining a reference temperature in the energy loss audit is that heat above this temperature is considered as recoverable and thus accounted as loss if it is rejected to coolers. The typical reference temperature is 100 °F. The actual reference temperature to be selected for the heat loss calculation depends on the technical feasibility and economic viability for low-temperature heat recovery.

Table 8.1 gives the example heat loss calculations based on Table 3.1 in Chapter 3. As what can be observed, Table 8.1 shows both the amount of energy lost and the temperature at which the loss occurs. For example, the reaction effluent enters the air cooler at 394 °F with cooling duty of 192 MMBtu/h. This clearly reveals a significant opportunity of recovering reaction effluent heat. Other major energy losses include fractionation column overhead vapor cooling, diesel product cooling, and kerosene pump-around. These losses occur at relatively high temperatures.

On the other hand, other heat losses occur at relatively low temperatures. The question is what to do with such low-temperature heat. In general, there are a number of ways to recover low-temperature heat. One way is to use it for process heating such as feed preheating. In this case, a new heat exchanger may be installed to

TABLE 8.1. Heat Losses for the Example Process

Equipment	Services	Process Temperature, °F		Duty, MMBtu/h
		In	Out	
Water coolers °C				
1323	Heavy naphtha product	174	78.0	4.6
1317	Kerosene product	230	77.0	3.8
1320	Diesel product	289	110	11.4
1310	Reaction effluent	150	112	27.5
1314	Fractionation column kerosene pump-around	306	173	39.4
1305	H_2-make-up gas	290	110	30.0
1306	H_2 make-up compressor intercoolers	235	105	6.4
1348	Lean Amin	131	120	1.4
Air coolers °C				
1309	Reaction effluent	394	150	192
1312	Debutanizer column overhead	165	110	21.3
1313	Fractionation column overhead	290	116	73.3
1322	Naphtha splitter overhead	171	129	12.9

TABLE 8.2. Low-Grade Heat Recovery Options

Low Grade Heat Recovery Options	Heat Sink Temperature, °F	Heat Source Temperature, °F	End Use
Cold Feed Preheat	>100	>125	Process use; less fuel or steam used for feed preheat
BFW Preheat (deaerated)	200–250	>230–270	More HP steam gen in boilers or less fuel fired for the same steam gen
LP Steam Generation (50 psig)	298	>318	Process use; air preheat, power generation via a condensing turbine
LLP Steam Generation (30 psig)	274	>294	Air preheat
MP Steam Generation (150 psig)	365	>385	Process use, power generation via an extraction or condensing turbine
Thermocompressors	265–300	>285–320	Raise LLP steam to LP or MP steam using MP or HP steam as motive steam

provide the service. Other ways include generation of low-pressure steam or for boiler feed water preheating. It is also possible to generate hot water for heating office buildings, tank farms, and residential houses if they are located nearby. Table 8.2 provides some general guidelines for the options of recovering low-temperature heat.

8.2.2 Power Loss Assessment for Rotating Equipment

It is common that pumps and compressors are used in the process industry to transport process liquid and gas streams. The power requirement for pumping liquid can be estimated via

$$\text{HP} = M \times \Delta P \times \eta/1715, \tag{8.1}$$

where HP is shaftwork in break horse power (BHP); M is mass flow in GPM (gallons per minute); and ΔP is pressure drop through expander in psi. Typical pump efficiency is 40–60%.

The power requirement for a compressor can be estimated (Branan, 2005) via

$$\text{HP} = \frac{(144/33{,}000) \times (k/k - 1) \times P_i M_i \times \left(r^{(k-1)/k} - 1\right)}{\eta}, \tag{8.2}$$

where M_i is actual mass flow measured at inlet conditions, CFM (cubic feet per minute); k is ratio of specific heat $= C_p/C_v$; r is compression ratio $= P_o/P_i$; P_i is suction pressure, psia; P_o is discharge pressure, psia; and η is overall efficiency for a compressor, percentage; typical compressor efficiency is 60–80%.

Steam turbines are commonly used as process drivers. The theoretical rate in lb/kWh for steam turbine is determined from Mollier charts following a constant entropy path. The actual steam rate is calculated by theoretical steam rate divided by turbine efficiency. The discussions on Mollier charts and calculation examples are presented in Chapter 15.

The assessment for pumps/compressors and steam turbines together can be done by first estimating the process power requirement for running pumps or compressors based on equations (8.1) and (8.2). Then the current steam rate can be measured online and specific steam rate is estimated from the theoretical steam table. The ratio of process power requirement and the theoretical power delivered from the steam turbine indicates the overall efficiency of pump/compressors and turbines combined. If the efficiency is much lower than design, extra steam must be required to run the same pump or compressor.

An example is that a large pump is driven by a steam turbine operating between 150 psig and atmospheric condition. 3800 lb/h steam is measured online. The pump power requirement is estimated by equation (8.1) as 72 hp. On the other hand, the theoretical power supplied by the steam turbine can be estimated from the theoretical steam table. The steam table indicates the specific steam rate of 18.22 lb/kWh, which indicates the steam turbine can deliver 208.6 kW (3800 lb/h/18.22 lb/kWh) or 279.6 hp. Thus, the overall efficiency of the turbine and pump is only 26% (72/279.6). This efficiency is much lower than design. It was found that steam turbine blades were severely eroded over time and no inspection and maintenance work was ever done on the turbine for a long time.

8.2.3   Power Loss Assessment for Letdown Valve

The power loss from a process liquid stream going through a pressure letdown valve could be significant. If a power expander is considered to replace the letdown valve, equation (8.1) could be used to estimate the power recovery potential, and a typical power generation efficiency for a liquid expander is 75%.

For the process shown in Figure 4.1, a process liquid stream from the hot separator with mass rate of 1817.2 gpm and pressure of 2635.3 psig reduces its pressure through a pressure letdown valve to the hot flash drum at 285.3 psig. Equation (8.1) gives an estimate of power recovery potential of 1.38 MW if a liquid expander is installed.

For the steam letdown valve, the power recovery potential can be estimated via

$$W = M \times (h_1 - h_2) \times \eta_{is}/3413, \tag{8.3}$$

where W is power, kWh; $(h_1 - h_2) =$ steam enthalpy change for isentropic expansion, Btu/lb; and $M =$ letdown steam rate, lb/h.

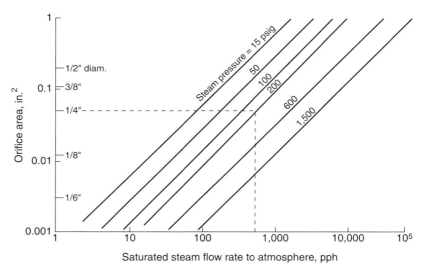

FIGURE 8.1. Steam loss rates at varying leak sizes and steam pressure. (From Kenney (1984), reprinted with permission by Elsevier.)

8.2.4 Motor Electricity Loss Assessment

The motor efficiency degradation could happen to old motors as motor windings deteriorate with age and will pull more amps. The horsepower output from a motor can be estimated (Branan, 2005) as

$$HP = E \times \eta/0.746. \tag{8.4}$$

E is electricity input in kW, which is calculated as

$$E = I \times V \times 1.73/1000 \times PF \text{ (three phase)}, \tag{8.5}$$

where I is electrical current in amps; V is volts (line-to-line), PF is power factor (fraction); and η is motor efficiency (fraction).

8.2.5 Heat Loss Assessment for Fired Heaters

Increasing heater efficiency can reduce operating costs significantly. For example, for a fired heater with a heat release of 200 MMBtu/h, improving the efficiency by 1% results in savings of $100,000 per year at a fuel cost of $6/MMBtu. The cost savings from heater improvement could be very significant because there are many fired heaters in a large processing complex.

To determine heater improvement opportunity, evaluation of heater efficiency must be conducted as a major part of an energy audit. The method for heater efficiency calculations is explained next.

The heat efficiency indicates how much of fuel fired (Q_f) is absorbed by the process (Q_p), which can be expressed as

$$\eta = \frac{Q_p}{Q_f} \times 100\%. \tag{8.6}$$

Since the amount of heat absorbed from the process as a percentage of vaporization is unknown, we have to turn our attention to heater heat losses (Q_L). The relationship of heater efficiency and heater losses can be described as

$$\eta = \frac{Q_f - Q_L}{Q_f} \times 100\%. \tag{8.7}$$

The amount of fuel fired, Q_f, can be readily obtained by measuring the flow rate of the fuel fired (lb/h) and the fuel's heating value (BTU/lb). Thus, the heater efficiency can be determined by estimating the heat losses in a heater. Three major heat losses occur to a heater:

- Stack loss due to high temperature of flue gas leaving the stack,
- Excess air loss due to high O_2 content used,
- Casing loss due to radiation loss from heater casing. This amount is usually constant.

Applying the method above can generate heat curves with stack temperature and oxygen content against efficiency. Thus, the heater efficiency can be readily obtained from measured stack temperature and oxygen content.

8.2.6 Heat Loss Assessment for Towers

The heat loss due to higher process energy use than necessary occurs mainly in over-reboiling for separation towers. Over-reboiling is at the expense of extra energy consumed in the reboiler. It is not uncommon for operators to operate towers in higher reflux rates than necessary. This is particularly true in turndown operation when reflux rate is not reduced accordingly, which results in over-reboiling. It is also found that some operators like to use high reflux rate to improve the tower operation stability.

Steam reboiler duty and efficiency can be estimated directly from steam rates and conditions. We can also estimate furnace reboiler duty and efficiency. The amount of fuel fired in the reboiler can be measured from the meter. The heat delivered by fuel combustion can be estimated from both fuel rate and heating value of the fuel.

8.2.7 Heat Loss Assessment for Steam Leakage

This loss can be a major component of overall energy losses. Leaks out of the steam system are relatively easy to spot from steam plume. Leaks can also occur inside the steam system when steam leaks from steam traps directly to the condensate system, which is invisible. The question is how to estimate the amount of steam losses occurring from both visible and invisible leaks.

8.2.7.1 Steam Leaks from Steam Pipes (Visible)   The most common causes of steam leaks come from holes in steam pipes and threaded pipe connections in a steam and condensate system. Steam flow through a leak can be calculated using an orifice equation, which is based on the diameter of the leak, pressure at the inlet of the orifice, and pressure at the outlet (atmosphere).

For simplicity, Figure 8.1 is used to give the steam loss rates in lb/h at varying leak sizes and steam pressures. Let us consider an application example. A steam leak survey found steam leaking from a gasket between two pipe flanges in a 200 psig steam pipeline located in a hydrocracking unit. The hole was measured as 0.2 in. wide and 1.25 in. long. The amount of steam lost can be estimated as shown next.

The area of the leak is calculated as

$$A = 0.2 \times 1.25 = 0.25 \text{ in.}^2.$$

From Figure 8.1, a hole of 1/4 in. for a 200 psig steam pipe would lose about 750 lb/h of steam. Since the process unit operates for 8,300 h per year, the annual savings from fixing the leak could be around $100,000 per year for a given 200 psig steam price at $16/klb.

8.2.7.2 Steam Leaks from Steam Straps (Invisible)   The purpose of a steam trap is to separate steam and condensate. It is designed in such a way that it opens to allow condensate to go through, but closes when presented with steam (hence "trapped"). If operating properly, steam traps have negligible loss of steam. However, when steam traps stick in the open position, steam passes through the trap valve directly to the condensate system. There are many hundreds of steam traps all over the place in a site and many of them could operate in poor conditions if there is no proper maintenance in place.

A steam trap is essentially a control valve and the valve characteristic is described by the C_v factor. How can one estimate steam trap losses via C_v factor? Calculation steps are explained below

- *Step 1:* Pyrometer measurements must be taken from the inlet (steam) and outlet (condensate) of the steam trap to identify if the trap valve fails to close when steam is present. The temperature difference between inlet and outlet of the steam trap indicates the subcooled degree of condensate. For example, $0\,°F$ subcooled condensate indicates the trap valve sticks in a full open position.
- *Step 2:* The C_v factor can be estimated from valve characteristic plots provided by manufacturing. Figure 8.2 shows condensate flow plots at different trap inlet pressures. In one example given by Branan (2005), the pyrometer measurements detect a trap on 150 psi steam line blowing live steam. The catalog rating

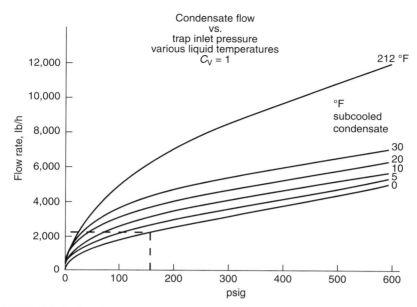

FIGURE 8.2. Estimating trap C_v. (From Branan (2005), reprinted with permission by Elsevier.)

of the trap is 5000 lb/h at saturation temperature at 150 psi. Figure 8.2 indicates the condensate flow is 2100 lb/h under 0 °F subcooled. Thus,

$$C_v = 5000/2100 = 2.38.$$

- *Step 3:* The steam rate (M_c) leaking through the trap under $C_v = 1$ is obtained from the plot of steam rate versus steam pressure, which is also provided by manufacturing. Figure 8.3 shows $M_c = 220$ lb/h at 150 psig with $C_v = 1$. Then the actual steam rate (M_{loss}) through steam trap is estimated as

$$M_{loss} = C_v \times M_c = 220 \times 2.38 = 523.6 \, \text{lb/h}.$$

- *Remarks:* The above steam loss from one steam trap could cost around $50,000/year if the 150 psig steam price is $11/klb. How much money could the plant lose if 40 steam traps malfunction at the same time?

8.2.8   Heat Loss Assessment for Condensate Loss

Loss of condensate is another major component of heat loss. For maximum efficiency, condensate should return to boilers as much as possible. Otherwise, cold fresh makeup water will be added for any amount of condensate lost. The boiler duty has to increase to make up for the sensible heat difference between makeup water and condensate. In addition, condensate return will reduce water treatment cost, which is expensive. The condensate is valued at the same price as boiler feed water (BFW). To achieve maximal condensate return, a condensate system is necessary, which collects condensate from various steam traps. Care must be taken,

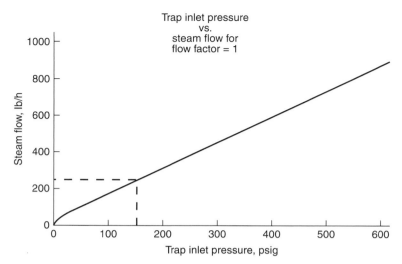

FIGURE 8.3. Estimating actual steam loss. (From Branan (2005), reprinted with permission by Elsevier.)

however, to ensure that the condensate is not contaminated. To do so, contamination level should be monitored if it can be readily detected. Even so, condensate from some traps may not be collected if the distance is too far and the amount of condensate is too small to justify the investment for condensate return. It is common to observe 20–30% of condensate loss, but 50% is unacceptable, which should be reduced to the acceptable level.

8.2.9 Heat Loss Assessment for Poor Turndown Operation

It is common that energy usage does not reduce accordingly when feed rate drops. Turndown assessment should address the following issues when operators fail to

- Close secondary air registers on furnaces
- Slow down steam-driven pumps and compressors
- Reduce tower reflux rate
- Cut-down tower stripping steam rates
- Reduce vacuum ejectors' steam rates, etc.

8.2.10 Heat Exchanger Fouling Assessment

Fouling reduces effective surface area and hence process heat transfer duty. The result is increased utility usage. Good fouling control could reduce total energy use significantly in the order of 5–10%. U value could be used to assess the level of heat exchanger fouling. Due to the fact that heat exchanger performance varies with flow rates, compositions, and fouling conditions, heat exchanger assessment must be conducted on a regular basis so that a performance trend over time can be obtained and faults can be detected at the early stage. A single point U value is good for getting a baseline data on its performance, but it must be done on a regular basis to define trends. The U value trend can tell the fouling levels as well as useful hints of problems in comparison with historical performance.

Heat exchange duty (Q) and LMTD (ΔT_{lm}) can be calculated first using equations (6.10) and (6.7) based on online data and then U value is determined using equation (6.1)

$$U = \frac{Q}{A F_t \, \Delta T_{\mathrm{lm}}}, \tag{6.1}$$

$$\Delta T_{\mathrm{lm}} = \frac{\Delta T_1 - \Delta T_2}{\ln\left(\Delta T_1 / \Delta T_2\right)}, \tag{6.7}$$

$$Q = MC(T_1 - T_2) = mc(t_2 - t_1), \tag{6.10}$$

where A is surface area, ft^2; C, c are fluid heat capacity of hot and cold streams, Btu/ (lb °F); T_1, t_1 are supply temperature of hot and cold streams, °F; T_2, t_2 are target temperature of hot and cold streams, °F; ΔT_1 is hot end temperature approach, °F; ΔT_2 is cold end temperature approach, °F; ΔT_{lm} is logarithmic mean temperature difference (LMTD), °F; F_t is correction factor to ΔT_{lm}, fraction; M, m is mass flow rate for hot and cold streams, lb/h; Q is heat duty, MMBtu/h; and U is overall heat transfer coefficient, Btu/(ft^2 °F h).

8.2.11 Miscellaneous Energy Losses

Although major energy losses are addressed as above, other miscellaneous energy losses could occur and some of the examples are:

- Ambient heat loss remains constant regardless of throughput.
- Cold air enters furnaces through leakage holes, which results in higher fuel use.
- Poor insulation could cause major heat losses from hot process equipment, hot flanges, and steam pipes.
- Pumps are pushed back on their performance curves, etc.

8.3 ENERGY LOSS AUDIT RESULTS

The energy loss assessment discussed above can be applied to a process unit. As an example, the hydrocracking unit with energy data shown in Table 3.1 is used for illustration of an energy loss audit. The energy loss audit for the unit is summarized in Table 3.2 and reproduced here. The reference for enthalpy calculation is 100 °F saturated liquid. For light gases that cannot exist as liquid at 100 °F, the reference state is that of an ideal gas at 100 °F. The fuel equivalent for purchased power is 9090 Btu/kWh. Power generation loss is thus assumed to be 5676 (9090−3414) Btu/kWh. Detailed discussions for calculations to produce Table 3.2 are given in Chapter 3.

Table 3.2 Tabulated Energy Loss Audit Summary for the Example of Table 3.1

Energy Input	FE MMBtu/h (above 100 °F)	Energy Output	FE MMBtu/h (above 100 °F)
		Energy Export	
Power	28.6	MP export	65.5
Fuel	337.1	Condensate return @141 °F	12.2
HP steam	292.3	Total	77.7
Heat of reaction	141.0	**Energy Lost**	**FE MMBtu/h (above 100 °F)**
Feed @170 °F	17.2	Power gen losses	17.9
Boiler feed water @250 °F	2.9	Air coolers	352.4
Total	819.2	Water coolers	148.9
		Furnaces stack loss	89
	FE MMBtu/h	Pumps & motors (mechanical loss)	2.86
Net Energy Input	741.4	Surface condensers	115.1
	FE kBtu/bbl	Condensate loss	0.9
Specific Energy Use	480.9	Unaccounted losses	14.6
		Total	741.5

Balance check: Energy input − Energy output = 819.2 − (77.7 + 741.5) = 0

The energy loss audit summary (Table 3.2) together with the energy loss assessment results discussed above reveals heat recovery opportunities and thus identifies several improvement ideas (Table 8.3). Listing potential ideas is often helpful in the early stage of idea discovery. From the table, three general directions for energy recovery for the example process unit can be suggested:

- *Direction 1:* Save power via liquid expander and/or steam turbine. Although it would be expensive to install a power recovery turbine, the capital cost needs to be justified from the amount of power recovered and price of power.
- *Direction 2:* Save feed furnace fuel via more feed preheating. The major part of waste heat for feed preheating could come from the reaction effluent as the reactor air cooler inlet temperature is 394 °F.
- *Direction 3:* Reduce steam via use of waste heat for steam generation and BFW preheating. The waste heat for this purpose could come from the diesel product run down, kerosene pump-around.

These ideas have distinctive merits that warrant further investigation, while some operation improvement ideas may be selected as quick hits for implementation. However, for improvement ideas requiring capital investment, it is too early to think about implementation at this stage. This is because the purpose of energy loss audit is to identify if there is a large amount of energy lost in operation and design. If yes, it triggers the need for a more detailed assessment discussed in the chapters later to determine specific improvement ideas. This energy loss audit is like doing an

TABLE 8.3. Idea List from Energy Loss Assessment

No.	Improvement Ideas	Services	Modifications (Qualitative)	Impacts
1	Recover more heat in reaction effluent	Frac feed preheating	Add heat exchangers	Reduce furnace duty
2	Recover diesel product heat	LP steam gen and BFW preheating	Add heat exchangers	Reduce steam import
3	Recover heat from fractionation column overhead vapor	Raw feed preheating	Add heat exchangers	Reduce steam import
4	Recover heat in kerosene pump-around	Reboiling service for other process units	Add heat exchangers	Reduce furnace duty
5	Power recovery from high-pressure liquid	Power generation	Install a liquid expander	Reduce power import

appraisal of energy performance from a high level. Otherwise, regretful investment may be the consequence of rushing into implementation when much better alternatives are identified later after implementation.

8.4 ENERGY LOSS EVALUATION

With ideas identified, one might wish to know the effects of capturing the energy loses from these ideas. Using the energy balance spreadsheet, the effects could be evaluated readily; however, it requires some assumptions based on experience. Table 8.4 shows such an example. The following crude but conservative assumptions are made in estimating the benefits as shown in Table 8.4:

- Half of the main fractionation tower overhead cooler duty can be captured for raw feed preheating.
- Half of the kerosene pump-around duty is used for providing reboiling services in other process units.
- Half of diesel rundown duty is captured for LP steam generation.
- Reaction cooler temperature is reduced to 350 °F from 394 °F and the heat recovered from the reaction effluent is used for feed preheat.

Potential benefits for each idea are incorporated in the energy balance and the "improved unit" is compared with the current performance in Table 8.4. The "improved unit" could reduce energy use close to 15%! This gives a very strong indication for the need of pursuing further energy study. However, it is too early to jump to conclusions as heat recovery assessment presented in Chapters 9 and 10 could provide deeper insights for the target and actual modifications required.

TABLE 8.4. Comparing Between Current and Improvement

(i) Energy In, MMBtu/h	Current	Improved Unit
Feed	17.2	17.2
Furnace	337.1	317.1 (1)
Heat of reaction	141.0	141.0
Electrical power	28.6	28.6
HP steam	292.3	292.3
Boiler feed water	2.9	2.9
Total in	819.15	799.2
(ii) Energy Out, MMBtu/h		
Energy recovered		
MP steam	65.5	71.2 (2)
Condensate return @141 °F	12.2	12.2
Power recover	–	5.2 (5)
Main frac tower overhead	–	36.7 (3)
Kero pump-around	–	39.4 (4)
Total recovered	77.74	164.7
(iii) Energy Lost		
Power generation losses	17.9	17.9
Water coolers	148.9	148.9
Air coolers	352.4	315.7
Furnaces	89	70.2
Pumps and motors	2.86	2.86
Surface condensers	115.1	115.1
Condensate	0.9	0.9
Unaccounted	14.6	14.6
Total lost	741.52	686.1
(iv) Net Energy Input	**480.9**	**411.6**

Note: (1), (2), (3), (4), (5) = ideas 1/2/3/4/5 in Table 8.3.

8.5 BRAINSTORMING

Brainstorming is a good vehicle for idea generation during the early stage of energy loss audit. It is used for identifying potential improvement ideas including operational, control, maintenance, as well as basic housekeeping such as leaking, insulation, and so on. The procedure and guidelines for brainstorming should be sent to the related people in advance given plenty of time for attendees to prepare. Attendees are encouraged to share one's own ideas and help others to expand their ideas. Meeting notes are made to record any ideas discussed and circulate them to attendees for feedback.

8.5.1 Expected Attendees

The most important factor for a successful brainstorming session is to have attendees who have sound technical and operational knowledge. The number of attendees should be restricted to less than 10 for focused discussions. To prevent discussions from being dominated by some big personality, the attendees could be invited to take turns in presenting their ideas.

- *Lead operators* are the operation owners of the process unit and are aware of almost everything that happened in the past and present. They must be invited for the brainstorming as they can provide valuable feedback for improvement ideas. In most cases, they are actually the main source of idea generation.
- *Process engineers* and *production engineers* for the process unit should be present as they are the technical owners of the unit. Process engineers know the overall process scheme and operating conditions. They determine key operating parameters for the unit operation. In today's environment, many process engineers are relatively young and enthusiastic; they can bring fresh and bold ideas on the table.
- *Control engineers* should be invited as control could be one of main areas for improvements. Operators may inform some erratic unit operation behavior, which could be the result of inappropriate control schemes and inconsistent operation, which become the lead for identifying ideas for control improvement.
- Other people could include the *asset manager* or *superintendent* of the unit if needed.
- The facilitator of the brainstorming session is preferably the *energy engineer* who has both energy and process knowledge and can ask out-of-the-box questions in the right context. *Consultants* from outside could be the alternative choice when they conduct energy optimization projects at the time.

8.5.2 Discussion Platform

A process flow diagram (PFD) provides a good basis for brainstorming. When a process engineer walks through the PFD line-by-line, attendees ask questions challenging the current operation and design from which ideas start to snow ball. There is a common saying: A figure is worth a thousand words. A PFD provides a visual aid to bring all discussions on the same page.

The procedure of walking through a PFD starts from the feeds entering the process unit. Opportunities for feed delivery include hot feed transfer directly from upstream processing instead of from feed tanks, as well as cold feed preheating using low-temperature heat available from process streams. Then the PFD review moves to reaction. Opportunities for a reaction circuit include changing of reaction conditions, changing of hydrogen partial pressure, recovery of reaction effluent heat, and minimization of power use in the recycle gas compressor etc.

The next step is the review of the fractionation system with the following opportunities in mind: change column conditions to maximize desirable products, avoid product specification give-away, and reduce reboiling duty. Other opportunities for a fractionation system is optimizing pump-arounds, changing feed temperature, recovery of heat from product rundowns, and changing of column internals to enhance throughput or fractionation efficiency.

The last but not least is identification of steam and power opportunities including optimizing reboiling and stripping steam in a fractionation tower and strippers, steam generation by use of process heat, and driver selection. Other issues worth discussing include control optimization for rotating machines and minimizing slop production. Here, only a few opportunities are mentioned among many potential ideas.

8.5.3 Time Duration

Sufficient time should be dedicated for brainstorming without interruption from other matters. Usually, 2 h could be good for one session and several sessions may be required for a process unit.

8.5.4 Ground Rules

Some ground rules must be set for brainstorming and attendees should be aware of these rules in advance. The first rule is that every attendee must not bring their ego into the meeting room. In other words, ego must be checked out of the meeting room and everyone is equal in discussions. The second rule is not killing any ideas during brainstorming. Remember the purpose of brainstorming is to generate ideas.

8.5.5 Nothing is Sacred

Although due merits for current operation practice and process design should be acknowledged, nothing should be regarded as sacred. Anything can be questioned and challenged when it comes to looking for improvement opportunities. Questions for "why" and "why not" should be asked to verify the rationales behind current operating practices and process configurations as they were developed based on different economic conditions and operation culture.

8.6 ENERGY AUDIT REPORT

The final step in the energy loss audit is to put things together in a report that details the findings and provides energy cost saving opportunities. The general guideline for the report is to keep the report simple and short. For each opportunity identified, three elements need to be reported: (i) the explanation of the opportunity; (ii) benefits and capital costs; and (iii) potential impacts on process operation and equipment reliability. To establish a baseline, the benchmarking assessment results should be conducted as a part of energy loss audit and included in the report.

The report may begin with an executive summary that provides the shareholders of the audited facility with a brief synopsis of the total savings available and the highlights of key opportunities with a roadmap of improving energy efficiency for the facility based on opportunities identified. An executive summary should be tailored to nontechnical personnel, and technical jargon should be minimized. The technical details can be discussed in the main body of the report. The report and related documents should be stored properly for future reference.

NOMENCLATURE

A	area
BHP	break horse power
C_p, c_p	fluid heat capacity of hot (cold) stream
C_v	Control valve characteristic factor
E	electricity
F_t	correction factor to ΔT_{lm}, fraction
h	enthalpy
HP	horse power
I	electrical current
k	ratio of specific heat $= C_p/C_v$
m, M	mass flow
ΔP	pressure drop
P	pressure
PE	power factor
Q	heat content
r	compression ratio $= P_2/P_1$
T_1, t_1	supply temperature of hot (cold) stream
T_2, t_1	target temperature of hot (cold) stream
ΔT_1	hot end temperature approach
ΔT_2	cold end temperature approach
ΔT_{lm}	logarithmic mean temperature difference (LMTD)
U	overall heat transfer coefficient
V	volts
W	power

Greek Letters

η efficiency

REFERENCES

Branan C (2005) *Rules of Thumb for Chemical Engineering*, Elsevier.
Kenney WF (1984) *Energy Conservation in the Process Industry*, Academic Press, Inc.

9

PROCESS HEAT RECOVERY TARGETING ASSESSMENT

Identifying savings opportunities for process heat recovery is an important step for process energy optimization. If the opportunity is large enough, it will warrant a more detailed study to determine what modifications are required at what capital investment. The benefit and cost will form the central part of the business case to be presented to management for approval of the capital budget.

9.1 INTRODUCTION

Although the energy loss audit discussed in Chapter 8 can reveal heat recovery opportunities, it does not give the overall view of process heat recovery options and does not address the two key questions for energy retrofitting. What is the maximal heat recovery target for a process unit? What modifications are required to close the gap against the target? The energy targeting assessment, which is the focus of this chapter, will seek answers to the first question and thus reveal the overall opportunity gap available on the table. If the opportunity is determined to be large, the attention will turn to retrofit analysis to search for specific modifications to answer the second question, which will be discussed in Chapter 10.

Let us consider the example problem in Figure 9.1. The process involves two reactors and one separation column and the current design features heat recovery between hot and cold process streams. The questions to the process engineer are: Is it

Energy and Process Optimization for the Process Industries, First Edition. Frank (Xin X.) Zhu.
© 2014 by the American Institute of Chemical Engineers, Inc. Published 2014 by John Wiley & Sons, Inc.

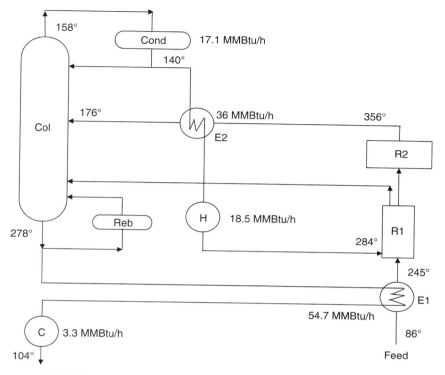

FIGURE 9.1. Process flow diagram for the example energy targeting.

possible to reduce the heating duty that is currently at 18.5 MMBtu/h and how much can be saved? This question will be answered in the following sections.

9.2 DATA EXTRACTION

A heat recovery targeting assessment starts with data collection for process streams based on existing process design and operation conditions. When extracting operating data, the questions are as follows: What time periods of data should be extracted? What feed rate and compositions should be used as the basis? The idea is to select the operating conditions that reflect the most common operation in terms of feed rate and product yields. On this basis alone, the benefit calculation has meaningful representation. Therefore, two criteria could be satisfied for data extractions:

- Feed rate and compositions should be most frequently used.
- Middle of run operation.

There are two kinds of process streams in the context of the energy targeting, namely, hot and cold streams. Hot streams are those whose temperature reduces and

TABLE 9.1. Stream Data Extracted from Figure 9.1

Stream Name	Inlet Temperature (°F)	Outlet Temperature (°F)	Flow Rate (lb/h)	CP (Btu/ (lb °F))
Column bottom product (A)	278.0	104.0	512,681	0.650
Column ovhd condenser (B)	158.0	140.0	102,536	9.249
Reactor 2 effluent to column (C)	356.0	176.0	256,341	0.780
Raw feed to reactor 1 (D)	86.0	245.0	512,681	0.668
Column ovhd liquid to reactor 1 (E)	140.0	284.0	51,268	7.382

thus heat is rejected. The hot streams in a process unit include reaction effluents, separation column overhead vapors, column bottom products, product rundowns, and so on. In contrast, cold streams are those whose temperature increases when heat is received. The cold streams are the feeds to reactors and columns, column reboilers, and so on. The data for each stream include the starting and terminal temperatures, flow rate, specific heat capacity, and heat load or enthalpy change. The stream data for the example process in Figure 9.1 are extracted as shown in Table 9.1. The column reboiler is not included as it is served by HP steam.

9.3 COMPOSITE CURVES

Composite curves were developed for heat recovery targeting (Linnhoff et al., 1982). The word "composite" reveals the basic concept behind the composite curves method: A system view of the overall heat recovery system. One hot composite stream represents all the hot process streams, while one cold composite stream represents all the cold process streams. In this manner, the problem of assessing a complex heat recovery system involving multiple hot and cold streams is simplified as a problem of two composite streams. In essence, the hot composite stream represents a single process heat source, while the cold composite stream represents a single process heat sink.

Consider the stream data of Table 9.1, which include three hot streams and two cold streams. Let us start building the hot composite curve first. The three hot streams are plotted on the T/H axis with their starting and terminal temperatures corresponding to the enthalpy changes (heat loads), respectively (see Figure 9.2).

What characteristics should a hot composite stream possess to represent the three hot streams? The composite stream should have two features: (i) It should go through the exact same temperature changes as the three streams do, and (ii) it should have the same total heat load as the summation of the heat loads of three streams.

For visualization of how these two conditions are met, let us inspect these three streams on the T/H axis as shown in Figure 9.2, where five temperature intervals are created from the starting and terminal temperatures of each stream. In interval 1, only stream C exists and thus the enthalpy change $\Delta H_1 = (T_1 - T_2) \times CP_C$. However, in interval 2, streams A and C contribute to it and thus the enthalpy change is the summation of two individual enthalpy changes, that is, $\Delta H_2 = (T_2 - T_3) \times (CP_A + $

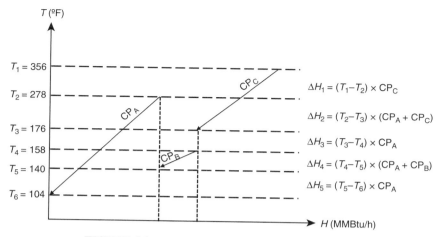

FIGURE 9.2. *T/H* representation of three hot streams.

CP_C). In interval 3, only stream A exists, $\Delta H_3 = (T_3 - T_4) \times CP_A$. For interval 4, streams A and B coexist and then $\Delta H_4 = (T_4 - T_5) \times (CP_A + CP_B)$. Finally, in interval 5, only stream A exists, so $\Delta H_5 = (T_5 - T_6) \times CP_A$.

By plotting the five temperature changes and corresponding enthalpy changes in sequence on the *T/H* axis, a hot composite stream is obtained and the resulting *T/H* diagram (Figure 9.3) is the hot composite curve, which satisfies the aforementioned two conditions. Clearly, the composite hot curve represents the three hot streams in terms of temperature and enthalpy changes.

Similarly, the *T/H* representation for the two cold streams can be shown in Figure 9.4 and the composite curve for the two cold streams can be constructed in Figure 9.5. By plotting both hot and cold composite curves on the *T/H* axis, we obtain the *T/H* diagram, which are so-called composite curves (Figure 9.6).

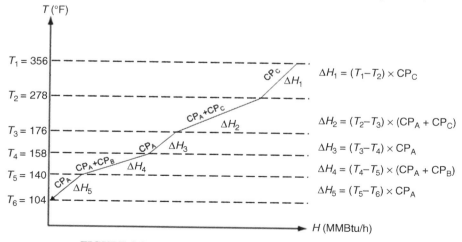

FIGURE 9.3. *T/H* representation of a hot composite stream.

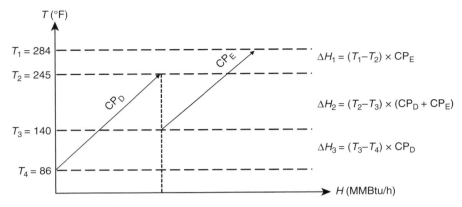

FIGURE 9.4. *T/H* representation of two cold streams.

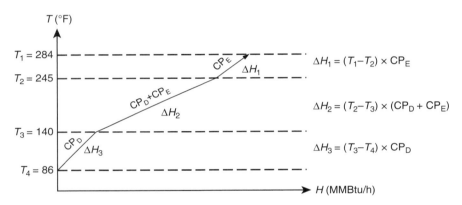

FIGURE 9.5. *T/H* representation of a cold composite stream.

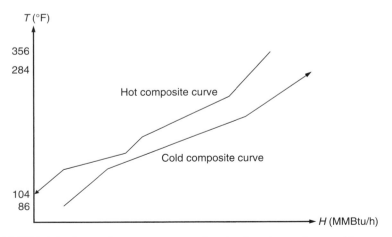

FIGURE 9.6. Composite curves representing the three hot and two cold streams.

9.4   BASIC CONCEPTS

What can be learned from the composite curves? Briefly, the curves can reveal very important insights for a heat recovery problem: maximal process heat recovery, pinch point, and hot and cold utility targets, which can be visualized in Figure 9.7. The explanations for these basic concepts are given as follows:

- *Minimum Temperature Approach* ΔT_{min}: For a feasible heat transfer between the hot and cold composite streams, a minimum temperature approach must be specified, which corresponds to the closest temperature difference between the two composite curves on the T/H axis. This minimum temperature approach is termed as the network temperature approach and defined as ΔT_{min}.
- *Maximal Process Heat Recovery:* The overlap between the hot and cold composite curves represents the maximal amount of heat recovery for a given ΔT_{min}. In other words, the heat available from the hot streams in the hot composite curve can be heat-exchanged with the cold streams in the cold composite curve in the overlap region.
- *Hot and Cold Utility Requirement:* The overshoot at the top of the cold composite represents the minimum amount of external heating (Q_h), while the overshoot at the bottom of the hot composite represents the minimum amount of external cooling (Q_c).
- *Pinch Point:* The location of ΔT_{min} is called the process pinch. In other words, the pinch point occurs at ΔT_{min}. When the hot and cold composite curves move closer to a specified ΔT_{min}, the heat recovery reaches the maximum and the hot and cold utilities reach the minimum. Thus, the pinch point becomes the bottleneck for further reduction of hot and cold utilities. Process changes must be made if further utility reduction is pursued.

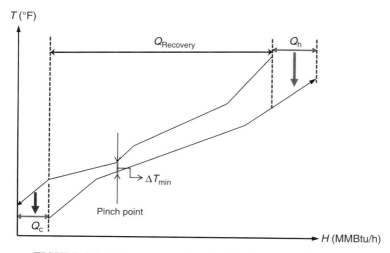

FIGURE 9.7. Basic concepts described in the composite curves.

9.5   ENERGY TARGETING

By assuming a practical ΔT_{min}, the composite curves can indicate targets for both hot and cold utility duties. For example, let the minimum approach temperature ΔT_{min} be 25 °F for the example problem in Table 9.1, the composite curves (Figure 9.8) give the targets for the minimum heating of 17 MMBtu/h versus the current duty of 18.5 MMBtu/h. At the same time, the minimum cooling requirement of 18.9 MMBtu/h is obtained in comparison with the current duty of 20.4 MMBtu/h.

The procedure of obtaining the composite curves and energy targets can be summarized. First, the stream data (Table 9.1) are collected and heating and cooling requirements of the process streams in terms of temperatures and enthalpies are determined. Next, the hot streams are plotted on the temperature–enthalpy axes (Figure 9.2) and then individual stream profiles are combined to give a hot composite curve (Figure 9.3). This step is repeated for the cold streams (Figure 9.4) to generate the cold composite curve (Figure 9.5). Finally, the two composite curves are plotted together to obtain the composite curves (Figure 9.8) for a given ΔT_{min}.

The general observation is that a large ΔT_{min} corresponds to higher energy utility but lower heat transfer area and thus lower capital cost, and vice versa (Figure 9.9). The calculations for heat transfer area based on the composite curves for a heat recovery system will be discussed in Section 9.7.

9.6   PINCH GOLDEN RULES

Once the pinch is identified, the overall heat recovery system can be divided into two separate systems: one above and one below the pinch, as shown in Figure 9.10a. The system above the pinch requires a heat input and is therefore a net heat sink.

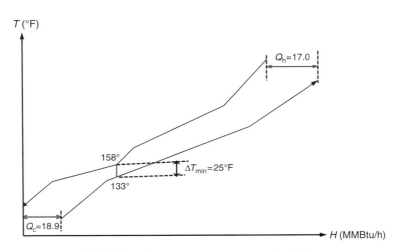

FIGURE 9.8. Energy targets for a specified ΔT_{min}.

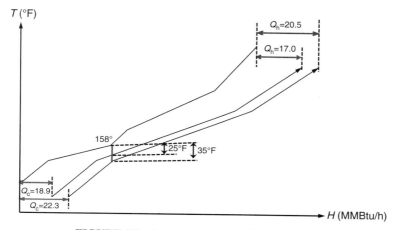

FIGURE 9.9. Energy targets for different ΔT_{min}.

Below the pinch, the system rejects heat and so is a net heat source. When a heat recovery system design does not have cross-pinch heat transfer, that is, from above to below the pinch, the design achieves the minimum hot and cold utility requirement under a given ΔT_{min}.

On the other hand, if cross-pinch heat transfer is allowed, what could happen? In Figure 9.10b, cross-pinch (XP) amount of heat is transferred from above to below the pinch. The system above the pinch now loses XP units of heat to the system below the pinch. To restore the heat balance, the hot utility must be increased by the same amount, that is, XP units. Below the pinch, XP units of heat are added to the system; therefore, the cold utility requirement also increases by XP units. The consequence of a cross-pinch heat transfer (XP) is that both the hot and cold utilities will increase by the XP amount, the cross-pinch duty.

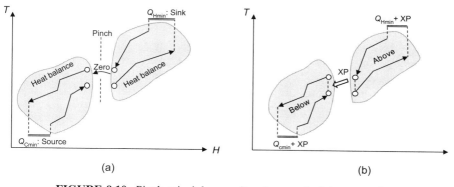

FIGURE 9.10. Pinch principle: penalty of cross-pinch heat transfer.

For the example process design in Figure 9.1, the cross-pinch heat transfer in the current design is equal to 1.5 MMBtu/h (18.5 MMBtu/h in current design versus 17 MMBtu/h as a target). Thus, for $\Delta T_{min} = 25$ °F, XP = 1.5 MMBtu/h, which is cross-pinch heat transfer.

Based on the cross-pinch heat transfer principle, if external cooling is used for hot streams above the pinch, it increases the hot utility demand by the same amount. Similarly, external heating below the pinch increases the overall hot and cold utility requirement by the same amount.

To summarize, there are three basic pinch golden rules (Linnhoff et al., 1982) that must be obeyed to achieve the minimum energy targets for a process:

- Heat must not be transferred across the pinch.
- There must be no external cooling above the pinch.
- There must be no external heating below the pinch.

Violating any of these rules will lead to penalty in both hot and cold utilities beyond the target.

9.7   COST TARGETING: DETERMINE OPTIMAL ΔT_{min}

The optimal ΔT_{min} is determined based on the trade-off between energy and capital such that the total cost for the heat recovery system is at a minimum.

For a grassroots design of a heat recovery system, the total annual cost consists of two parts, namely, the capital cost and the energy operating cost:

- The energy operating cost includes energy expenses for both hot and cold utilities, which is billed annually in $/year.
- The capital cost of the network is the summation of installed costs for all individual heat transfer equipment, including heat exchangers, furnaces, steam heaters, water coolers, and air cooler and refrigeration, and other related costs, including foundation, piping, instrumentation, control, and so on. Thus, it is a total investment ($) required to build the entire heat transfer system.

The surface area calculation for heat transfer equipment is the most importent part of cost targeting. For simplicity, the area model (Townsend and Linnhoff, 1983a, 1983b) is used to explain how surface area is calculated directly from composite curves. To do this, utilities are added to composite curves to make heat balance between hot and cold composites. Then the balanced composite curves are divided into several enthalpy intervals, and each enthalpy interval must feature straight temperature profiles (Figure 9.11).

There could be several hot and cold streams within an enthalpy interval. For each heat exchanger involving hot stream i and cold stream j in kth interval, the surface area and cost can be calculated via equations (9.1) and (9.2):

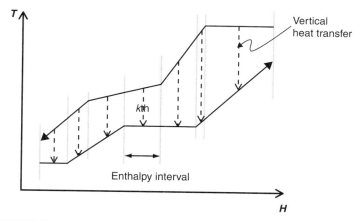

FIGURE 9.11. Calculation of surface area from the composite curves.

$$A_{i,j,k} = \frac{Q_{i,j,k}}{U_{i,j,k}\text{LMTD}_{i,j,k}}, \qquad (9.1)$$

$$C_{i,j,k} = a_{i,j,k} + b_{i,j,k}A_{i,j,k} = a_{i,j,k} + b_{i,j,k}\frac{Q_{i,j,k}}{U_{i,j,k}\text{LMTD}_{i,j,k}}, \qquad (9.2)$$

where A is the surface area, ft^2; Q is the heat load, MMBtu/h; LMTD is the logarithmic mean temperature difference, °F; U is the overall heat transfer coefficient, MMBtu/h/(ft^2 °F); a is the fixed cost for heat transfer equipment, \$; and b is the surface area cost, \$/ft^2.

Thus, total surface area and purchase cost for all heat transfer equipment in kth interval can be calculated via equations (9.3) and (9.4):

$$A_k = \sum_{i,j} A_{i,j,k} = \sum_{i,j} \frac{Q_{i,j,k}}{U_{i,j,k}\text{LMTD}_{i,j,k}}, \qquad (9.3)$$

$$C_k = \sum_{i,j} \left[a_{i,j,k} + b_{i,j,k}\frac{Q_{i,j,k}}{U_{i,j,k}\text{LMTD}_{i,j,k}} \right], \qquad (9.4)$$

where A_k is the total surface area for kth interval, ft^2, and C_k is the total purchase cost for all heat transfer equipment in kth interval, \$.

The overall surface area and capital cost for the network can be calculated via equations (9.5) and (9.6). The exchanger equipment costs must be converted to installed costs, which include purchase cost, foundation, piping, instrumentation, control, erection, and so on.

$$A_{\text{Network}} = \sum_k A_k = \sum_{i,j,k} \frac{Q_{i,j,k}}{U_{i,j,k}\text{LMTD}_{i,j,k}}, \qquad (9.5)$$

$$C_{\text{Network}}^{\text{Cap}} = \sum_k \sum_{i,j} I_{i,j,k} \left[a_{i,j,k} + b_{i,j,k}\frac{Q_{i,j,k}}{U_{i,j,k}\text{LMTD}_{i,j,k}} \right], \qquad (9.6)$$

where A_{Network} is the total surface area, ft^2; $C^{\text{Cap}}_{\text{Network}}$ is the total installed cost, \$; and $I_{i,j,k}$ is the installation factor for exchanger between streams i and j in kth interval.

On the other hand, utility consumption and costs can be calculated for both hot and cold utilities, respectively, as:

$$Q_{\text{h,Network}} = \sum_{\text{l}} Q_{\text{h,l}}, \tag{9.7}$$

$$Q_{\text{c,Network}} = \sum_{\text{p}} Q_{\text{c,p}}, \tag{9.8}$$

$$C^{Q}_{\text{Network}} = \sum_{\text{l}} c_{\text{h,l}} Q_{\text{h,l}} + \sum_{p} c_{\text{c,p}} Q_{\text{c,p}}, \tag{9.9}$$

where $Q_{\text{h,l}}$ is the heat load for hot utility l, MMBtu/h; $Q_{\text{c,p}}$ is the heat load for cold utility p, MMBtu/h; $c_{\text{h,l}}$ is the cost for hot utility l, \$/MMBtu; $c_{\text{c,p}}$ is the cost for cold utility p, \$/MMBtu; and C^{Q}_{Network} is the total utility cost, \$/h.

The installed cost for each ΔT_{min} can be calculated based on equation (9.6), while the energy cost by equation (9.9). By calculating both energy and capital costs for different ΔT_{min} values, two cost curves, namely, the energy and costs curves, for a range of ΔT_{min} can be obtained, as shown in Figure 9.12.

The total annualized cost for the entire heat recovery system can then be defined as:

$$C^{\text{Total}}_{\text{Network}} = FC^{\text{Cap}}_{\text{Network}} + KC^{Q}_{\text{Network}}, \tag{9.10}$$

where F is the capital annualized factor, 1/year; K is the time annualized factor = 24 h/day * operating days/year; and $C^{\text{Total}}_{\text{Network}}$ is the total annualized cost, \$/year.

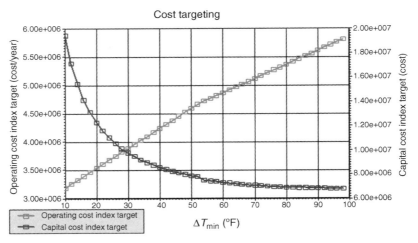

FIGURE 9.12. Capital and energy trade-off.

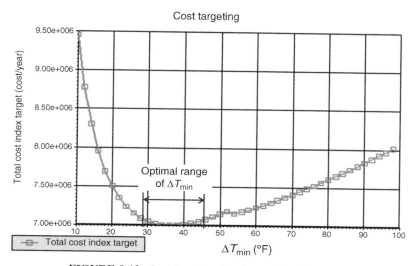

FIGURE 9.13. Cost targeting for determining $\Delta T_{\min,\text{opt}}$.

For different $\Delta T_{\min}$, it will be expected to have different capital cost and utility cost. When $\Delta T_{\min}$ increases, capital costs drop as the LMTD increases and thus surface area reduces, while utility consumption and then energy costs go up. When $\Delta T_{\min}$ reduces, capital costs go up as the LMTD reduces, while utility consumption and then energy costs go down (Figure 9.12). Thus, there is a trade-off between utility cost and capital cost. This trade-off can be visualized by plotting the total annualized cost on the graph as shown in Figure 9.13. The optimal network approach, denoted as $\Delta T_{\min,\text{opt}}$, is the one corresponding to the lowest total annualized cost. In many cases, there is a range of $\Delta T_{\min}$ values in which total costs are similar and thus selection of $\Delta T_{\min,\text{opt}}$ in this range should be made toward design simplicity.

The significance of $\Delta T_{\min,\text{opt}}$ is in setting the design basis for where the heat recovery design should start and what to expect for the utility and capital costs to result from design. If the design comes up with much higher costs than the targeted costs, design evaluation must be conducted to find out why and figure out a measured correction.

Several software including Aspen HxNet and Honeywell UniSim ExchangerNet can be applied to build composite curves and conduct energy targeting and total cost targeting to determine $\Delta T_{\min,\text{opt}}$.

9.8 CASE STUDY

The case study is a diesel hydrotreating process and the flowsheet is shown in Figure 9.14. The combined feed is preheated by reaction effluent via E-100 and furnace and then goes to the reactor. After transferring heat to the combined feed, the reaction effluent is cooled in the air cooler and then goes to the high-pressure

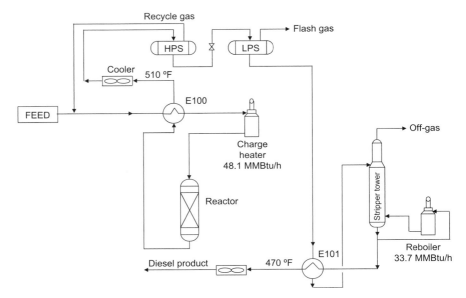

FIGURE 9.14. A schematic of the diesel hydrotreating process.

separator (HPS) where hydrogen-rich gas is separated and recycled back to be mixed with the fresh feed to form the combined feed. The HPS liquid reduces its pressure through the letdown valve and goes to the lower-pressure separator (LPS) where the LPS liquid goes to the stripper column through the stripper bottom-feed exchanger (E-101). The stripper bottom is cooled by E-101 first and then water-cooled and goes to the storage tank as a diesel product.

The energy loss audit detected two major energy loss items: Reaction effluent heat is rejected to the air cooler at 510 °F and stripper bottom product runs down in the water cooler at 470 °F, respectively. However, the energy loss audit could not determine the fuel-saving benefits of recovering these two heat losses and this question can only be answered by the heat recovery targeting assessment discussed later.

9.8.1 Selection of a Representative Operation Basis for Targeting Assessment

This question may deserve an explanation because it is common that people would consider using the most constrained operating scenario, that is, maximum throughput and end of run data, as the basis for revamp designs. This is understandable as sufficient design margins must be provided to allow the revamped unit to deal with production variations. However, for the heat recovery targeting purpose, the data to be used should be based on the most common operating scenario, which is different from the most constrained operating scenario, to avoid underestimation or over estimation of benefits. The most constrained scenario is used as the basis for equipment sizing in the revamp study.

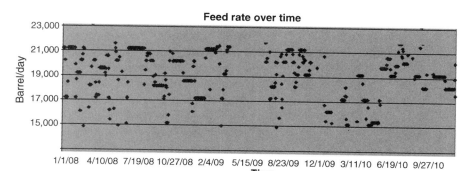

FIGURE 9.15. Process feed rate over time.

Let us use an example to show what is meant by the "most common" operating scenario. Figure 9.15 shows the process feed rate for the process operation in the past 3 years, which varies widely from 15,000 to 21,000 bpd with the arithmetic feed rate average over the entire time period as 18,000 bpd and maximal operating rate at 21,000 bpd. In determining the most common feed rate, feed rate frequency was calculated in Table 9.2, which shows that the most common feed rate is in the range of 19,000–21,000 bpd. The average in this range is calculated as 20,500 bpd and this average could be used as the basis for heat recovery assessment. Clearly, 20,500 bpd is very different from the arithmetic average of 18,000 bpd. If the arithmetic average was used as the basis, the benefit would be underestimated. On the other hand, if the maximal feed rate was used, the benefit would be overestimated. Thus, the average of feed rate in the most frequent operation is the best representation of the common operation. The arithmetic average cannot represent the most common operation because the average can be strongly affected by either too large or too small feed rates which infrequently occur.

With the most common operating feed rate as the typical feed rate selected, the time period with the feed rate close to the typical feed rate and in the middle of the run was selected as the data basis for the heat recovery assessment, which is given in Table 9.3. Specific heat capacity data for each stream are obtained from the physical property database. At the same time, the fuel consumption for both charge heater and stripper reboiling heater are obtained from the historian and then converted to absorbed duty based on heater efficiency, which are 48.1 and 33.7 MMBtu/h, respectively.

TABLE 9.2. Operating Frequency

Feed Rate (bpd)	Frequency (%)
19,000–21,000	68
17,000–19,000	22
15,000–17,000	10

TABLE 9.3. Process Stream Data for the Diesel Hydrotreating Unit

Stream Name	Inlet Temperature (°F)	Outlet Temperature (°F)	Flow Rate (lb/h)	C_p (Btu/ (lb °F))
Combined feed to reactor	165.8	700.0	256,341	0.762
Total reaction effluent to HPS	737.1	115.0	256,341	0.765
Stripper bottom to storage	513.0	144.0	232,287	0.588
LPS liquid as stripper feed	117.2	355.0	254,886	0.544
Stripper condenser	253.0	115.0	29,968	1.275
Stripper reboiler	460.7	513.0	364,956	1.394

9.8.2 Benefit Estimation

Based on the process stream data, the composite curves were constructed as shown in Figure 9.16. The ΔT_{min} for the existing process is determined as 250 °F from the composite curves based on the total absorbed duty of 81.8 (48.1 + 33.7) MMBtu/h. Clearly, a large ΔT_{min} at 250 °F manifests a very poor heat recovery design, which has a simple reason: The design was a 1950s vintage design when energy was very cheap and the design effort was to minimize capital investment. If the same process was designed today, the optimal ΔT_{min} could be determined as 70 °F indicated by the minimum total cost, which corresponds to 34.8 MMBtu/h of total absorbed duty.

Therefore, the potential benefit for an energy retrofit of this old process design could be calculated as follows.

- Energy savings—the maximal scope for energy savings in absorbed duty could be 47 (81.8 − 34.8) MMBtu/h or fired duty of 55 MMBtu/h assuming a heater efficiency of 85%.
- Energy cost savings could be $2.9 MM/year assuming fuel price at $6 / MMBtu and operating time of 360 days/year.

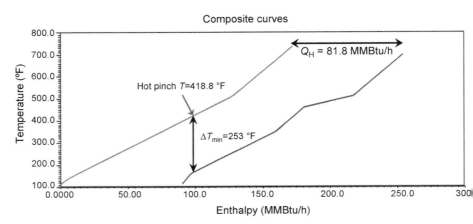

FIGURE 9.16. Composite curves for example diesel hydrotreating process.

The heat recovery targeting assessment uncovers a very significant cost-saving opportunity and thus the significance of the heat recovery targeting is in creating the financial incentive to justify further investigations. However, the heat recovery targeting cannot address the question: What specific modifications are required to capture the savings opportunity? This question will be answered by applying the network pinch method, which will be discussed in Chapter 10.

9.9 AVOID SUBOPTIMAL SOLUTIONS

The common mistake in determining $\Delta T_{min,opt}$ is optimizing subsystems. To avoid this, the battery boundary for a process unit should be determined from the feed to products. The following issues should be addressed to define the process boundary:

- What sections should be included in the boundary? In general, a process includes feed preparation, reaction, fractionation and separation, heat recovery, and product delivery sections. To avoid suboptimal solutions for ΔT_{min}, these sections should be included when determining $\Delta T_{min,opt}$.
- What is the optimal feed temperature? It is not always true that the higher the feed temperature is better because hotter feed is at the sacrifice of the upstream unit(s), which produces the feed. The counter example occurs when an upstream unit is badly in need of the heat and the process under design has plenty of heat from reaction effluent available. Therefore, the optimal feed temperature should be determined based on net effect from considerations of both upstream units and the unit under design. The same argument applies to determining product temperatures when the products become the feeds for downstream units.
- Is the energy pricing realistic? The bottom line is that energy pricing should be directly connected to the marginal utility, which the process plant pays for purchasing or making it.

To illustrate possible occurance of suboptimal solutions, let us consider an example of the hydrocracking process, which makes transportation products. The data were extracted from process simulation and it was assumed that the feed preheating and reaction sections are designed first followed by the fractionation section.

By applying the cost targeting procedure, the total cost curve for the feed preheating and reaction sections is obtained and $\Delta T_{min,opt}$ occurs at 60 °F as shown in Figure 9.17. The occurrence of $\Delta T_{min,opt}$ at lower values is due to the availability of a large amount of reaction effluent heat and high cost of fuel.

Now let us turn our attention to the product fractionation and gas concentration sections. The cost targeting curve for these sections is shown in Figure 9.18, which indicates $\Delta T_{min,opt}$ in the range of 110–120 °F. The reason for the high $\Delta T_{min,opt}$ is simple: There is not enough process heat available at a high temperature resulting in a large fuel consumption.

However, if all these sections are considered simultaneously in one design boundary, the cost targeting indicates the $\Delta T_{min,opt}$ range of 65–70 °F. By integrating

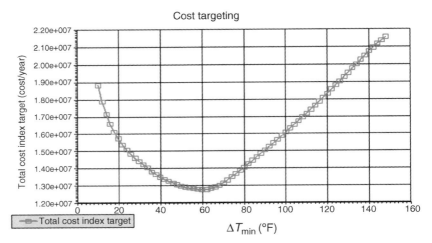

FIGURE 9.17. $\Delta T_{\min,\text{opt}}$ for the reaction section.

both the reaction and fractionation sections, some of the reaction effluent heat at relatively high temperatures can be utilized for the fractionation section, while some of the product heat at relatively low temperatures could be recovered for preheating of fresh feed coming into the unit at a low temperature. With this integration, the furnaces in the reaction may become slightly larger than designs in isolation, but the fractionation furnace is much smaller. The net consequence is that the overall fuel consumption and furnace capital costs reduce significantly.

Let us see the effect on total cost. If the heat recovery system for reaction and fractionation sections are designed separately, $\Delta T_{\min,\text{opt}}$ will be at around 60 °F for the reaction section resulting in the total annualized cost at around $13 MM/year

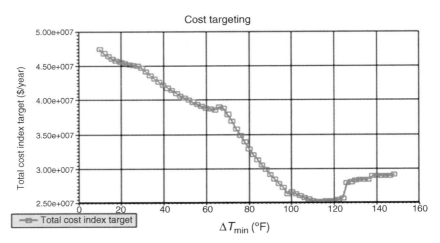

FIGURE 9.18. $\Delta T_{\min,\text{opt}}$ for the fractionation section.

(Figure 9.17), while $\Delta T_{min,opt}$ for the fractionation section would be around 115 °F corresponding to the total annualized cost of $25 MM/year (Figure 9.18). However, the simultaneous design of heat recovery for both sections could give a total annualized cost of around $35 MM/year, which is about 8% lower than $38 (13 + 25) MM/year by designing the heat recovery systems for the reaction and fractionation sections separately. It is not uncommon to observe that a complex process is designed in sequence and this kind of design approach results in heavy penalty in total investment cost.

9.10 INTEGRATED COST TARGETING AND PROCESS DESIGN

9.10.1 Conventional Design Approach

The conventional process design starts from the process flowsheeting without address-ing energy use and generation. Once the process flow schemes are designed and process conditions are determined, process utility demands are summed to obtain overall utility balances in terms of cooling water, steam, power, and fuel. The deficits in these utilities are then satisfied by the central steam and power system together with the import of electricity and natural gas. The unit design approach outlined previously often results in insufficient consideration of two important opportunities: (i) heat integration within the process unit of interest and across the process units, and (ii) heat integration between process units and utility systems.

9.10.2 Systematic Design Approach

To explore all opportunities and identify the true optimum energy solution, a systematic design procedure is discussed (Zhu and Martindale, 2007), which is depicted in Figure 9.19 and explained as follows.

First, several design cases are established including both start of run and end of run for process units with significant variance in energy utilization and catalyst performance. A base case is then determined. The preliminary heat and weight balances for the base case are then generated to determine the process configuration and critical process conditions such as reaction temperature and pressure, separation temperatures and pressures, conditions of feeds, products and recycle streams, and so on. This initial process design information provides the basis for conducting heat integration targeting analysis.

During the targeting stage, major opportunities in what if scenarios are explored, which include heat integration between different process sections. Heat integration between process units is addressed mainly by steam use and generation and by optimizing the temperatures of intermediate products connecting upstream and downstream process units. Opportunities to integrate process conditions with heat recovery are also explored. For example, increasing feed temperature to a separator could reduce reboiling duty and thus hot utility demand; however, the effects on separation need to be verified because an increased feed temperature may adversely affect vapor–liquid flows in the separation section. The results of the targeting

FIGURE 9.19. A systematic design procedure embedding energy optimization.

analysis give the optimal flow schemes and levels of heat recovery based on trade-offs between capital and utility costs. Realistic capital costs should be provided in the targeting stage for the entire system—including fired heaters, heat exchangers, fin fans, as well as utility systems costs for steam, fuel, and cooling media (e.g., air, cooling water, and refrigeration).

The opportunities identified during the targeting stage are thoroughly reviewed by process designers, operating technical service staff and clients to verify technical feasibility and assess any impacts on operating flexibility and safety. Additional process simulation is sometimes needed at this stage to fully assess technical, economic, and operational feasibility. Once the design options are selected based on optimum economics and confirmed feasibility, the process configurations and conditions are then finalized and incorporated in updated heat and weight balances. Further heat integration studies are then completed to finalize the specifics of the heat exchange schemes in the final process design.

9.11 CHALLENGES FOR APPLYING THE SYSTEMATIC DESIGN APPROACH

The concept of targeting sounds very powerful, but there are challenges in application. One challenge is the selection of energy costs to be used as the basis—including fuel, steam at different levels, power, and cooling media (air, cooling water, chilled water, and refrigeration). We will use natural gas as an example to illustrate the challenge and importance of selection of the energy prices for optimization. In the late 1990s, the price for natural gas was about $3/MMBtu—prices in early 2010 are about twice as high. The use of different energy prices could lead to very different energy–capital trade-off. It is advisable to use a more conservative estimate of fuel price as the basis for targeting, and it is wise to include the legislative penalty on gases emissions.

Proper steam pricing is also challenging. For most process plants, pricing for medium-pressure and low-pressure steam is the most confusing as they are intermediate products from the steam system. Realistic steam pricing should be based on incremental fuel cost required instead of based on enthalpy (account for energy quantity), which could be grossly misleading.

Another key factor for successful cost targeting is capital cost estimation. Capital costs for heat recovery systems include costs for heat exchangers, heaters, and coolers. Improper estimates of capital cost can lead to suboptimal trade-offs between capital and utility costs. One important aspect is good estimates of heat transfer coefficients for individual process streams since heat transfer coefficients determine the size and surface area of heat exchangers. A second consideration for accurate capital costs is realistic cost calculation for process furnaces and cooling towers in the targeting phase of work—these are much more expensive than heat exchangers.

There is a way to deal with these uncertainties, which is to explore the flat cost range in the cost range targeting analysis. In this flat cost range, the total costs for different heat recovery levels are similar and close to the minimum total cost. A heat recovery level within this flat cost range with expected design simplicity is then selected as the basis for the design of the heat recovery system. This method provides effective mitigation of the risk of a design being rendered suboptimum due to changes in energy costs.

Yet another key factor is keeping plant layout and safety in mind throughout the energy optimization work. Application of energy optimization tools based only on calculated heat recoveries and cost trade-offs can lead to impractical designs. Often a good heat integration design is expected to be highly integrated among different processing sections and process units. However, a highly integrated design is not desirable in terms of operating flexibility, start-up, and control as well as piping infrastructure. Another issue that is often not adequately considered is the amount and routing of piping—a long piping run may be required to bring a hot stream from one location to transfer heat in a heat exchanger resulting in costly and complex piping connections.

Application of this systematic design approach needs to address plant layout and safety from the start of the energy optimization effort—in the targeting phase and throughout the remainder of the work. A key principle applied to develop optimum solutions is simplicity of design. Heat exchange between process units is only recommended when it provides a significant improvement in economics and does not impact operation flexibility versus the alternatives. Also, to optimize the operability of cross-unit heat integration as well as economics, feed temperatures for downstream process units must be optimized, and steam integration across process units is considered. Another strategy applied to improve operability is limiting direct heat integration across units to applications where they routinely operate in the same pattern.

Last, but certainly not least, a design task schedule needs to be developed carefully to minimize the time required to develop a process design. Allocation of design activities should be proper such that the energy optimization work is done and completed in parallel with other process design work; thus, no extra project time is required.

NOMENCLATURE

a fixed cost for exchanger, \$
A surface area, ft^2
b surface area cost, $\$/ft^2$
C cost, \$
F capital annualized factor, 1/year
$I_{i,j,k}$ installation factor for exchanger between streams i and j in kth interval
K time annualized factor $= 24$ h/day* operating days/year
LMTD   logarithmic mean temperature difference, °F
Q heat load, MMBtu/h
U overall heat transfer coefficient, MMBtu/h/(ft^2 °F)

REFERENCES

Linnhoff B, Townsend DW, Boland D et al. (1982) *A User Guide on Process Integration for the Efficient Use of Energy*, IChemE, UK.

Townsend DW, Linnhoff B (1983a) Heat and power networks in process design. Part 1: criteria for placement of heat engines and heat pumps in process networks, *AIChE Journal*, **29**(5), 742–748.

Townsend DW, Linnhoff B (1983b) Heat and power networks in process design. Part II: design procedure for equipment selection and process matching, *AIChE Journal*, **29**(5), 748–771.

Zhu XX (Frank), Asante NDK (1999) Diagnosis and optimization approach for heat exchanger network retrofit, *AIChE Journal*, **45**(7), 1488–1503.

Zhu XX (Frank), Martindale D (2007) Sustainable energy improvements: reduce energy costs for both new and revamped units, *Hydrocarbon Engineering Journal*, September, pp. 81–87.

10

PROCESS HEAT RECOVERY MODIFICATION ASSESSMENT

It is not trivial to determine the best modification options that could feature minimal capital cost and the smallest effect on existing infrastructure and thus achieve maximum energy savings. The search for such options could be very time consuming for a complex heat recovery system. The incentives to find practical yet optimal modifications have been discussed by many other researchers and practitioners. The network pinch method developed by Zhu and Asante (1999) has been proven successful for practical applications, which will be discussed in detail in this chapter.

10.1 INTRODUCTION

Most process plants undergo at least one major revamp in their lifetime to take advantage of advances in process technology, to improve energy efficiency, or to increase the plant capacity. During such revamps, retrofit of the heat recovery system is normally required to ensure that required processing temperatures and other conditions (e.g., pressure) are attained under the new operating conditions. Alternative designs with varying operating and capital costs need to be generated to satisfy new operating requirements, and final retrofit design must be selected from these alternative designs. The alternative designs may also vary, to different degrees, in operability, flexibility, and inherent safety, and these factors must also be considered in selecting the final retrofit design.

Energy and Process Optimization for the Process Industries, First Edition. Frank (Xin X.) Zhu.
© 2014 by the American Institute of Chemical Engineers, Inc. Published 2014 by John Wiley & Sons, Inc.

In the past, the retrofit methods based on the process pinch concept was widely used (Tjoe and Linnhoff, 1986; Polley et al., 1990; Shokoya and Kotjabasakis, 1991). However, the process pinch is developed for grassroots design and is fundamentally irrelevant to the retrofit scenario. The method (Zhu and Asante, 1999) to be discussed here is a two-stage approach: identify the network pinch, the true bottleneck of an existing heat recovery system, and determine modifications with minimum capital costs to overcome the network pinch—that is the reason why this method is called the network pinch method.

10.2 NETWORK PINCH—THE BOTTLENECK OF EXISTING HEAT RECOVERY SYSTEM

In the example problem shown in Figure 9.1 in Chapter 9, the targeting assessment indicates the minimum energy use for $\Delta T_{min} = 25\,°F$ is 17 MMBtu/h, which is 9.2% lower than the current design. One may ask: Is it possible to increase the heat recovery by adding exchanger surface area to some of the existing exchangers? To answer this question, let us return to the five-stream example in Figure 9.1. The heat recovery in Figure 9.1 is represented by a grid diagram in Figure 10.1 for better visualization of the heat exchange system. In the grid diagram, the hot streams are from left to right and cold streams right to left. A heat exchanger is represented by two circles and one line connecting two process streams.

If we want to reduce the reaction heater duty in Figure 10.1, the simplest way is to increase the feed temperature to reactor 1. To do this, additional surface area must be inserted to Exchanger E1 (the feed–column bottom exchanger), while process heat recovery from this exchanger is increased. For achieving the same reaction temperature, the overhead liquid temperature to reactor 1 must be reduced, which results in lower heating duty. However, there is a limit for heat recovery increase from E1 no

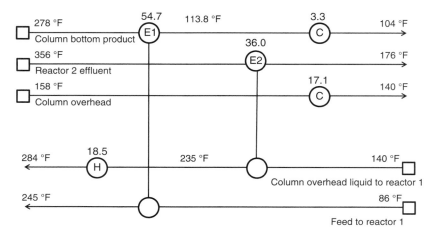

FIGURE 10.1. Grid representation for heat exchanger network in Figure 9.1.

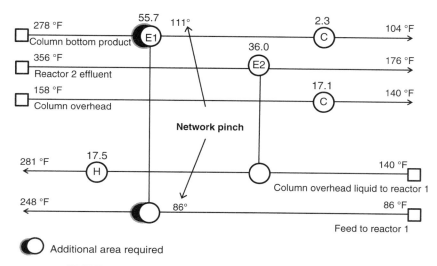

FIGURE 10.2. Network pinch for the network in Figure 10.1.

matter how much surface area is added to it. In this case, the exchanger E1 that limits the heat recovery is called the *pinching exchanger*, and the point at which this occurs is termed the *network pinch*, which is the bottleneck of the existing network. This can be identified as the cold end of E1 where the existing exchanger E1 reaches ΔT_{min} of 25 °F (Figure 10.2). The minimum approach temperature of 25 °F is used for dealing with process variation. The reaction heating duty for this improved design is 17.5 MMBtu/h, which is a 5% savings from the current heat duty of 18.5 MMBtu/h, but still 3% higher than the target of 17 MMBtu/h.

What do these discussions teach us? To simply put it, with the fixed network structure, we are unable to achieve the target of 17 MMBtu/h heating without violating ΔT_{min}. Heat recovery can be increased to a certain limit (E1 as the pinching match) when surface area is added to the existing exchangers in a heat exchanger network (HEN) without making changes to the structure. Beyond this limit, any additional surface area will not increase the heat recovery. Clearly, the chosen network configuration is imposing a constraint, which is independent of surface area and stops us from achieving the target.

10.2.1 Difference Between the Process Pinch and the Network Pinch

It must be noted that the process pinch is different from the network pinch. For the network in Figure 10.2, the process pinch occurs at a hot pinch temperature of 158 °F and cold pinch temperature of 133 °F, respectively, as shown in Figure 9.8, while the network pinch occurs at a hot pinch temperature of 111 °F and cold pinch temperature of 86 °F, respectively, as displayed in Figure 10.2. The reason for the difference is that the process pinch is defined by the process conditions such as temperatures and heat capacities, and the process pinch is mainly used for energy targeting in grassroots design. However, for an existing heat recovery network, heat exchangers are already in place and the network configuration will impose a limit for increased heat recovery. These limits (pinching matches) can only be identified by the network pinch method.

10.2.2 Uniqueness of the Network Pinch Location

Figure 10.3a shows a simple heat exchanger network with temperatures and heat duties. By maximizing heat recovery or minimizing heater duty by only adding surface area to process–process exchangers, this network will reach the conditions as given in Figure 10.3b. Clearly, exchanger 1 is the pinching match and the limit for the minimum heating duty is 100 MMBtu/h. For simplicity, a zero value of exchanger minimum approach temperature (EMAT) is used in this example. In reality, a practical EMAT should be used for identification of pinching matches.

One may argue for the uniqueness of the pinch matches because heat load could be shifted among heat exchangers and thus temperatures could change as a result. For example, exchangers 2 and 4 form a loop around which heat can be shifted without altering the network heat recovery. Consequently, either exchangers 1 and 2 can be pinched (as shown in Figure 10.3a) or exchanger 1 will be pinched (see Figure 10.3b). In both cases, the network remains at minimum heating duty of 100 MMBtu/h.

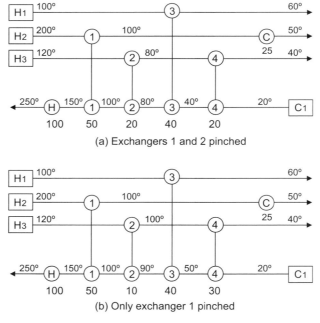

FIGURE 10.3. Example network at the heat recovery limits.

Which exchanger 1 or 2 or both, is the pinching match? The answer is exchanger 1 because the definition of pinching matches excludes those "pinched" exchangers whose limiting driving forces can be relaxed by shifting heat around a loop. Therefore, it is important to remember that pinch matches for a given network structure are unique and they can be determined mathematically by minimizing the number of pinch matches.

10.2.3 Significance of the Network Pinch Concept

The network pinch identifies the heat recovery limit inherent in an existing heat exchanger network and therefore indicates the requirement of structural changes to the HEN. In other words, when a network reaches its network pinch limit, the only way to overcome it or increase heat recovery is to make structural changes if no process changes are considered.

However, not all network structural changes are able to overcome the network pinch, but only those that can move heat from below to above the network pinch can. This principle provides a valuable guideline for screening out nonbeneficial modification options, which will be discussed next.

10.3 IDENTIFICATION OF MODIFICATIONS

Having identified the network pinch or the bottleneck, what can we do about it? The answer is that only changes to the network structure can overcome the bottleneck. In the stage of selecting structural changes, the maximization of heat recovery is used as the selection criteria. Four types of modifications can be considered for the heat exchanger network, which are:

- *Resequencing:* The order of two exchangers can be reversed, and this sometimes allows better heat recovery. Usually, this option involves relative minor piping changes and hence the least cost required.
- *Repiping:* This is similar to resequencing, but one or both of the matched streams can be different to the current situation.
- *Adding a New Match:* This can be used to change the load on one of the streams in the pinching match. This option could be relatively expensive as it requires a new foundation, piping, and control system.
- *Splitting:* Split a stream, again reducing the load on a stream involved in the pinching match. In practice, the stream split could be very asymmetric in both flow and temperature, and a special configuration would be needed.

In most cases, many structural changes capable of overcoming the network pinch can be identified, and usually they produce different energy savings and capital costs. To select the best modification(s), among many alternatives, it is necessary to define a measure of optimality. Ideally, the cost-based objective should be employed.

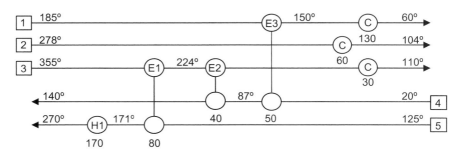

FIGURE 10.4. Retrofit example: existing network.

However, it would be a very daunting task to make it ready with the cost information for piping, labor, foundation, and installation for all potential modifications prior to design.

As an alternative to the cost-based objective, the maximization of heat recovery is used as a selection criterion of modifications in the identification stage. The modification options selected will be evaluated for further selection in capital costs, and in effects on operation and safety to make sure the modifications selected will justify implementation costs as well as operation and safety criteria.

For illustration, we will apply the network pinch method to the network retrofit problem as shown in Figure 10.4. The network pinch method identifies the modification of repiping exchanger 2 as shown in Figure 10.5 and exchanger 2 is pinched at $\Delta T_{min} = 25\,°F$. This modification could reduce the heating duty from 170 to 124.5 MMBtu/h, which is a 27% reduction from the base case of Figure 10.4. Alternatively, the resequence modification to E2 as shown in Figure 10.6 could achieve the same heating duty as the repiping design in Figure 10.5 with exchanger 1 as pinching exchanger, but the resequence design can eliminate the cooler for stream 3. Alternatively, a new exchanger could be added to the network, which can reduce the heating duty to 126 MMBtu/h (Figure 10.7). The new exchanger is the pinching exchanger. Evaluation of these three retrofit designs will determine the best one based on capital costs and operation aspects.

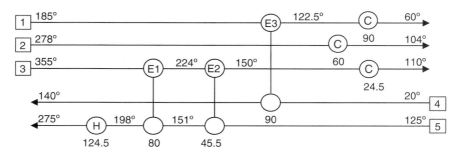

FIGURE 10.5. Retrofit example: repipe exchanger 2.

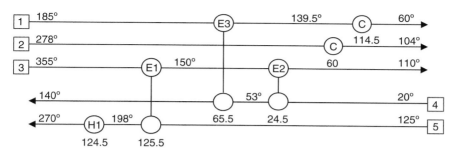

FIGURE 10.6. Retrofit example: resequence exchanger 2.

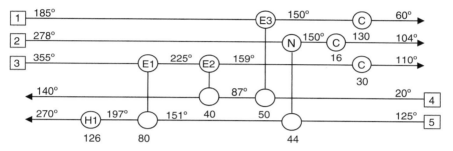

FIGURE 10.7. Retrofit example: adding a new exchanger.

10.4 AUTOMATED NETWORK PINCH RETROFIT APPROACH

The previous discussions provided an insight into the nature of the network retrofit problem. In particular, it has been demonstrated that the concept of the network pinch can be used to identify modifications that can increase heat recovery. This forms the basis of the network pinch approach. The network pinch retrofit approach consists of three stages: a modification identification stage, an evaluation stage, and an optimization stage. This network pinch approach is described in Figure 10.8, which has been implemented into several software including Aspen HxNet and Honeywell UniSim ExchangerNet, which are available in the software market.

The identification stage is developed to identify promising modification options, which include resequence and repiping of existing exchangers, addition of new exchangers, and stream splitting. Two linear models must be used in the identification stage. A linear programming model is used for identification of the heat recovery limits in the existing network by pushing the network to its maximum heat recovery limit without altering the network structure. This linear programming (LP) model identifies the pinching matches and the network pinch that initiates the search for structural changes. The mixed integer linear programming (MILP) model is then used repeatedly to identify structural changes as being capable of overcoming the network pinch. The new network pinch is determined at the same time when structural changes are found.

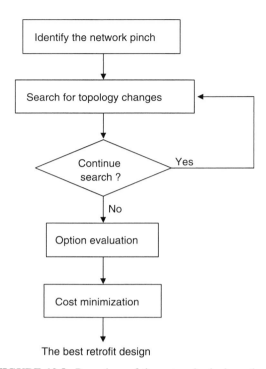

FIGURE 10.8. Procedure of the network pinch method.

Those modification options identified will be further assessed in the evaluation stage. This stage provides opportunities for meaningful user interaction. In this stage, the costs for structural changes identified are estimated (such as costs for piping, foundation, and area) while safety and operability issues can be considered based on the user's experience and knowledge. As a result, impractical options are screened out. After the identification and evaluation, only promising options are selected. Cost estimates are only required for selected options that show benefit and feasibility. This makes the design evaluation task much simpler than what would be if the cost estimates for all possible modifications are required prior to design stage.

In the final stage of optimization, the options selected are optimized in terms of the trade-off between capital investment and utility saved. During this optimization, the main structural features of a network remain the same, but exchanger heat loads are allowed to vary. In some cases, several different topologies can be obtained, and they are considered separately in the optimization stage. The details of mathematic formulations for the approach can be seen in the work of Zhu and Asante (1999).

The rationale of the network pinch method is based on the effort and difficulty required for each assessment task. The effort for conducting a revamp study is demanding as it could involve weeks or up to months depending on the complexity of the process. During the revamp evaluation stage, not only itemized costs need to be determined, but effects on piping, pressure drops, and operation will be evaluated as well. To reduce the manpower effort, the network pinch method has been implemented into mathematical models, which can generate modification options automatically with users being in a control position in making decisions for rejecting or selecting modification options based on impacts on layout, infrastructure, and operation.

Recent experience provides new insights and guidance on process revamp and areas that will often represent good opportunities for improved energy efficiency with attractive economics. Optimization efforts for existing facilities often focus on improved heat recovery. However, recent work has identified other areas that may provide significant opportunities, but are not commonly explored. Also, some innovative work has been completed to identify some new process changes that may be of interest. These aspects will be discussed in Chapter 11.

10.4.1   About Utility Modifications

In most cases, individual processes are serviced by a central utility system, and any modifications to the utility system has a significant effect on other processes and consequently on the performance of the site as a whole. For this reason, the optimization of the utility system must generally be considered over a whole operating site. A technique for a site-wide energy optimization will be discussed in Chapter 17, in which the concept of marginal economical values for stream, power and fuel consumptions is introduced. The marginal values determined from the utility system are implemented into the mathematical models of the network pinch method as the economic values for energy savings, which becomes the basis for utility selection and optimization in the context of process energy retrofit projects.

10.4.2   About Nonlinear Behavior of Stream Heat Capacity

The heat capacity values of a process stream vary as the temperature changes. This variation could be very significant for process streams spanning a large temperature change, in particular for reaction effluent and other process streams undergoing preheating to vaporization. If a single value is used to represent the heat capacity for a whole stream, it could give a very misleading result in terms of temperature driving forces for related heat exchangers and thus surface area required. When applying the network pinch method, if a stream has a large variation in heat capacity, it should be treated as multiple segments each of which is modeled with a specific heat capacity. This modeling treatment allows more precise estimates of temperature driving forces and thus surface areas for heat exchangers.

10.5   CASE STUDIES FOR APPLYING THE NETWORK PINCH RETROFIT APPROACH

10.5.1   Case Study 1: Increase Feed Rate for the Crude Unit

The retrofit objective in this example was to debottleneck the heat exchanger network to accommodate a 10% increase in crude throughput for the crude distillation unit as shown in Figure 10.9a. The main retrofit constraint is the maximum furnace duty of

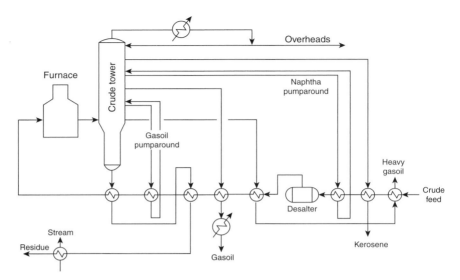

FIGURE 10.9a. Process flow diagram for the crude distillation unit. (From Zhu and Asante (1999), reprinted with permission by AIChE.)

100 MW. Figure 10.9b shows the grid diagram for the existing network design. The minimum temperature considered for retrofit was at 10 °C.

In revising the base case in Figure 10.9b, the crude feed rate (stream C1) is increased by 10% and the product streams H1–H7 are increased accordingly. Then

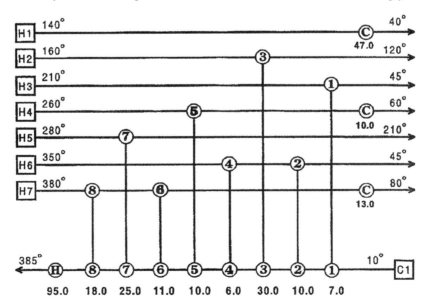

FIGURE 10.9b. Base case heat exchanger network. (From Zhu and Asante (1999), reprinted with permission by AIChE.)

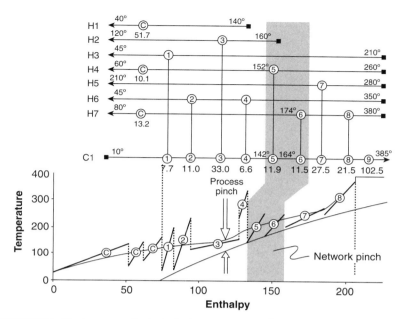

FIGURE 10.10. Heat recovery limits in the base case network. (From Zhu and Asante (1999), reprinted with permission by AIChE.)

the heat recovery limits for the increased feed rate case with the original network topology are established, which are shown in Figure 10.10. The furnace duty at the maximum heat recovery for the given topology is a 102.5 MW and exchangers 5 and 6 are pinched. Because the furnace duty at the maximum heat recovery (R_{max}) for the given topology is at 102.5 MW, which is above the maximum allowable design duty of 100 MW, changes to the topology are needed to reduce the furnace duty below the design limit.

The first modification option considered is exchanger resequence, and the option selected by the method is the resequence of exchanger 4 as illustrated in Figure 10.11. This modification produces a 4.4 MW increase in heat recovery, and reduces the minimum furnace duty to 98.1 MW. Although the minimum furnace duty after this resequence modification is below the design limit of 100 MW, it is quite close to it. It can be expected that a retrofit design close to the topology R_{max} would require excessive exchanger surface area. Thus, another topology change is sought to further increase heat recovery, and consequently, reduce the exchanger areas required for the retrofit.

The topology produced after the resequence of exchanger 4 features three adjacent pinch matches (exchangers 4/5/6) as shown in Figure 10.12. As the first of these pinching matches occurs at the process pinch, the process and network pinches become coincidental, and this fulfils the conditions of stream split heuristic. Thus, the stream split is implemented. Exchangers 4/5/6 being the three adjacent pinching matches are placed in parallel with each other to produce a 1.8 MW increase

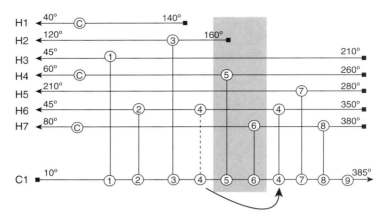

FIGURE 10.11. Resequence of exchanger 4: min QH = 98.08; ΔQRec = 4.4. (From Zhu and Asante (1999), reprinted with permission by AIChE.)

in heat recovery. Although the three exchangers are initially placed in parallel with each other (Figure 10.13), the optimal slit configuration will be determined during the optimization stage.

The search for modifications could be stopped at this point, and the resulting topology is submitted to the optimization stage. If, however, the search is continued for

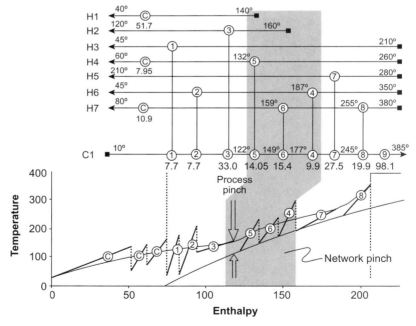

FIGURE 10.12. Heat recovery limits for the network after resequence of exchanger 4. (From Zhu and Asante (1999), reprinted with permission by AIChE.)

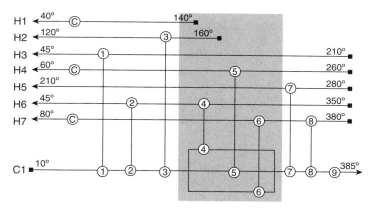

FIGURE 10.13. Stream split: min QH = 96.3; ΔQRec = 1.8. (From Zhu and Asante (1999), reprinted with permission by AIChE.)

another modification and it will be the addition of a new exchanger, which further increases the heat recovery by 3.9 MW to give a minimum furnace duty of 92.4 MW. It is not possible at this stage to determine whether the addition of the new exchanger can be economically justified. The optimization stage will provide the necessary evaluation.

After the optimization at a fixed furnace duty of 99.6 MW the same as the design published in the literature, the retrofit design without the new exchanger requires 1974 m^2 of additional surface area in total, while the design with the new exchanger requires only 1265 m^2 (Figure 10.14). With this information, the retrofit designs can be effectively compared to identify the best retrofit design. The installation cost of the new exchanger can be fairly accurately estimated at this stage to assist in the decision process.

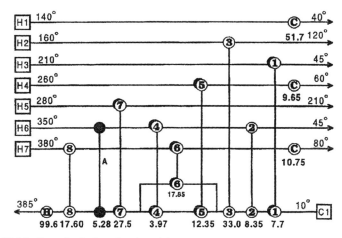

FIGURE 10.14. Retrofit design developed by the network pinch method. (From Zhu and Asante (1999), reprinted with permission by AIChE.)

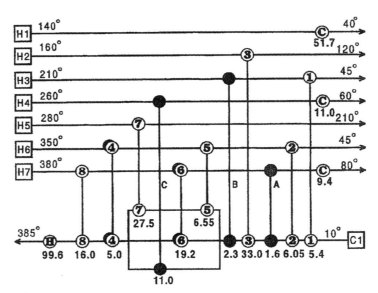

FIGURE 10.15. Retrofit design developed by Ahmad et al. (1989). (From Zhu and Asante (1999), reprinted with permission by AIChE.)

The design produced by using the network pinch method (Figure 10.14) is compared well with the designs by Ahmad et al. (1989) and Shokoya and Kotjabasakis (1991) based on the process pinch methods, which are shown in Figures 10.15 and 10.16, respectively. The comparison is given in Table 10.1, which reveals that the design by using the network pinch method requires the least number of modifications and minimum exchanger area.

10.5.2 Case Study 2: Reduce Energy Cost for the DHT Unit

The diesel hydrotreating (DHT) process design is shown in Figure 9.14. The existing design is not energy efficient and thus requires relatively high fuel consumption in

TABLE 10.1. Design Comparison

	Network Pinch Figure 10.14	Ahmad et al. (1989) Figure 10.15	Shokoya and Kotjabasakis (1991) Figure 10.16
Number of new exchangers	1	3	2
Number of resequenced exchangers	1	1	2
Number of repiped exchangers	0	1	0
Number of stream splits	1 (1 × 2)	1 (1 × 3)	1 (1 × 2)
Number of existing exchangers requiring additional area	5	2	4
Total additional area (m²)	1265	1990	1257

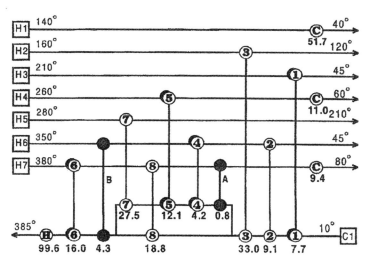

FIGURE 10.16. Retrofit design by Shokoya and Kotjabasakis (1991). (From Zhu and Asante (1999), reprinted with permission by AIChE.)

the charger heater and stripper reboiler. The scope for energy savings is in recovering more heat wasted in the reaction cooler and the stripper bottom product cooler. The recovered heat will be used for preheating both the combined feed and stripper feed with the purpose of reducing firing duty. In addition, medium-pressure (MP) steam generation is considered if possible.

The targeting analysis in Chapter 9 was applied to this process and indicated overall energy-saving opportunity of 55 MMBtu/h. Following the targeting analysis, the network pinch method was applied to the base heat exchanger network and several modification options were generated. These modifications are shown in Figure 10.17 and are summarized here:

- *Modification 1:* Installing reaction effluent–feed exchanger, which can reduce feed heater absorbed duty by 22.6 MMBtu/h.
- *Modification 2:* Installing reaction effluent–stripper feed exchanger, which can reduce stripper reboiler absorbed duty by 10.7 MMBtu/h.
- *Modification 3:* Installing stripper bottom–MP steam generator with duty of 16.2 MMBtu.

In selecting modification options, practical constraints were considered. For example, the reaction effluent stream was not used for MP steam generation. This constraint was to protect the steam system because in case of exchanger tube leaks, reaction effluent could leak into the steam system.

To give conservative estimates of capital costs, the retrofit design was simulated based on the controlling operating scenario, which is the maximal feed rate under

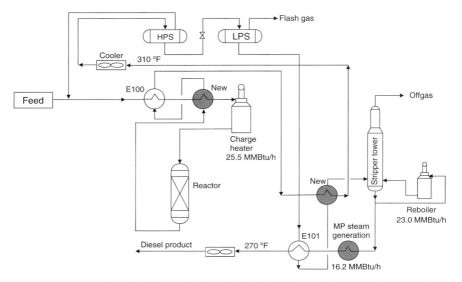

FIGURE 10.17. Modifications to the diesel hydrotreating unit (the base case design as shown in Figure 9.14).

end-of-run conditions. Use of the control case allowed the new exchangers to be properly sized to handle all possible operating scenarios.

The modifications 1 and 2 reduce heater absorbed duty by 33.3 MMBtu/h (39.2 fired duty), while modification 3 generates MP steam by 16.2 MMBtu/h. The total energy savings is 55.4 MMBtu/h. The overall payback for the above modifications is less than 2 years. The composite curves for the retrofit case are shown in Figure 10.18.

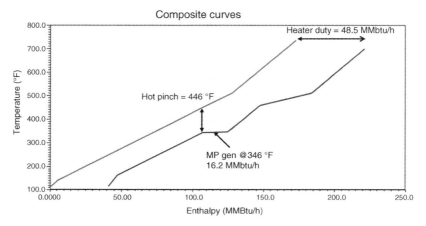

FIGURE 10.18. Composite curves for the diesel hydrotreating unit after retrofit.

10.5.3   Case Study 3: Overcome Limitations for Implementing Energy Modification Projects

It is possible that energy modifications could impact processes. Thus, it is an important part of the energy study to uncover the major limiting factors and identifying solutions so that energy modifications could be implemented. In an energy retrofit study for the preheat train in a crude distillation unit, several modifications were identified by applying the network pinch method. These modifications have significant energy-saving potential. However, when historical data were gathered, a severe limitation was identified: the feed furnace inlet temperatures or the preheat train outlet temperatures were quite close to the crude bubble points. In other words, the current preheat train cannot accommodate the large increase in heat recovery that can be achieved by implementing the modification options. The feed would be vaporized before the furnace and percentage of vaporization would be much more than 3%, which is the limit that the furnace flow control scheme could handle. This indicated that the crude vaporization before the furnace limits the increase in heat recovery. Only if the vaporization problem can be resolved, the identified modifications can then be implemented.

The question is: What could cause feed vaporization? First, hydraulic analysis was conducted for the exchangers. Comparing pressure drops between the clean condition and the measured pressure data in 1998 indicated a gap of 60 psi in pressure drops. With a gain in pressure by 60 psi in the preheat train, the feed would not vaporize even after the energy modification projects are implemented.

Second, the rating calculations were carried out to determine the root causes for high pressure drops in the preheat train. The purpose of rating calculations was to seek the possibility of inappropriate velocity in exchangers. This is because too low velocity could cause accumulation of deposits, which makes the flow channel narrower leading to high pressure drop. To find out, tubeside velocities were calculated for those exchangers, which consumed unusually large pressure drops, based on feed rates, physical properties, and exchanger geometry. It turned out that the velocities inside tubes of these exchangers are reasonable.

Third, bubble points for different crudes were calculated. In Figure 10.19, the historic data for heater pass control valve opening and the preheat train terminal temperatures were obtained and plotted together with bubble points versus the preheat train outlet pressure. As can be observed from the figure, the most common operating preheat train outlet pressure was between 20.5 and 22.5 barg. In this pressure range, the heater pass control valve is almost 100% open indicating no room for pressure control. At the same time, the preheat train outlet temperature is very close to the bubble points of crudes 1 and 4 with the base preheat train design. Figure 10.20 gives a focused view of preheat temperature compared with the bubble points of crude 1 against preheat outlet pressure. If the modifications were implemented, the preheat train outlet temperature would have increased by 45 °C. This would make heater operation unsafe as the preheat train outlet temperature would be much higher than the bubble points of crude. From the crude scheduling data, it was found that crude 1 was used most often recently due to the cheap price.

To understand the nature of fouling, fouling analysis was conducted as the fourth task. Samples were taken from the deposits in some of the exchangers. The lab tests

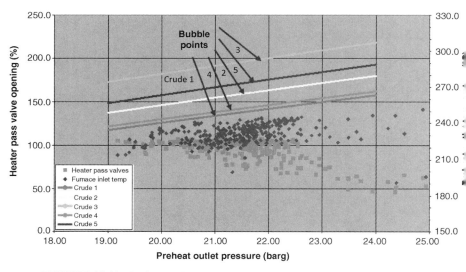

FIGURE 10.19. Preheat train outlet temperature versus crude bubble points.

indicated that the deposits are mainly wax type. At this point, the fact was revealed that crude 1 contains wax materials. In operation, wax materials could precipitate, stick to the tube walls, and make the tube flow channel narrower leading to high pressure drops on the tube side. The plant faced a dilemma: Face the consequence of processing wax crude (crude 1) or use different crudes if energy modifications were considered to implement. However, the choice was clear: It was more economical to use crude 1 due to its very low price, which easily outweighs the benefit of energy savings. What could be done to have both: Use the cheap crude and implement energy-saving projects?

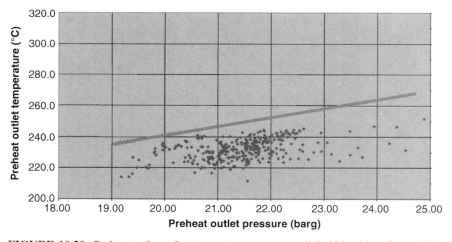

FIGURE 10.20. Preheat train outlet temperature versus crude bubble points for crude 1.

To overcome the vaporization while allowing the wax crude to be processed, three major options have been identified. The first option is to increase the discharge pressure for the second discharge pump by using a new impeller. This option could have the least impact on the current infrastructure with relatively cheap cost. However, this option was restricted with the current pressure rating in some of the flanges.

The second option is cleaning the most fouled heat exchangers on a regular basis. This option aims to squeeze current exchangers to make the best performance out of them. The advantage of this option is that it has almost no impact on the existing infrastructure and requires relatively low capital. However, this option is not effective because the wax materials could build up in a couple of weeks. Cleaning could be too frequent and production loss could be incurred during cleaning.

The third option is to use preflash so that vaporized feed is routed to a preflash drum. The drum bottom liquid is then sent to the feed furnace, while the drum vapor is sent to the top section of the distillation column. This option alone can completely resolve the vaporization problem and allow energy modifications to be implemented. The payback for this option would be less than 1 year if the benefit of using the cheap crude is also included.

To determine the effective percentage of flash, a simulation was done for different percentages of preflash for crude 1. For 5% preflash, the bubble point temperature range for crude 1 is between 251 and 286 °C corresponding to the pressure range of 19 to 24 bar. The minimum bubble point temperature was found to be still too small since the modifications mentioned above can achieve furnace inlet temperature higher than 251 °C. At 8 and 10% preflash, the bubble point ranges increase to 261–296 °C and 267–300 °C, respectively. If considering to have a safe margin for suppressing vaporization, 8–10% preflash should be selected.

For conservative design, 10% preflash was selected for the preflash drum, which was evaluated in simulation together with the modifications. The bubble point for the preflash drum bottom, that is, the feed to the furnace, is much higher than the preheat outlet temperature. Therefore, the preflash drum of 10% preflash provided a solution for suppressing the vaporization.

10.5.4 Case Study 4: Applying the Retrofit Method to a Total Site

The objective of this study was to provide a road map for improving energy efficiency throughout the refinery on the basis that capital investment would be considered for opportunities with rapid payback. The design basis for energy optimization was established based on a good understanding of design conditions, nominal operation conditions, throughput, and operational variations. Data were gathered via different sources based on online operation, engineering design, models, and engineers' experience.

In the initial phase of the study, the benchmarking assessment (see Chapter 3) was applied leading to the selection of 17 major process units for a more detailed study. Application of the network pinch method enabled the identification of many opportunities within a 3 month period of the study. The study identified 73 energy modifications, with a cumulative operating cost reduction of $61.3 MM/year. The total capital cost required for these projects was estimated as $75.5 MM. Capital cost and NPV (net present value) estimates were completed consistent with previous experience at this particular refinery. Also, technical and operational feasibility was assessed for each option. As a result, 40 of the 73 opportunities were determined to be economically viable and technically feasible, and 31 of those 40 opportunities were finally selected for implementation over the next 6 years with a cumulative operating cost reduction of $33 MM/year resulting in a payback of less than 2 years.

REFERENCES

Ahmad S, Polley GT, Petela EA (1989) Retrofit of heat exchanger networks subject to pressure drop considerations, Paper No. 34a, AICHE Meeting, Houston, April.

Polley GT, Panjeh Shahi MH, Jegede FO (1990) Pressure drop considerations in the retrofit of heat exchanger networks, *Transactions of IChemE*, Part A, **68**, 211.

Shokoya CG, Kotjabasakis E (1991) A new targeting procedure for the retrofit of heat exchanger networks, International Conference, Athens, Greece, June 265–279.

Tjoe TN, Linnhoff B (1986) Using pinch technology for process retrofit, *Chemical Engineering*, **93**(8), 47–60.

Zhu XX, Asante NDK (1999) Diagnosis and optimization approach for heat exchanger network retrofit, *AIChE Journal*, **45**(7), 1488–1503.

11

PROCESS INTEGRATION OPPORTUNITY ASSESSMENT

In many applications, the benefits from process changes outweigh those from energy modification projects. The challenge is how to identify process changes that generate significant process benefits. The methods and tools for idea discovery are discussed here.

11.1 INTRODUCTION

Improved heat recovery is the most common for improved energy efficiency. However, a recent work (Zhu et al., 2011) has pointed to the areas of process and equipment innovations, which are less commonly explored, that may provide significant opportunities for both process and energy improvements. Implementing process innovations often results in combined benefits in process yields, throughput, and energy efficiency. Many of these areas include process condition, flowsheet optimization, as well as use of advanced equipment. Some examples can be listed as below

(1) Optimize reactor system
   - Selection of reaction catalysts
   - Reaction temperature and pressure conditions
   - Optimizing hydrogen to hydrocarbon (H_2:HC) ratio
   - Recovery of reaction effluent heat

Energy and Process Optimization for the Process Industries, First Edition. Frank (Xin X.) Zhu.
© 2014 by the American Institute of Chemical Engineers, Inc. Published 2014 by John Wiley & Sons, Inc.

(2) Optimize complex fractionation/separation systems
  – Column sequence for a separation system involving multiple columns
  – Column temperature and pressure conditions
  – Column reflux ratio
  – Adding a pump-around, intermediate reboiler, and condenser
  – Column internal design
(3) Use of advanced equipment/control
  – Enhanced heat exchange technology
  – Advanced fractionation technology such as dividing wall columns, high capacity fractionator internals
  – Power recovery turbines
  – Advanced process control

However, there are major challenges for finding the best process changes when a new process is designed or an existing process is revamped. The challenge is to find what process changes could generate the most significant benefits.

11.2 DEFINITION OF PROCESS INTEGRATION

The traditional process design approach can be described by the so called "Onion Diagram" as shown in Figure 11.1, which provides an overall view of energy considerations throughout the traditional design procedure.

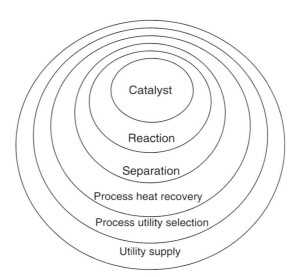

FIGURE 11.1. Sequential process design: traditional design approach.

The design of a process complex starts from defining a design basis. This step consists of defining physical and chemical conditions for feeds, products, and utilities. The design then concentrates on the chemical reaction system. The reaction system is the core of a process complex where the conversion of feeds to products takes place. The goal of a reaction system design is to achieve a desirable product yield structure via selection of a catalyst and design of reactors. Because reaction effluents contain a large amount of heat at high temperatures, the heat recovery of the reaction effluent is a major consideration for process energy efficiency.

After the reaction, the reaction effluent goes through a separation system to separate desirable products from by-products and wastes. For separating multiple products, a separation system involving several columns are required. Heat recovery from products makes significant contributions to process energy efficiency. At the same time, there could be a large amount of excess heat available in fractionation columns where multiple products are made. It is essential that this excess heat is removed from pump-around and used for process heating purposes.

Process heat recovery design comes after the design of reaction and separation systems. The latter defines the basis for process energy demand via the selection of process conditions for reaction and separation. In the heat recovery design step, the goal is to minimize overall process energy use (fuel, steam, and power) for a given process energy demand. This is achieved by heat recovery between those process streams with heat available (such as reaction effluent, separation products, column over head vapor), and process streams with the need for heat (such as reaction feed, separation feed, and reboiling).

After the process heat recovery is done, the next design step is to determine the utility supply in terms of heating and cooling, and power based on the needs and characteristics of process energy demand. In this step, the means of heat supply for the reaction and separation system will be addressed. For example, a choice for the reboiling mechanism must be made for a separation column between a fired heater and steam heater. Similarly, a choice of process driver between steam turbine and motor will be determined. Selection is made based on operation considerations, reliability and safety limits, and capital cost. Selection of process utility supply defines the basis for the design of a steam and power system.

The above steps complete the process design in the process battery limit. The last design step is to design the utility system, which is mainly the steam and power system. The main design consideration of the steam and power system is technology selection in terms of combined cycle (gas turbine plus steam turbine) or steam ranking cycle (steam turbine) for power generation. At the same time, fuel selection, system configuration, and load optimization of the steam and power system need to be determined. Furthermore, offsite utility demand should be addressed and this involves feed and product tank farm design with proper insulation and heating.

In short, the traditional design approach adopts a sequential design approach. In contrast, the process integration methodology for process design takes a different approach in that process design aspects in the inner part of the onion diagram are allowed to change, which may enhance the possibility of heat recovery and enable more energy savings in the utility system in the outer part of the onion. The effects of process changes can be evaluated effectively using the pinch analysis method (Linnhoff et al., 1982) in the early stage of design without waiting for the completion of process design. Otherwise, it would require major rework, which could result in nonoptimal solutions as well as lead to waste of engineering hours and delay in schedule.

11.3   PLUS AND MINUS (+/−) PRINCIPLE

Consider Figure 9.1. The question for the process engineers was: How to reduce the hot and cold utility for the process? Qualitatively, the answer is: increase (plus) the heat load for the hot streams above pinch and the cold streams below the pinch. Conversely, decrease (minus) the heat load of the cold streams above the pinch and the hot streams below the pinch. Obviously, the composite curves (CC) for the process will be modified accordingly if any of such process changes occur.

Following this general principle, several ideas for changes to this process can be proposed. For example, raising the temperature of reactor 2 could increase the heat load for the effluent from reactor 2, a hot stream above the pinch, which could reduce the heater duty for reactor 1. On the other hand, if the reactor 1 temperature could be reduced, the heat load of the reactor 1 feed stream, a cold stream above the pinch, could be decreased and thus the heater duty would be reduced as a result. For the separation column, if the top temperature of the column could be reduced via reducing the column pressure, the duty of the overhead condenser (a hot stream below the pinch) could be reduced. Of course, the effects of these changes have to be evaluated in the overall context of yields, product quality, energy use, and equipment cost.

Thus, the plus and minus (+/−) principle (Linnhoff and Vredeveld, 1984) can be stated as the following:

A. (+) Increase the heat duty of hot streams above pinch
B. (−) Decrease the heat duty of cold streams above pinch
C. (−) Decrease the heat duty of hot streams below pinch
D. (+) Increase the heat duty of cold streams below pinch

Process changes based on principle A and principle B will reduce the hot utility, while principle C and principle D will reduce the cold utility. This simple (+/−) principle provides a definite reference for any adjustment in process heat duties, such as vaporization of a recycle, pump-around condensing, and so on, and indicates which modifications would be beneficial and which would be detrimental.

Often it is possible to change temperatures rather than heat duties. It is clear from Figure 11.2a that temperature changes that are confined to one side of the pinch will not have any effect on the energy targets. Figure 11.2b illustrates how temperature changes across the pinch can change the energy targets. Due to the reduction in feed

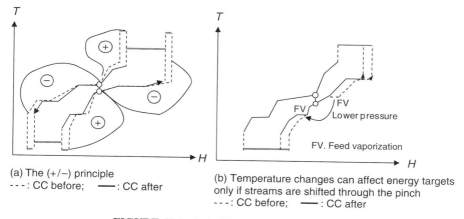

(a) The (+/−) principle
- - - : CC before;   ——— : CC after

(b) Temperature changes can affect energy targets only if streams are shifted through the pinch
- - - : CC before;   ——— : CC after

FIGURE 11.2. (+/−) Process change principle.

vaporization pressure, the feed vaporization duty has moved from above to below the pinch. As a result both hot and cold utility are reduced by the vaporization duty.

Thus, the beneficial pattern for shifting process temperatures can be summarized as:

E. Shift hot streams from below the pinch to above the pinch
F. Shift cold streams from above the pinch to below the pinch

The (+/−) principle is in line with the general idea that it ought to be beneficial to increase the temperature of hot streams (this must make it easier to extract heat from them) and that likewise reduce the temperature of the cold streams (except of subambient conditions). Changing the temperature of streams in this fashion will improve the driving forces in the heat exchanger network, but can also decrease the energy targets if the temperature changes extend across the pinch. With the help of composite curves, the designer can predict which modifications would be beneficial, detrimental, or inconsequential ahead of design. Applications of the composite curves for evaluating process changes prior to design will be discussed later in this chapter.

11.4   GRAND COMPOSITE CURVES

Beside the composite curve, another pinch tool that can be used for assessing process changes is called the grand composite curve (GCC), which can be constructed based on the composite curves. The first step is to make adjustments in the temperatures of the composite curves in Figure 11.3a to derive the shifted composite curves of Figure 11.3b. This involves increasing the cold composite temperature by one half ΔT_{min} and decreasing the hot composite temperature by one half ΔT_{min}.

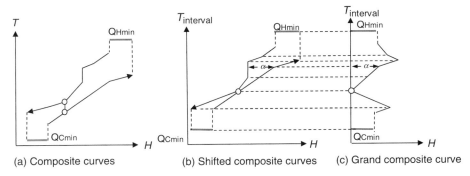

(a) Composite curves   (b) Shifted composite curves   (c) Grand composite curve

FIGURE 11.3. Construction of grand composite curve.

As a result of this temperature shift, the hot composite curve moves down vertically by one half ΔT_{min}, while the cold composite curve moves up by one half ΔT_{min}. As a result, shifted hot and cold composite curves touch each other at the pinch (see Figure 11.3b). In doing so, the minimum approach (ΔT_{min}) condition is built in for the shifted composite curves, which makes the task easier for utility selection on GCC (this will become self-evident later).

The grand composite curve is then constructed from the enthalpy (horizontal) differences between the shifted composite curves at different temperatures (shown by distance α in Figures 11.3b and c). The grand composite curve provides the same overall energy target as the composite curves, that is, targets are identical in Figures 11.3a and c. Furthermore, GCC represents the difference between the heat available from the hot streams and the heat required by the cold streams, relative to the pinch, at a given shifted temperature. Thus, the GCC is a plot of the net heat flow for any given shifted temperature, which can be used as the basis for assessing process changes and intermediate utility placement.

11.5 APPROPRIATE PLACEMENT PRINCIPLE FOR PROCESS CHANGES

11.5.1 General Principle for Appropriate Placement

Assume there is a hot utility that can be used for process heating at any temperature level. Where should we place it for process heating? Of course, we do not want to use it below the pinch according to the pinch golden rule (Chapter 10): Do not use hot utility below the pinch. To be smart, we should consider minimizing its use since the hottest utility is the most expensive. If intermediate utilities are available, we should consider maximizing the use of the utility at the lowest temperature first (e.g., low-pressure steam) and then the second lowest temperature (e.g., medium-pressure steam), and so on (e.g., high-pressure steam) above the pinch prior to the hottest utility (e.g., furnace heating).

Similarly, the cooling utility at the highest temperature should be used first (e.g., air cooling) and then second highest temperature (e.g., cooling water) and so on (e.g., chilled water) below the pinch prior to the coldest utility (e.g., refrigeration).

The above discussions point to the general principle for appropriate placement, originally introduced by Townsend and Linnhoff (1983a, 1983b). The penalty of violating the appropriate placement principle is that the costs of hot and cold utility go up and the process no longer achieves its utility cost targets.

This general principle was developed for utility selection in terms of the correct levels and loads. However, it is much less obvious that the principle also applies to process changes. For better illustration, application of this principle for utility selection will be discussed first and then discussions will cover unit operations such as reactors, separation columns, feed preheating, and so on.

11.5.2   Utility Selection

When there are multiple utility options available, the question is which utility is to be selected to reduce overall utility costs. This involves setting appropriate loads for the various utility levels by maximizing cheaper utility loads prior to use of more expensive utilities. The grand composite curve is an elegant tool for accomplishing this purpose.

Consider a process that requires heating and cooling. High-pressure (HP) steam is sufficient for heating at any temperature level and likewise, refrigeration is sufficient for cooling at any temperature level. The simplest way of utility selection is to use HP steam everywhere for heating, and refrigeration everywhere for cooling (Figure 11.4a). However, this could be a very costly option as HP steam and refrigeration are expensive. However, there exist intermediate utilities for use. If medium-pressure (MP) steam and cooling water (CW) can be used, a grand composite curve can be constructed as shown in Figure 11.4b. The target for MP

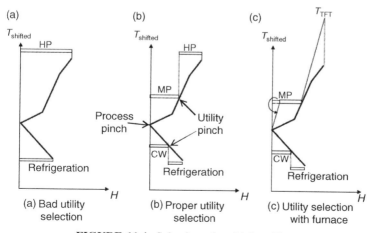

FIGURE 11.4. Selection of multiple utility.

steam is set by simply drawing a horizontal line at the MP steam temperature level starting from the vertical (shifted temperature) axis until it touches the grand composite curve. Remember that the minimum approach temperature is built in when constructing GCC via shifting hot composite curve as explained previously. The remaining heating duty is then satisfied by the HP steam. This maximizes the use of MP steam before HP steam and therefore minimizes the total hot utility cost because MP steam is cheaper than HP steam. The additional benefit from using MP steam versus HP steam is that higher latent heat is available in MP steam, which reduces the MP steam rate to meet the same duty requirement. Similarly, maximal use of cooling water before refrigeration reduces the total cold utility costs. The points where the MP and CW levels touch the grand composite curve are called the "Utility Pinches."

If the process requires furnace heating at higher temperature than HP steam, how can the furnace duty be reduced in design because furnace heating is more expensive? Figure 11.4c shows the possible design solution where the use of MP steam is maximized. In the temperature range above the MP steam level, the heating duty has to be supplied by the furnace flue gas. The flue gas flowrate is set as shown in Figure 11.4c by drawing a sloping line starting from the MP steam temperature to theoretical flame temperature (T_{TFT}).

The above discussions lead to the principles for utility selection as follows:

- Minimize furnace heating or high-pressure steam via maximizing the use of lower-pressure steam first.
- Minimize refrigeration or chilled water by maximizing the use of air and water cooling first.
- Maximizing generation of higher quality utility first.

11.5.3 Appropriate Placement for Reaction Process

Reaction placement implies appropriate heat integration of the reaction effluent. The reactor integration can be evaluated by the process GCC, which is constructed without the reaction effluent stream and then the reaction effluent stream is placed on top of the GCC. The general appropriate placement principle states that the heat of the reaction effluent should be released above the process pinch.

With guidance provided from the GCC, the reactor integration for new and existing process designs can be assessed. For existing processes, the process GCC is fixed, but reaction temperature might be adjusted to a small degree. Also, the integration of the reaction effluent can be modified by retrofitting the existing heat exchange scheme. The general guideline is to maximize heat recovery of the reaction effluent heat above the pinch.

For the new design, if reaction effluent stream does not fit well with the background process (Figure 11.5a), the reaction conditions such as temperature may be required to vary. However, only small changes in reaction conditions may be tolerated because any significant change would impact on conversion and product

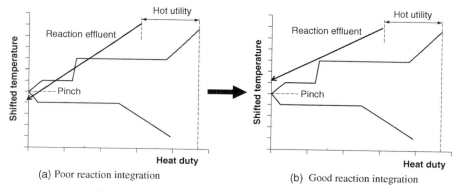

(a) Poor reaction integration

(b) Good reaction integration

FIGURE 11.5. Reaction integration against process.

yields, which usually outweighs energy costs. Thus, in grassroots design, there is little opportunity to change the desired reaction temperature, which is determined based on yield. However, instead of changing the reaction temperature, can we modify the background process to have better reaction integration (Figure 11.5b)? This topic is discussed in more detail by Glavic et al. (1988).

11.5.4 Appropriate Placement for Distillation Column

There are several key opportunities for column optimization, which include reflux ratio improvement, pressure changes, feed preheating, side reboiling/condensing, and feed stage location. A pinch tool called a column grand composite curve (CGCC) (Dhole and Linnhoff, 1993) was developed to provide an aid for the evaluation of these improvements.

11.5.4.1 The Column Grand Composite Curve
The column grand composite curve can be constructed based on a converged column simulation as shown in Figure 11.6a. From the simulation, the column stage-wise data are extracted and these data are then organized to generate the CGCC in Figure 11.6b. The stage-wise data relates to the "Ideal Column" design. For the ideal column design, the column requires an infinite number of stages and infinite number of side reboilers and condensers as shown in Figure 11.6c, which represents the minimum thermodynamic loss in the column. In this limiting condition, the energy can be supplied to the column along the temperature profile of the CGCC instead of supplying it at extreme reboiling and condensing temperatures. The CGCC is plotted in either $T–H$ (T, temperature; H, enthalpy) or Stage-H diagrams. The pinch point on the CGCC is usually caused by the feed.

Similar to the GCC for utility selection for a process, the CGCC provides a thermal profile for evaluating heat integration ideas for a column such as side condensing and reboiling (Figure 11.6b). In a practical column, energy is supplied to the column at feasible reboiling and condensing temperatures.

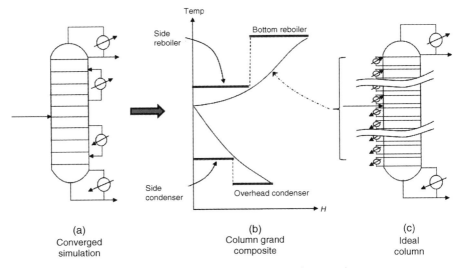

FIGURE 11.6. Construction of column grand composite curve.

11.5.4.2 Column Integration with Process Column integration implies heat exchange of the column heating/cooling duties against a background process or the external utility available. The principles of appropriate placement of columns against a background process are explained next.

Let us look at Figure 11.7a where the reboiler receives heat above the pinch of the background process, while the condenser rejects heat below the pinch. The background process is represented by its grand composite curve. Therefore, this distillation column is working across the pinch. In this case, Figure 11.7a represents a case of no integration of the column against the background process. The column is therefore inappropriately placed with regard to its integration with the background process.

Assume the pressure of the distillation column is raised and the condenser and reboiler temperatures can increase accordingly. As a result, the column can fit entirely above the pinch. This case represents a complete integration between the column and the background process via the column condenser as shown in Figure 11.7c. The column is now on one side of the pinch (not across the pinch). The overall energy consumption (column plus background process) equals the energy consumption of the background process. Energy-wise, the column is running effectively for free. The column is therefore appropriately placed with regard to its integration with the background process. Alternatively, lowering the column pressure so that its temperature drops will make the column fit below the pinch. Placing the column above or below the pinch is another application of the appropriate placement principle.

In reality, a big change to the operating pressure of the distillation column is rarely possible due to the process limits such as product specifications, capital costs, safety, or other considerations. However, there are other ways of reducing heat transferred across the pinch. One option is to install an intermediate condenser so that it works at a higher temperature than the main condenser at the top of the column. Figure 11.7b

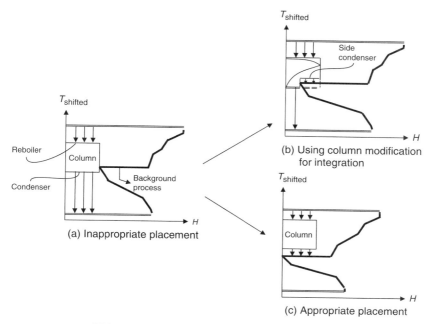

FIGURE 11.7. Column integration with process.

shows the CGCC of the column. The CGCC indicates a potential for side condensing. The side condenser opens up an opportunity for integration between the column and the background process. Compared to Figure 11.7a the overall energy consumption (column plus background process) has been reduced due to the integration of the side condenser. Alternatively, use of intermediate reboiler or pump-around can be considered.

In summary, the column is inappropriately placed if it is located completely across the pinch because the column has no heat integration with the background process. On the other hand, the column is appropriately placed if it is placed on one side of the pinch and can be integrated against the background process. Although appropriate column integration can provide substantial energy benefits, these benefits must be compared against associated capital investment and difficulties in operation. In some cases it is possible to integrate the columns indirectly via the utility system, which may reduce operational difficulties.

11.6 EXAMPLES OF PROCESS CHANGES

The process integration principles outlined above provide the guidelines for process changes in general. It is clear that both composite curve and grand composite curve are very powerful tools that allow us to optimize process integration for the entire process.

By applying the process integration principles, the design is no longer confined to working in one direction in design as suggested in the onion diagram (Figure 11.1), from reaction system to separation system, heat exchanger network, and site heat and power systems. Instead, we can have a two-way interaction such that the effects of process changes on the heat exchange and utility systems can be assessed prior to design and vice versa. In many studies, the savings from process changes far outweigh those from heat recovery projects. It is critical to get the overall plant design right in the early design stage and produce the most elegant and efficient solutions. Example case studies will be discussed in depth below.

11.6.1 Catalyst Improvement

Catalysts with high distillate selectivity can also improve energy efficiency at the same time they are improving yield. These kinds of catalysts reduce chemical H_2 consumption, lowers natural gas consumed, lowers makeup gas power, and reduces the need for quench. In this example as shown in Figure 11.8 (Zhu et al., 2011), natural gas is reduced by 80,000 Btu/bbl feed due to lower consumption in the H_2 plant. Makeup gas compression power is reduced by 4000 Btu/bbl on a fuel equivalent basis. Recycle gas compression power is reduced by 500 Btu/bbl on a fuel equivalent basis. The total energy savings are approximately 85,000 Btu/bbl on a fuel equivalent basis translating to energy savings of approximately \$10 million per year for a 50,000 bpd hydrocracking unit.

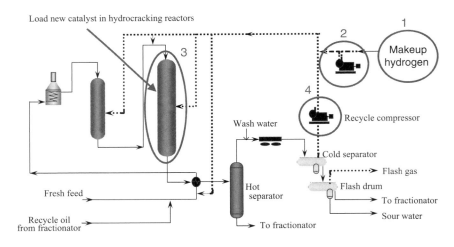

Benefits of catalyst change:
1. Lower chemical H_2 consumption reduces natural gas consumed at H_2 plant.
2. Lower chemical H_2 consumption reduces makeup gas compression energy..
3. Higher distillate selectivity reduces heat release in the new distillate catalyst; less quench is required
4. Lower quench requirement reduces the recycle gas compressor utilities.

FIGURE 11.8. Effects of catalyst improvements on process energy efficiency.

11.6.2   Process Flowsheet Improvement

The traditional hydrocracking process design features one common stripper, which receives two feeds containing very different compositions, one from the cold flash drum and the other from the hot flash drum; a typical design is shown in Figure 11.9. The processing objective of the stripper is to remove H_2S from the feeds. These two feeds originate from the reaction effluent, which first comes to the hot separator. The overhead vapor containing light products of the hot separator goes to the cold flash drum, while the bottom containing relatively heavy products of the hot separator goes to the hot flash drum. Eventually, the liquids of both hot and cold drums are fed to the same stripper. Then the stripper bottoms become the feed for the main fractionator. The shortcoming of this process sequence can be summarized as: separation and then mixing and separation again.

Thus, the inefficiency of this single stripper design is rooted in mixing of the hot flash drum and cold flash drum liquids, which undo the separations upstream. To avoid this inefficient mixing, it is proposed to use two strippers (Hoehn et al., 2013), namely, a hot stripper that receives the hot flash drum liquid as the feed, and a cold stripper to which the cold flash drum liquid is used as the feed. Furthermore, the cold stripper bottom does not pass through the main fractionator feed heater but goes directly to the main fractionators. The cold stripper bottom was part of the liquid in the single stripper bottom (Figure 11.10). Because only the hot stripper bottom goes to the main fractionator feed heater, the heater duty is reduced accordingly. This proposed design is shown in Figure 11.10. The idea sounds promising; however, how do we evaluate the effect of this idea?

Let us turn to composite curves for getting the answer. The stream data for both the single-stripper and two-stripper designs are obtained separately and the data include both the reaction and fractionation sections. For the relative comparison purpose, the composite curves are plotted based on zero degree ΔT_{min}, absolute pinch point. The single-stripper design is described by Figure 11.11, which indicates that the net total hot utility (fuel in this example) will be 110 MMBtu/h. In contrast, the two-stripper design depicted by Figure 11.12 shows zero net hot utility required at zero ΔT_{min}. The net difference between two designs is 110 MMBtu/h by the proposed two stripper design, which is 42% reduction of the fractionator heater duty or 23% of total energy reduction for the hydrocracking unit. This improvement is significant.

Furthermore, the process GCC curves are generated, one for the single-stripper design (Figure 11.13) and the other for the two-stripper design (Figure 11.14). The pinch point is the zero enthalpy point on GCC. As can be observed from the single stripper GCC (Figure 11.13), the process heat curve (above the pinch point) moves sharply away from zero enthalpy, which indicates the need for a large amount of external heating. In contrast, the GCC for the two-stripper design moves closer to zero enthalpy implying the reduction of external heating and cooling demand.

Although further design effort is required to complete the design details for the two-stripper design, the composite curve method is able to show the significance of the two-stripper design at the early stage. Since the benefit is revealed to be

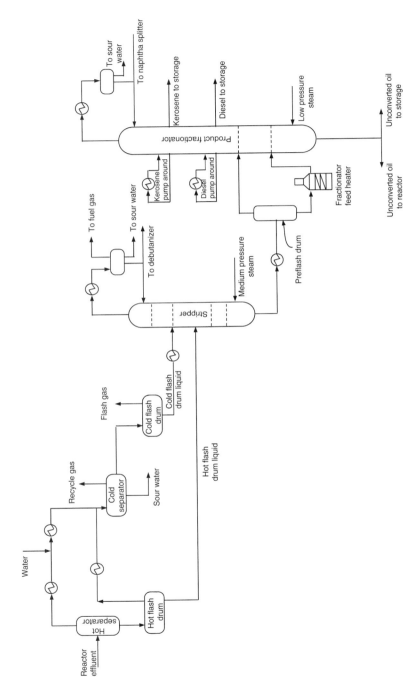

FIGURE 11.9. Single-stripper fractionation scheme.

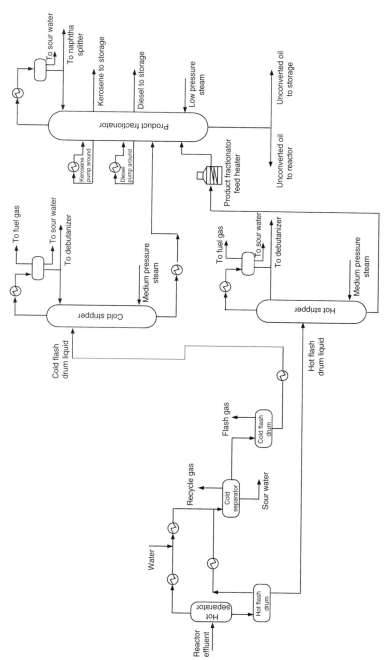

FIGURE 11.10. Proposed two-stripper fractionation scheme.

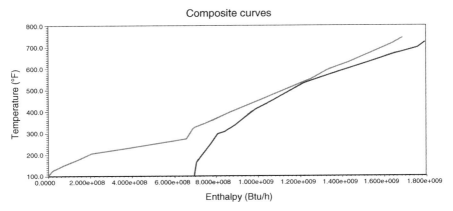

FIGURE 11.11. Composite curves for the single-stripper hydrocracking unit.

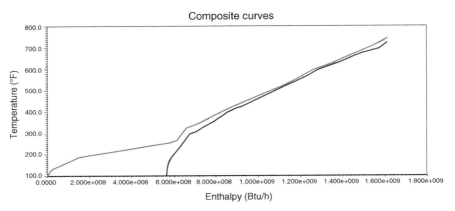

FIGURE 11.12. Composite curves for the two-stripper hydrocracking unit.

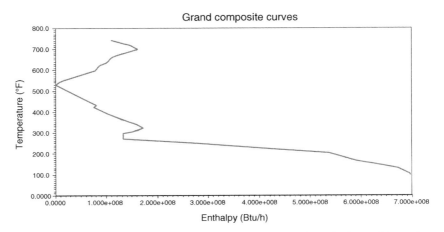

FIGURE 11.13. Grand composite curve for the single-stripper hydrocracking unit.

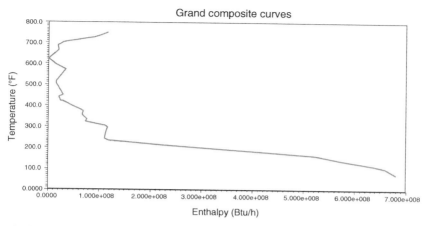

FIGURE 11.14. Grand composite curve for the two-stripper hydrocracking unit.

significant, further investigation of details is warranted. To make the story short, the more detailed cost–benefit analysis indicates operating cost savings of $2.5 MM/year as well as equipment cost savings of $2.0 MM. The capital cost savings are obtained from much reduced size of the feed heater for the main fractionators. Although there is a capital cost increase due to the addition of the second stripper, it is insignificant in comparison with the capital cost savings from the feed heater. The capital cost can be further reduced from other equipment innovations, such as having the common overhead system for two stripers and having one vessel to accommodate two columns via stacking two strippers on top of each other.

The composite and grand composite curves will remain the same for fixed process flowsheet and conditions. The energy targeting, in this case, will indicate the pinch point and minimum hot and cold utilities required for a given ΔT_{min}. However, the process design engineer may feel constrained with fixed stream flow rates and temperatures, which restrict his/her ability to design an integrated heat recovery system. This desire motivates him/her to explore process changes to improve the current process design since a large benefit could be gained if good process changes can be found.

Obviously, process modifications will have a profound impact on stream data such as flow rates, temperatures, and so on. Consequently, composite and grand composite curves will be very different from the base case. The above example has shown that these curves could become powerful tools to demonstrate the benefit of process changes.

11.6.3 Use of Advanced Technology

This example is about the use of a dividing wall column (DWC) for saving both energy and capital required in product fractionation. Due to market opportunity, an existing refinery wants to make four naphtha products from the two naphtha products it currently makes. Using the conventional design, two new columns are required. The showstopper for the revamp is the plot space, and a major concern is the prohibitive capital required for new equipment and related supporting infrastructure. The innovations discussed in this case not only resolve the plot space issue but reduce capital and energy use in a very significant way. Let us see how these improvements could be achieved.

11.6.3.1 Description of the Base Case

Fluid catalytic cracking (FCC) is widely employed to convert straight-run atmospheric gas oils, vacuum gas oils, certain atmospheric residues, and heavy stocks into high-octane gasoline, light fuel oils, and olefin-rich light gases. The reactor effluent from the reaction section of the FCC unit is sent to the main fractionator. The overhead of the main fractionator, which includes unstabilized gasoline and lighter material, are processed in the FCC unsaturated gas plant. A detailed description of the FCC process is given by Meyers (2004).

The existing unsaturated gas plant is shown in Figure 11.15, which consists of absorbers and fractionators to separate the main fractionator overhead into gasoline and other desired light products. According to this figure, overhead gas from the FCC main fractionator receiver is compressed and mixed with the primary absorber bottom and stripper overhead gas and directed to the high-pressure separator. Gas from this separator is sent to the primary absorber, where it is contacted by unstabilized gasoline from the main fractionator overhead receiver to recover C_3 and C_4 fractions. The primary absorber off gas is directed to a secondary or "sponge" absorber, where a circulating stream of light cycle oil from the main column is used to absorb C_5+ material and the remaining C_3 and C_4 fractions in the feed. The sponge absorbed rich oil is returned to the FCC main

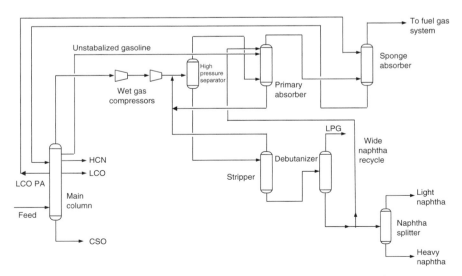

FIGURE 11.15. Existing naphtha separation for two-naphtha products.

fractionator. The sponge absorber overhead is sent to the fuel gas or other processing.

Liquid from the high-pressure separator is sent to a stripper column where C_2 material is removed overhead and recycled back to the high-pressure separator. The bottom liquid from the stripper is sent to the debutanizer, where an olefinic C_3–C_4 product is separated. Typically, 25–50% of the debutanized gasoline (full range naphtha) is recycled back to the primary absorber to control the recovery of light hydrocarbons. Currently, two gasoline cuts are produced. To achieve this, the debutanizer bottom, which is stabilized gasoline, is sent to naphtha splitters to make light and heavy naphtha products.

With increased market demand, the refinery wants to make four naphtha products from two naphtha products. If using the conventional design, two new columns are required with a direct sequence of three columns to make four naphtha products as shown in Figure 11.16. However, the showstopper for this design is the plot space for installing two new columns. The question then becomes: Are there any advanced separation technologies available that can resolve the plot space issue?

11.6.3.2 Dividing Wall Technology The dividing wall technology caught the attention of the refinery because multiple products can be made in one column with a dividing wall inside the column. Applications of dividing wall columns are discussed in detail by Schultz et al. (2002), Schultz and Zhu (2012), and Alves et al. (2013).

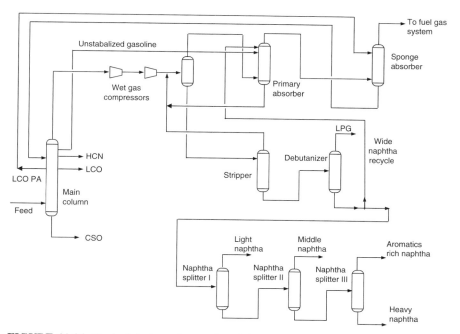

FIGURE 11.16. Typical design scheme of naphtha separation for four-naphtha products.

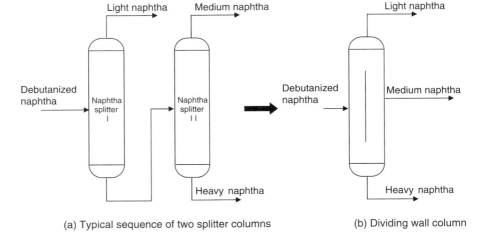

(a) Typical sequence of two splitter columns   (b) Dividing wall column

FIGURE 11.17. Separation of three naphtha products.

To verify the benefit of using a dividing wall column, a simulation model was conducted for two naphtha splitters (Figure 11.17a). Concentration profiles were generated for medium naphtha (Figure 11.18). This profile shows inefficient separation in splitter 1 for the medium naphtha; the concentration of medium naphtha increases from top down and then peaks in the middle of the column, but reduces sharply toward the bottom, which is an indication of undone separation for the medium naphtha. The reason for the concentration drop is the occurrence of remixing of medium and heavy naphtha in the middle toward the bottom section of the column. Consequently, separation between medium and heavy naphtha must take place again and accomplished in the splitter 2 column where energy has to be provided via bottom reboiling. Thus, the separation inefficiency due to remixing in the two-column system must be overcome with additional stages and reboiling.

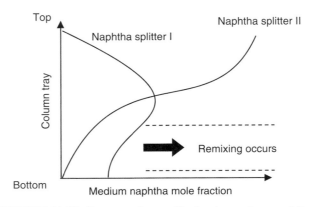

FIGURE 11.18. Concentration profile for the medium naphtha.

On the other hand, a dividing wall column where a vertical partition separates the column sections is simulated for making three naphtha products (Figure 11.17b). With the partition of the column into two sections, separation of light naphtha from the rest occurs in the partition on the left-hand side of the column similar to what happens in splitter 1 while the partition on the right-hand side acting similar to splitter 2 mainly deals with separation between medium and heavy naphtha. With medium naphtha withdrawn in the middle of the column where its concentration peaks, there is no remixing occurring. The remaining light naphtha separated travels up to the top and joins that on the left side to go to the overhead condenser. In this case, reboiling duty is reduced by 25% with a single dividing wall column versus two simple columns. This proof of principle based on a simulation is demonstrated by the benefit of applying a dividing wall technology for naphtha separation in general. However, it was not certain if it was a clear cut case for applying DWC for this existing naphtha system. It was found that there are criteria for using a dividing wall, which are: (i) the side cut must be relatively large in flow rate compared with other products, and (ii) there is no requirement for high-purity product specifications for the medium naphtha product.

By assessing the naphtha separation from the FCC process, it was found that these conditions could be satisfied readily because there is no strict specifications for medium and aromatics rich naphtha, which are the two middle cuts naphtha products, and the quantity of these two middle cuts are relatively large. Since dividing wall technology has been successfully applied to many processes, this boosted the confidence of the refinery to apply the technology. Therefore, it was concluded that a dividing wall column was the most promising for the naphtha separation.

11.6.3.3 Application of Dividing Wall Technology

Now the question becomes more specific: How to apply the technology to the existing gas plant?

Due to the restriction of plot space, the revamp team was looking into the revamp of the existing debutanizer column for a dividing wall column (Figure 11.19). By retrofitting the debutanizer column with dividing wall technology, LPG can be recovered from the overhead and light naphtha withdrawn as side cut. Part of the light naphtha is recycled and used as sponge liquid in the primary absorber instead of full range naphtha as used in the current design (Figure 11.15). Light naphtha has much better absorbing ability than the full range naphtha; consequently, the recycle can be significantly reduced. The debutanizer bottom is directed to the downstream separation. Naturally, the existing naphtha splitter was also revamped for a dividing column by inserting a metal wall to make a partition of the column. With the naphtha splitter using dividing wall technology, three more naphtha products are made, that is, medium naphtha, aromatics grade naphtha, and heavy naphtha.

The dividing wall column strategy reduces the overall reboiler duty by 30%. At the same time, considerable cost savings are realized. With the revamped naphtha separation scheme by applying the dividing wall technology, four naphtha products can be made without the need for installing two new columns. This resolves the plot space issue. Consequently, capital cost was reduced significantly by around 25% although a major revamp for the existing two columns was required.

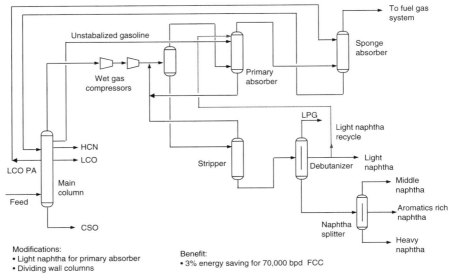

FIGURE 11.19. Applying dividing wall to naphtha separation.

11.6.4 Integrate Power Recovery with Existing Steam System

This discussion for the case study comes from Couch and Leonard (2006). The purpose of using this case study here is to show the benefits of integrating power generation turbines with the existing steam system.

11.6.4.1 Description of Power Recovery Turbine   The hot flue gas from the catalyst regenerator, which operates at approximately 1350° F, is passed through a turbine to generate power. The benefit of installing such a power recovery turbine (PRT) is around $6.5 MM/year savings in operating cost for a 70,000 bpd FCC. However, due to high capital cost, installation of the FCC flue gas power recovery turbine has been treated as an "accessory," which is installed only for higher-capacity, higher-pressure FCC units in areas of high electrical cost. To make this technology applicable for a wider range of FCC operators, innovative improvements have been developed for the way power recovery systems are incorporated into the FCC unit. These innovations significantly increase the power production and reduce the capital cost per unit of energy recovered.

A typical PRT design is of a five-body train, consisting of a hot gas expander, main air blower, steam turbine, motor/generator, and gear box as necessary. A five-body train is shown in Figure 11.20. In this configuration, the expander is coupled to the main air blower and provides a direct transfer of energy to the shaft. The direct transfer of energy to the main air blower minimizes power transfer losses, and is the most energy efficient configuration. The steam turbine is used to start up the train,

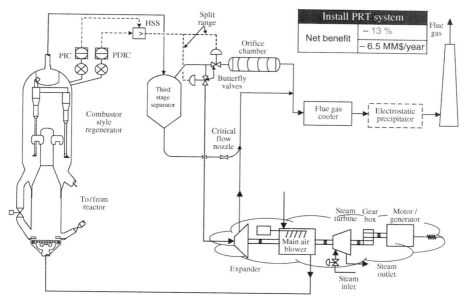

FIGURE 11.20. Traditional five-body power recovery train.

and the combination of a steam turbine and motor can provide the required power to operate the air blower at design conditions, with the PRT expander out of service. After the start up, when enough flue gas is present, and process conditions are stable, the PRT expander can be commissioned, while the steam turbine will be in hot standby operation. At this point, the steam turbine becomes a marginally utilized asset until the next FCC shutdown.

11.6.4.2 Steam System   If we look at the steam system around the FCC unit, we can identify opportunities of energy recovery from steam letdown. A typical FCC can generate HP steam from a number of places such as the flue gas cooler, waste heat boiler, as well as the main fractionator bottom. The FCC process also uses steam as feed dispersion steam, lifts steam, and spent catalyst stripping steam. Often the HP steam is let down to MP and LP to satisfy these process needs. The opportunity here is that power can be generated if a steam turbine can be installed to replace steam letdown valves (Figure 11.21). A steam turbine can be incorporated to recover power from the HP steam letdown. For our example, power production can be increased, which results in $2.6 MM/year for a 70,000 bpd FCC. More power can be generated if additional HP steam is brought in from the refinery steam main header. However, it is too often that a capital project of installing a steam turbine in such applications could not pass the capital review process as the cost for installing a steam turbine and related supporting infrastructure is too expensive to justify the power generated.

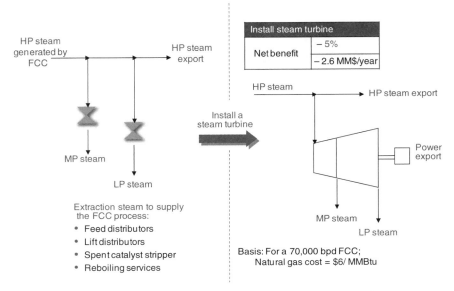

FIGURE 11.21. Install a steam turbine to replace letdown valves.

11.6.4.3 Integration of PRT with the Steam Turbine

There are two separate opportunities identified so far. On one hand, there is a hot flow gas PRT in FCC, while there is a letdown steam turbine on the other hand. However, if these two opportunities are treated separately, both turbines will require a separate shaft, gear box, and generator, as well as a separate foundation, piping, electrical infrastructure, and control system. The capital cost could be prohibitive for either opportunity.

What about integrating these two turbines? The idea is to make better use of the steam turbine helper for start up in the PRT five-box train design in Figure 11.20. In other words, the steam turbine helper becomes a letdown turbine and the normal FCC process steam can be routed through this letdown turbine. Thus, the letdown turbine is integrated with the PRT as shown in Figure 11.22. The integrated power recovery system can produce electricity while the main air blower is run by motor.

When installing an FCC flue gas power recovery system, most of the auxiliary equipment is already required; that is, the generator, 13.8 kVa cable, switches gear, foundation, electrical controls, and substation. The incremental cost of adding the steam letdown turbine to the power recovery train is low compared to the potential energy recovered, and the integration can significantly increase the return on investment for installation of a power recovery system in the FCC unit. This idea will create synergies for implementing both the PRT and letdown steam turbine at the same time. The total benefit will be $9.1 MM/year in operating cost savings for a 70,000 bpd FCC.

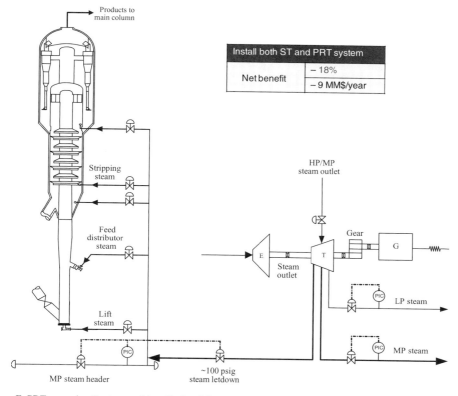

Install both ST and PRT system	
Net benefit	– 18%
	– 9 MM$/year

E: PRT expander; T: steam turbine; G: electricity generator

FIGURE 11.22. Integration of letdown turbine with PRT.

11.6.5 Integrated Energy and Process Optimization

This example is about how energy retrofitting can support capacity expansion and yield improvement (Zhu et al., 2011). A North America refinery wishes to increase hydrocracking capacity by 15% to meet new diesel demand in the region; the existing process design is shown in Figure 11.23. A screen study indicated several major bottlenecks that could require significant capital investment and make the expansion financially infeasible. In the screen study, the following equipment would not be able to handle 15% expansion because

- The heaters for the reaction and fractionation will be too small;
- The space velocity will be too higher for the existing reactors;
- The fractionation tower and debutanizer tower will have severe flooding.

However, the refinery management wanted a more detailed assessment by a technology company to seek a possibility of reducing capital costs. The results from this assessment can be summarized as follows.

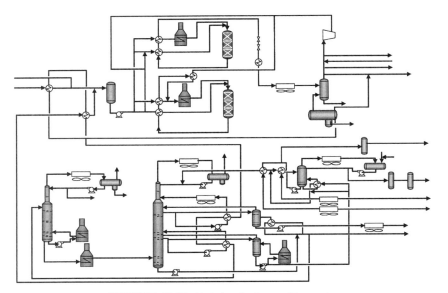

FIGURE 11.23. Existing hydrocracking unit.

11.6.5.1 Removing the Heater Bottlenecks   To do this, the network pinch method (Zhu and Asante, 1999) was applied to the existing heat exchanger network. The modifications for reducing the reactor charge heaters and the fractionator heater were identified. As a result, installation of four heat exchangers (A–D) was required that use reactor effluent heat for preheating the reactor feed and the fractionation feed (Figure 11.24). These new exchangers can reduce total heater duty by 20% with a

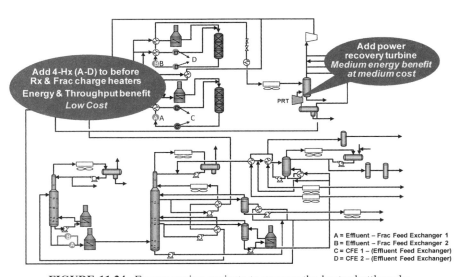

FIGURE 11.24. Energy-saving projects to remove the heater bottlenecks.

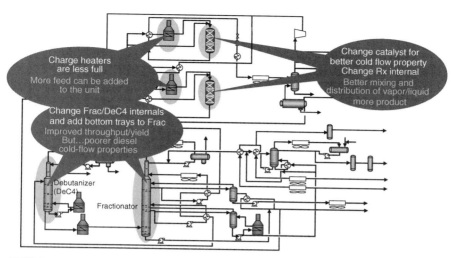

FIGURE 11.25. Reaction and fractionation projects to remove the process bottlenecks.

payback of less than 2 years based on energy savings alone. Thus, the need for revamping existing heaters was avoided. At the same time, an opportunity of using a liquid expander for power recovery was identified. In the existing design, high-pressure liquid at 2200 psig was throttled to around 450 psig through a valve.

11.6.5.2 Removing the Reactor Bottlenecks The hydrocraking unit was designed in the 1970s. It was not surprising to find that the existing reactor internals experienced poor distribution. By installing better mixing and distribution devices, gas and liquid distribution was improved, and the existing reactors could handle high space velocity from feed rate increase (Figure 11.25). A new catalyst was also used to deal with diesel cold flow property issues.

11.6.5.3 Removing the Fractionation Bottlenecks In evaluating both the fractionation tower and debutanizer tower, the simulation model predicted too high liquid loading and thus a downcomer backup flooding. To avoid this, UOP ECMD trays were considered for replacement because the ECMD trays feature multiple downcomers and equalized loadings for downcomers. This feature could mitigate downcomer backup flooding and thus allow towers to accommodate more than 15% feed rate increase.

During evaluation, a new opportunity was identified. In the current operation of the fractionation tower, a 58 °C overlap exists between the recycled oil and the diesel product, which corresponds to a 340 °C TBP cut point. Newly designed units will typically have a TBP cut point of 380–410 °C and a gap of 10–30 °C. The poor separation in the fractionation column results in 13 wt% of diesel range material slumping into recycle oil (fractionation column bottom). Due to the large amount of diesel range material that is being recycled, the diesel yield is reduced from over-conversion of the gas oil and excess hydrogen is consumed. Additionally, a higher reactor severity is needed, which leads to a higher activity catalyst (or shorter catalyst life) and higher temperatures when compared to a unit with better separation. Also, more heavy poly nuclear aromatics (HPNA) components are made at the higher reactor severity. Recovery of diesel from the recycled oil could be worth $20 MM/ year for the refinery margin. A possible solution was to add several trays in the bottom. The good news is that there was enough space available to accommodate a few trays in the column bottom section (Figure 11.25).

11.6.5.4 Modification Summary   In summary, the modifications for the process unit include the following:

- Four new heat exchangers to reduce both the reaction fractionation heaters' duty more than 20%. Thus, the heater bottlenecks were resolved.
- Reactor internals with mixing and distribution devices installed to avoid addition of new reactor or major revamp of the existing two reactors.
- New catalyst to deal with diesel cold flow property issues.
- Replacement of part of the tower internals for fractionators and debutanizer to mitigate liquid loading to avoid severe downcomer flooding.
- Adding several trays in the bottom section of the column to reduce the diesel slumping into the recycled oil from 13 wt% to less than 4 wt%.
- The power recovery turbine was not selected in this revamp package as it was not required for feed expansion. However, it can be considered as a major energy-saving project in the future.

The overall study identified energy savings and allowed the desired throughput increase with minimal capital costs because expensive modifications were avoided. The synergy between energy savings and process technology know-how was critical to achieving these results. The refinery management was very pleased with the results and got approval from the board of directors for investment of these projects.

REFERENCES

Alves J, Ulas Acikgoz S, Zhu XX (2013) Process for providing one or more streams, US Patent No. 8,414,763, April 9.

Couch KA, Leonard EB (2006) Concepts for an overall refinery energy solution through novel integration of FCC flue gas power recovery, NPRA Conference, San Antonia.

Dhole VR, Linnhoff B (1993) Distillation column targets, *Computers and Chemical Engineering*, **17**(5/6), 549–560.

Glavic P, Kravanja Z, Homsak M (1988) Heat integration of reactors. 1. Criteria for the placement of reactors into process flowsheet, *Chemical Engineering Science*, **43**, 593.

Hoehn RK, Bowman DM, Zhu XX (2013) Process for recovering hydroprocessed hydrocarbons with two strippers, Patent Publication No.: US 2013/0,046,125 A1.

Linnhoff B, Townsend DW, Boland D, Hewitt GF, Thomas BEA, Guy AR, Marsland RH (1982) *A User Guide on Process Integration for the Efficient Use of Energy*, IChemE, Rugby, UK.

Linnhoff B, Vredeveld DR (1984) Pinch technology has come of age, *Chemical Engineering Progress (CEP)*, **80**(7), 33–34.

Meyers RA (2004), *Handbook of Petroleum Refining Processes*, Chapter 3, 3rd edition, McGraw-Hill.

Schultz, MA, Stewart DG, Harris, JM, Rosenblum, SP, Shakur, MS, O'Brien, DF (2002) Reduce costs with dividing-wall columns, *Chemical Engineering Progress (CEP)*, **98**(5), 64–71.

Schultz MA, Zhu XX (2012) Zone or system for providing one or more streams, US Patent No. 8,246,816, August 21.

Townsend DW, Linnhoff B (1983a) Heat and power networks in process design. Part 1: Criteria for placement of heat engines and heat pumps in process networks, *AIChE Journal*, **29**(5), 742–748.

Townsend DW, Linnhoff B (1983b) Heat and power networks in process design. Part II: Design procedure for equipment selection and process matching, *AIChE Journal*, **29**(5), 748–771.

Zhu XX, Asante NDK (1999) Diagnosis and optimization approach for heat exchanger network retrofit, *AIChE Journal*, **45**(7), 1488–1503.

Zhu XX, Maher G, Werba G (2011) Spend money to make money. *Hydrocarbon Engineering Journal*, p. 34–38, September issue.

PART 3

PROCESS SYSTEM ASSESSMENT AND OPTIMIZATION

12

DISTILLATION OPERATING WINDOW

The operating window defines a feasible operating region for a distillation column. If the region can be determined and made visible, engineers can know where the column stands in performance and how close it is to operating limits. This chapter provides practical insights for the characteristics, calculations, and criteria for the distillation design and operation.

12.1 INTRODUCTION

Separation is widely applied in the process industry, and typically follows reaction processes to separate desirable products and by-products from reaction effluents. There are many ways of separation. In this chapter, we mainly deal with product distillation or fractionation, which forms the majority of the separation processes in the process industry.

What criteria should be used for assessing a distillation tower? How does one know if a tower operates under stable operation or near the best performance? What turnkey options are available to operators to simultaneously optimize product separation and energy use? How can the best performance be sustained? These are the questions that engineers constantly ask and these are the focus of this chapter.

Because distillation is a complex topic, this chapter intends to cut through the maze straight to the most important matters, such as the concept of feasible operating region, key operating parameters, and optimizing distillation from design and

Energy and Process Optimization for the Process Industries, First Edition. Frank (Xin X.) Zhu.
© 2014 by the American Institute of Chemical Engineers, Inc. Published 2014 by John Wiley & Sons, Inc.

operation. For effective understanding, simplified explanations and symbolic terms are adopted without sacrificing fundamentals. This chapter is designed to focus on these aspects supported by comprehensive examples. The detailed and excellent discussions for all aspects related to column design and operations can be found in Kister (1990, 1992).

12.2   WHAT IS DISTILLATION?

A complex distillation column (Figure 12.1) is designed and operated for separating products from feed(s) and some of the products may need to go through further separation downstream to meet product specifications. Energy is provided in the form of reboiling or feed heating or steam injection as the driving force for desirable distillation. Pump-arounds are used to remove excess heat within the column to achieve required vapor and liquid loading balances locally in the column. Although every distillation column is different in feeds/products hence resulting in different design and operation, the basic elements are common. A good understanding of the major features of these elements will provide a sound basis to know what to watch for (key indicators), what knobs to turn (operating parameters), and directional changes for improvements (effects on process and equipment performance) within operating limits (process and equipment constraints).

Distillation occurs in a tower due to the relative volatility between light and heavy components. Vapor flows upward at a relatively higher temperature, meets liquid that flows downward at a relatively lower temperature. On the tray, heat transfer takes place between vapor and liquid. As a result, vapor becomes cold, which makes heavy components separate out from the vapor and join the liquid traveling down. In contrast, liquid is heated up and thus light components are flashed out and join the vapor traveling upward. This is what happens in mass and

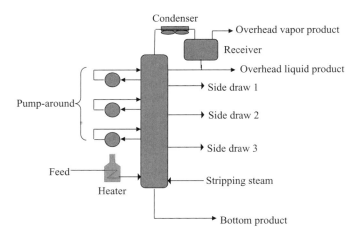

FIGURE 12.1. A complex configuration of a distillation column.

heat transfer between vapor and liquid on each tray. Consequently, two things happen. First, vapor and liquid approach equilibrium on each tray. Second, vapor becomes lighter and colder traveling upward, while liquid becomes heavier and hotter traveling downward. The limit for the mass transfer between vapor and liquid on each tray is set by the composition equilibrium, which depends on the relative volatility. With a certain number of equilibrium stages, a distillation column accomplishes its operating objective that light and heavy products are separated and produced from the overhead and the bottoms, respectively. The ease or difficulty of using distillation to separate the more volatile components from the less volatile components in a mixture is denoted by the relative volatility. For a mixture with low (high) relative volatility, a tall (short) tower with many (fewer) number of stages is required.

It is often that distillation columns do not perform as they should and operation performance deviates from design. Energy usage is often not optimized during operation when feed conditions and product specifications change due to operational requirements. In some cases, product quality could be better than specifications. We call this as product quality giveaway, because it consumes more energy than necessary. In other circumstances, distillation columns are operated in abnormal conditions such as flooding. Under abnormal operation, distillation efficiency suddenly drops leading to little or virtually no distillation taking place. Therefore, the topic of distillation efficiency needs to be well understood before discussing the operating window.

12.3 DISTILLATION EFFICIENCY

The ultimate goal of a distillation tower is separation of products, and naturally distillation efficiency becomes the key performance metric. A tower, if properly designed, can achieve 10% higher distillation efficiency. In operation, the tower operated with better distillation efficiency requires less energy use.

Many different measures of efficiency have been developed. Let us look at the two commonly used measures, which are the stage-based Murphree tray efficiency (Murphree, 1925) and overall efficiency.

The Murphree tray efficiency is defined as

$$\eta_M = \frac{\text{change in vapor for actual stage}}{\text{change in vapor for equilibrium stage}}. \tag{12.1}$$

The above expression can be better explained using Figure 12.2. The denominator in equation (12.1) is the vertical distance or vapor composition difference from the operating line to the equilibrium curve $(\overline{AC})$, while the numerator is the vertical distance or vapor composition difference from the operating line to the actual concentration curve $(\overline{AB})$. For example, $\eta_{M2} = \overline{AB}/\overline{AC}$ represents the efficiency for stage 2. The efficiency for other stages can be determined similarly.

The overall tower efficiency is defined as the ratio between the number of theoretical stages and the actual number of stages required for the

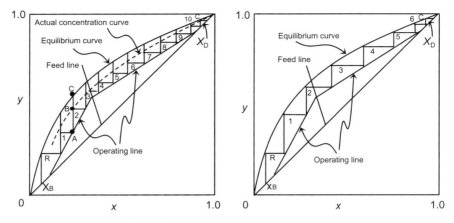

FIGURE 12.2. McCabe–Thiele diagram.

separation as

$$\eta_o = \frac{N_{eq}}{N_{act}}. \tag{12.2}$$

As an example for illustration, the McCabe–Thiele diagram (McCabe and Thiele, 1925) in Figure 12.2 indicates 12 actual stages required in comparison with 8 theoretical stages in the tower. Partial condensers and partial reboilers are counted in both the theoretical stages and actual stages. Thus, the overall tower efficiency is 67% (= 8/12).

The overall efficiency lumps everything that happens in the column into one value. Based on the assumptions of constant molar overflow and constant value of η_M for trays in each distinctive section in a tower, Lewis (1936) developed a relationship between the Murphree tray efficiency and overall efficiency, which is expressed as:

$$\eta_o = \frac{\ln[1 + (\lambda - 1)\eta_M]}{\ln \lambda}, \tag{12.3}$$

where

$$\lambda = m\frac{V}{L}. \tag{12.4}$$

Equation (12.3) applies separately to rectifying and stripping sections as the V/L ratio is different between these two sections. In the rectifying section, vapor rate is higher while liquid rate is lower compared with those in the stripping section. However, in each section, equation (12.3) is based on the assumptions of straight operating and equilibrium lines and constant V/L ratio and η_M from tray to tray.

For complex towers with pump-arounds, side condensers or side reboilers, and side product draws, a tower can be divided more than two sections, and the efficiency defined in equation (12.3) is then applied to each section. This efficiency calculation method requires a tower simulation to give internal V/L distributions from stage to stage, which becomes the basis to determine the sections each of which features near-constant molar flows between stages. For the example in Figure 12.2, three actual stages (including the feed stage but not the reboiler) are required in the stripping section versus two theoretical stages. Thus, the efficiency for this stripping section is 67%. In contrast, seven actual stages (not including the condenser) are needed in the rectifying section versus four theoretical stages, which results in the section efficiency of 57%. The overall tower efficiency is 67% (including the reboiler and condenser). For industrial towers, stage efficiency is typically around 60–75%; it is uncommon to have stage efficiency less than 50%.

Many studies of efficiency have been conducted on both sieve and valve trays in industrial and academic laboratories. Most academic studies are performed in towers far too small in size to be useful for industrial applications; but the industrial efficiency data are proprietary. The best example is the FRI data (Fractionation Research Inc.), which are only available to FRI consortium members.

The simplest approach is to use a correlation to determine overall tower efficiency. The O'Connell (1946) correlation shown in Figure 12.3 is the standard of the industry for industrial tower efficiency (Kister, 1992).

Figure 12.4 shows a typical trend of tower efficiency dependent on the balance of vapor and liquid rates. In the middle of the efficiency curve corresponding to stable operation, there is a relatively flat region although with marginal variation. Trays with good turndown features such as the valve tray compared with sieve tray have a wider flat or stable operating region. On the either side of the curve,

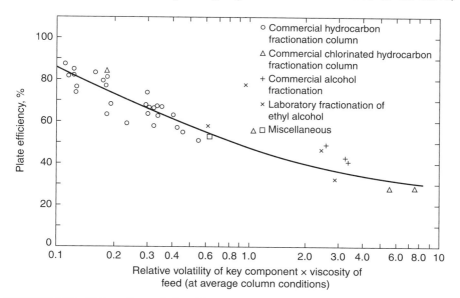

FIGURE 12.3. O'Connell correlation. (From O'Connell (1946), reprinted with permission by AIChE.)

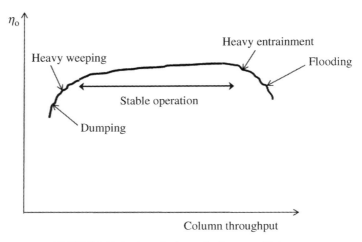

FIGURE 12.4. A typical trend of tower efficiency.

efficiency drops off dramatically. Efficiency declines under low feed rate corresponding to turndown operation and falls off the cliff when liquid dumping occurs. On the other hand, efficiency reduces at excessive entrainment and plummets when flooding occurs. Optimization in design and operation tends to push the tower toward the boundary of stable operation. An understanding of these controlling mechanisms can shed insights into how to optimize tower design and operation while achieving stable operation.

12.4 DEFINITION OF FEASIBLE OPERATING WINDOW

How can we be sure if a tower is designed and can be operated to give a feasible operation with a satisfactory distillation? A valuable tool for this purpose is the capability diagram or feasible operation window. This operation window is bounded by tray capability limits (Figure 12.5). Any specific operating scenario represents an operating point within this operating window. Any variation in tower feed rate, product rates, and quality and operating conditions such as feed temperature, tower pressure, and reflux rate will move to a different operating point in the window. From the simplest terms, it can be said that the design and operation of a tower is all about *balancing between vapor and liquid loadings* with an utmost goal of achieving desirable separation at minimal operating cost.

12.5 UNDERSTANDING OPERATING WINDOW

The mechanisms of capability are described qualitatively here based on the concept of relative momentum between vapor and liquid while quantitative discussions will be given afterward. To understand the main modeling parameters

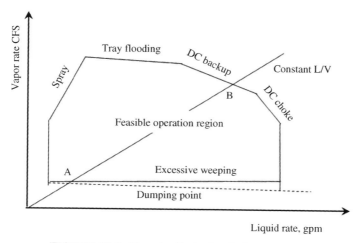

FIGURE 12.5. Capacity limits for distillation tower.

of interest, Figure 12.6 is used for illustration. Adjacent trays are apart with tray spacing H_S. Vapor flows upward through the perforations with hole diameter d_h. Froth is generated due to the momentum exchange between vapor and liquid with the froth height h_f. To measure the equivalent liquid height for the froth, the clear liquid height h_c is defined for modeling purposes, which is the height that the froth level h_f could collapse in the absence of vapor. From the froth level, vapor continuously flows upward and creates a two-phase layer with height of h_V. In this vapor dominant region, vapor is in a continuous-phase, but containing a small portion of liquid in the form of fine droplets. The concept of *relative momentum* between vapor and liquid is the key to understand the operating window (Lockett, 1986; Bennett and Kovak, 2000).

The liquid momentum is mainly generated by gravitational force represented by liquid height or liquid hold up, h_c, on the tray deck, while the vapor momentum is generated by the buoyant force due to vapor flowing upward through perforations, which is indicated by $u_h d_h$. The relative momentum is largely defined by the ratio of $h_c/u_h d_h$. In the case of higher ratio of $h_c/u_h d_h$, small u_h and/or small d_h, vapor exchanges momentum sufficiently with liquid, which slows down the vapor momentum and prolongs the vapor residence time within the liquid zone. The result is better interactions between vapor and liquid leading to better distillation. Because liquid loading is responsible for liquid height h_c, while vapor loading for vapor velocity u_h, a combination of high vapor and liquid loading with small perforation size d_h promotes mass transfer between vapor and liquid and enhances distillation. On the other hand, in cases when the combination of (h_c, u_h, d_h) is inappropriate, the tower could be under unstable operation.

The first unstable operation of interest is *spray* flow, which occurs under a high vapor but low liquid loading on the tray. One could visualize what happens in spray operation: High vapor momentum causes liquid to blow off and reach the tray above, which contaminates the tray above and undoes separation. Under spray operation,

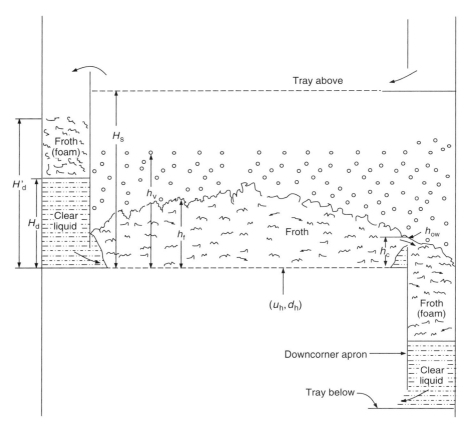

FIGURE 12.6. Vapor–liquid flow structure on tray deck.

with severe entrainment of liquid passing through the tray above, distillation efficiency reduces sharply and can drop half of the normal operation.

Spray flow occurs under very small $h_c/u_h d_h$ ratio when vapor momentum is much higher than liquid and there is a very thin layer of liquid h_c. In this case, vapor does not adequately exchange momentum with the liquid and maintains high momentum entering the two-phase region. The resulting spray flow carries fine liquid droplets to the tray above causing excessive entrainment (Figure 12.7). The spray flow can be analogous to fluidization or blowing effect. Spray flow should be avoided in design and operation. Because preventing spray is equivalent to increasing the ratio of $h_c/u_h d_h$, two effective ways of spray prevention can be devised. One is to increase liquid loading using a picket fence weir as an example. The other is to reduce hole diameter d_h, while increasing the number of perforations for the same hole area. If spray flow cannot be sufficiently avoided by using these two methods for the tray column, a packed column could be the choice. Under spray regime, valve trays can perform better than sieve trays in both capacity and efficiency.

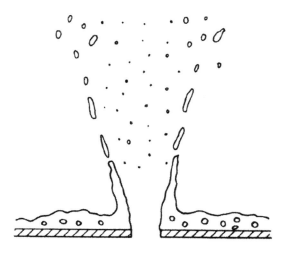

FIGURE 12.7. Spray.

The second abnormal operation is *tray flood* when the froth level reaches the tray above. The tray flood occurs when both liquid and vapor loadings on the tray are high and thus sufficient vapor momentum generates higher froth level h_f. When $h_V > H_S$, entrainment occurs as vapor carries liquid droplets to the tray above. The tower can still operate in a stable manner under significant amount of entrainment if the downcomer can handle additional liquid loading. However, tray distillation efficiency drops depending on the amount of entrainment. If vapor rate further goes up and increased vapor momentum pushes the froth level to reach the tray above ($h_f > H_S$), excessive entrainment occurs and thus distillation efficiency drops significantly. Because this happens on the tray deck, it is named tray flood in differentiation with downcomer flood. The tower may not be able to achieve stable operation under severe tray flood. In the literature, it is commonly called *jet flood*, which is a misconception as the name of jet flood is more suited for spray. To avoid misunderstanding, tray flood is used in place of jet flood in this book.

Clearly, tower diameter and tray spacing H_S are the two major parameters in design affecting tray flood. A large tower diameter reduces vapor velocity and liquid hold up (h_c), while high tray spacing increases liquid settling space. In operation, reboiling duty is the major parameter to be adjusted to avoid tray flood.

The third unstable operation is *downcomer flood* when the liquid stacks up in the downcomer and the liquid froth in the downcomer reaches the tray above ($H'_d > H_S$). Downcomer flood is caused by high hydraulic head loss along the liquid flow path. Downcomer flood could be *downcomer backup* and/or *choke*. Downcomer backup occurs when the downcomer cannot allow additional liquid to freely flow out of the downcomer when there is high hydraulic resistance along the liquid pathway (Figure 12.8). In a different mechanism, downcomer choke happens when large

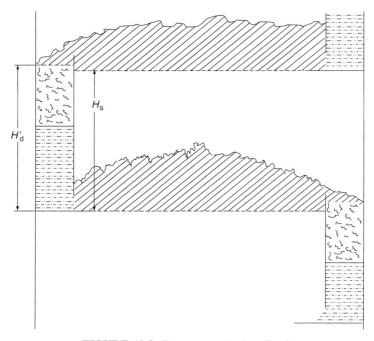

FIGURE 12.8. Downcomer backup flood.

froth crest gets stuck in the downcomer mouth and blocks the liquid to flow over the outlet weir (Figure 12.9).

In both downcomer back up and choke cases, downcomer liquid inventory increases and downcomer liquid backs up until the downcomer froth level reaches the tray above ($H'_d > H_S$). This phenomenon is called downcomer flood. When downcomer flood occurs to any tray, the whole tower will be flooded very quickly. A tower under downcomer flood provides virtually no distillation. In contrast, under tray flood, liquid can still leave the tower and the tower could still operate if the control system allows it although distillation efficiency suffers. Downcomer flood can be prevented in design by providing adequate downcomer area and clearance underneath the downcomer and minimizing tray pressure drop. Reducing reflux rate in operation could be effective in avoiding downcomer flood in operation.

The fourth unstable operation is *weeping*, which occurs at too low vapor loading when vapor momentum cannot hold liquid gravitational force. Under weeping, some of the liquid falls down through perforations and short-circuits the flow path; thus, distillation suffers. In the worst case, all liquid flows through perforations; this is called *dumping*. A tower can still operate under a certain amount of weeping although at reduced distillation efficiency; however, the operation is very unstable under dump.

Weeping usually happens at turndown operation when feed rate is reduced significantly from the nominal rate. For the need of high turndown operation,

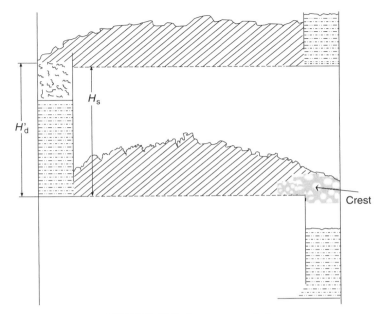

FIGURE 12.9. Downcomer choke.

selection of valve tray over sieve tray should be considered. Typically, valve tray can achieve 3:1 and higher turndown in comparison with 2:1 turndown by sieve tray.

Having discussed unstable operations, you may like to know what it takes to achieve *stable operation*. Under stable operation, vapor can travel upward without excessive liquid entrainment, while liquid travels down through the tray flow path effectively. This is achieved by two kinds of tower geometry arrangements. The first one is the tray dimension, which allows sufficient momentum exchange from vapor to liquid on the tray deck. The tray dimension includes tower diameter, tray spacing, tray flow path, hole diameter, and hole area. The other is the downcomer dimension, which includes downcomer flow area, clearance, weir type, and weir length. The tray and downcomer dimensions together are called tower dimensions, which are determined in design. In operation, a proper vapor and liquid loading balance is the key for ensuring stable operation for a given tower design. A balanced vapor and liquid loading is indicated by the liquid and vapor ratio (L/V). Feed rate and reboiling duty are the two major operating parameters affecting the L/V ratio. Reflux ratio is a dependent parameter of reboiling duty as reboiling generates reflux rate. A preferred stable operation is to locate the operating point near the upper right-hand side (RHS) of the operating window (Figure 12.5). This is because such an operation features high feed rate resulting in high vapor and liquid loadings and thus high distillation efficiency. A tower designed based on this upper RHS point could achieve a lower tower diameter and hence lower capital for a given separation objective.

However, how does one design a column away from unstable conditions in the first place and then operate it within the operating window when process conditions vary?

12.5.1 Spray

As mentioned above, spray occurs under large vapor momentum ($u_h d_h$) but small liquid gravitational force (h_c). By considering density difference in vapor and liquid, the equation for spray factor describing the relative momentum was developed by Lockett (1986) as

$$S_p = \frac{h_c}{u_h d_h} \left(\frac{\rho_L}{\rho_v} \right)^{0.5}. \tag{12.5}$$

Since the Lockett's correlation was developed for the sieve tray, Summers and Sloley (2007) extended it for the valve tray by introducing a constant K to the correlation:

$$S_p = K \frac{h_c}{u_h d_h} \left(\frac{\rho_L}{\rho_v} \right)^{0.5}. \tag{12.6}$$

$K = 1$ for sieve trays and 2.5 for movable and fixed valve trays. The introduction of K to the spray equation manifests that the valve tray has a better capability in suppressing entrainment than sieve trays. This is achieved by the mechanism of vapor entering the tray horizontally with valves, which reduces the entrainment significantly at low liquid loadings. According Lockett (1986) and Summers and Sloley (2007), spray factor S_p in equation (12.6) must be larger than 2.78 to avoid spray regime.

The relationship of vapor and liquid under spray was observed by Sakata and Yanagi (1979) for the sieve tray: As the liquid rate reduces beyond a certain amount corresponding to weir loading of 2 gpm/in. (gpm is gallons per minute), vapor rate must reduce to maintain the same entrainment rate. This reducing trend of both vapor and liquid rates under very small weir loading defines the spray phenomenon. This trend is different from the tray flood phenomenon under which vapor rate increases as liquid load reduces.

12.5.2 Tray Flooding

Because tray flooding occurs when vapor loading is too high, let us start by defining column vapor load as

$$V_L = \text{CFS} \sqrt{\frac{\rho_v}{\rho_L - \rho_v}}. \tag{12.7}$$

CFS is vapor flow in ft^3/s. ρ_v and ρ_L are densities for vapor and liquid, respectively. Vapor load represents the vapor flow under relative density difference of vapor and liquid, which contact each other on tray.

To describe the vapor load V_L over net area A_N ($=A_T - A_d$) on the tray, let us define

$$C = \frac{V_L}{A_N} = \frac{CFS}{A_N}\sqrt{\frac{\rho_v}{\rho_L - \rho_v}}. \tag{12.8}$$

Let the vapor velocity u be expressed as

$$u = \frac{CFS}{A_N}. \tag{12.9}$$

Thus,

$$C = u\sqrt{\frac{\rho_v}{\rho_L - \rho_v}}. \tag{12.10}$$

C defined in equation (12.10) is called the C-factor in literature, which has the meaning of vapor capacity factor.

However, how can we know the limit for vapor capacity at which the column starts to flood? This question can be answered by making a force balance between vapor and liquid on the tray. Assume the bulk of liquid with a volume v and the liquid gravitational force is $v \cdot (\rho_L - \rho_v)g$. At the same time, vapor flows upward with velocity of u and thus vapor momentum is $f \cdot (\rho_v u^2/2)$ where f is the friction coefficient between vapor and liquid.

When both forces are equal, the bulk of the liquid stays floating on the tray. If the vapor velocity is raised by any small amount above u, the bulk of the liquid will be carried away by the vapor. By denoting this threshold velocity as u_F, the force balance that defines the threshold condition for the occurrence of flooding can be expressed as

$$f\frac{\rho_v u_F^2}{2} = v(\rho_L - \rho_v)g. \tag{12.11}$$

Rearranging equation (12.11) gives

$$u_F = \sqrt{\frac{2vg}{f}}\sqrt{\frac{\rho_L - \rho_v}{\rho_v}}. \tag{12.12}$$

As it is difficult to measure bulk volume v and friction factor f in an actual column, C_F-factor (C-factor at flood) is used to represent $\sqrt{2vg/f}$. Let

$$C_F = \sqrt{\frac{2vg}{f}}. \tag{12.13}$$

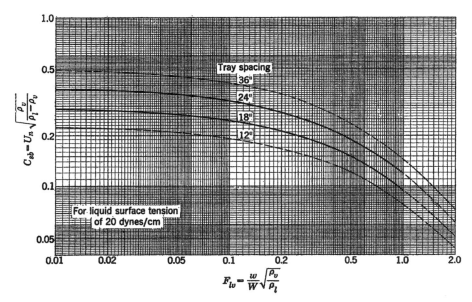

FIGURE 12.10. Fair's C_F correlation. (From Fair (1961), reprinted with permission by HART ENERGY PUBLISHING.)

Thus, equation (12.12) can be expressed as

$$u_F = C_F \sqrt{\frac{\rho_L - \rho_v}{\rho_v}}. \tag{12.14}$$

To predict C_F, Fair (1961) developed a C_F correlation, which can be plotted as Figure 12.10. Summers (2011) converted Fair's correlation into an equation form by ignoring the part with very low liquid loadings:

$$C_F = \left(\frac{H_S}{24}\right)^{0.5} \left(0.455 - 0.0055\rho_v^2\right). \tag{12.15}$$

For foaming materials, a derating factor or system factor (SF) may apply to C_F as a multiplier. SF is generally related to the foaming tendency of the material. The higher the foaming tendency, the lower the SF value and vice versa. From the point of view of practicality, the purpose that derating factors are used is to provide a design margin or overdesign to account for any inaccuracy associated with the empirical correlations. Excellent discussions related to derating factors can be found in the work of Kister (1992), and typical system factors are given in Table 12.1 based on Glitsch (1974).

TABLE 12.1. System Factors

Nonfoaming, regular system	1.00
Fluorine systems, e.g., BF3, Freon	0.90
Moderate foaming, e.g., oil absorbers, amin and glycol regenerators	0.85
Heavy foaming, e.g., amin and glycol absorbers	0.73
Severe foaming, e.g., MEK units	0.60
Foam-stable systems, e.g., caustic regenerators	0.30–0.60

Replacing the C_F term in equation (12.14) with equation (12.15) yields the expression for vapor velocity at flood as

$$u_F = \left(\frac{H_S}{24}\right)^{0.5} (0.455 - 0.0055\rho_v^2) \sqrt{\frac{\rho_L - \rho_v}{\rho_v}}. \tag{12.16}$$

With actual vapor velocity being calculated from equation (12.9), Fair's flooding limit is expressed as the ratio of actual vapor velocity and flooding vapor velocity, which defines the close proximity of actual operation from tray flooding:

$$\%\text{flooding} = \frac{u}{u_F} \times 100\%. \tag{12.17}$$

Alternative to the above equation for flood% based on vapor velocity, Glitsch's correlation (Glitsch, 1974) based on both vapor and liquid loadings is arguably the most widely employed:

$$\%\text{flooding} = \frac{V_L + (\text{GPM} \times \text{FPL})/13,000}{C_F A_a}. \tag{12.18}$$

FPL is the liquid flow path length and A_a is the active area. Assume $\text{FPL} = A_a/L_w$, Weiland and Resetarits (2002) rearranged the second term in equation (12.18) to yield

$$\%\text{flooding} = \frac{V_L}{C_F A_a} + \frac{W_L}{13,000 C_F}, \tag{12.19}$$

where

$$W_L = \frac{\text{GPM}}{L_W}, \tag{12.20}$$

W_L is weir liquid loading in gpm/in. while L_w is the outlet weir length. Equation (12.19) states that both vapor and liquid loading contribute to tray flood although vapor loading is much more dominant in tray flood. However, for downcomer flood, the weir loading plays a more dominant role, which will be discussed.

Clearly, a column is already in flooding when %flooding is near 100%. To prevent this from happening in the design phase, a certain design margin for avoiding tray flooding is built into the design. Hower and Kister (1991) recommends that large columns be typically designed with 80–85% flooding and 65–77% for small columns. For vacuum columns, %flooding is 75–77%.

The main factors affecting tray flooding include column diameter, tray spacing, and vapor and liquid loading. In operation, three key parameters can be used to control flooding, which are feed rate, column pressure, and reboiling duty (reflux rate depends on reboiling). When feed rate is raised or column pressure is lower or reboiling duty gets higher, more liquid is vaporized and hence vapor velocity is raised; thus, the column will work closer to flooding. In contrast, reducing feed rate or raising the column pressure or lowering reboiling duty could suppress tray flooding. However, they come at a cost. Obviously, reduced feed rate will immediately impact column throughput and affect economic margin. Increasing pressure could make the column more stable, but will have a negative effect on relative volatility and thus distillation efficiency. To compensate reduced distillation efficiency, higher reboiling duty is needed. On the other hand, increased reflux rate helps distillation because it reduces the theoretical stages required, and provides a means for improved mass transfer in the column. However, remember that reflux rate is made from reboiling. Higher reflux rate requires extra reboiling duty.

12.5.3   Downcomer Flooding

Downcomer flooding is caused by too much hydraulic resistance along tray flow path resulting in downcomer liquid backup. Figure 12.11 is used to show parameters of interest. The downcomer liquid level depends on tray hydraulic balance, which can be derived from the Bernoulli equation below based on two liquid levels as

$$H_d + \frac{p_1}{\rho_L g} = h_c + \frac{p_2}{\rho_L g} + h_{da}, \qquad (12.21)$$

where H_d is the liquid height in the downcomer, h_c is clear liquid height or liquid hold up on tray, and h_{da} is the equivalent clear liquid height for pressure drop through the downcomer apron. The unit for all these heights is inches.

We define tray pressure drop as

$$h_t = \frac{\Delta P}{\rho_L g} = \frac{p_2 - p_1}{\rho_L g}.$$

Thus, equation (12.21) can be converted to

$$H_d = h_c + h_t + h_{da}. \qquad (12.22)$$

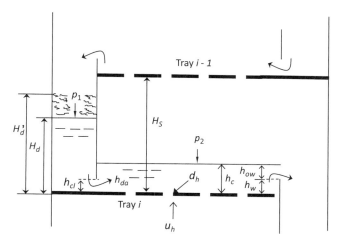

FIGURE 12.11. Key parameters for tray hydraulics (this graph is used for model illustration).

$h_t = \Delta p/(\rho_L g)$ is the equivalent clear liquid height to tray pressure drop due to vapor pressure drop through holes, which is called dry pressure drop, h_d, as well as liquid pressure drop through liquid on tray, which is named wet pressure drop, h_l. As defined in equation (12.22), H_d can also be called total hydraulic head of liquid going through downcomer and tray deck.

Does downcomer flooding occur when the downcomer clear liquid level H_d reaches the tray above? As a matter of fact, the flooding occurs earlier than the point of clear liquid reaching the tray above. This is because there is a layer of liquid froth on top of the clear liquid. As soon as the froth reaches the tray above, flooding happens as froth carries significant amount of liquid to the tray above. Thus, the total liquid height including froth level is $H'_d = H_d/\phi_d$ ($\phi_d \leq 1$), where φ_d is the downcomer froth density. To be conservative, we assume the downcomer backup limit is 80% and thus downcomer flood capacity is expressed as

$$H'_d = \frac{H_d}{\phi_d} = 80\%(H_S + h_w). \tag{12.23}$$

Combining equations (12.22) and (12.23) yields

$$\frac{h_c + h_t + h_{da}}{\phi_d} = 80\%(H_S + h_w), \tag{12.24}$$

where h_w is the outlet weir height. φ_d is downcomer froth density and is also considered as a derating factor by which a safety margin can be built into. φ_d is 0.3 to 0.4 for high foaming tendency, 0.6–0.7 for low foaming tendency, and 0.5 for average foaming tendency.

Clearly, downcomer backup can be estimated if we know the clear liquid height H_d. However, it is not straightforward in predicting H_d. As H_d consists of three components, namely, clear liquid height h_c, tray pressure drop h_t, and downcomer apron pressure drop h_{da}, let us look into each individual components one at a time.

12.5.3.1 Clear Liquid Height on Tray h_c can be calculated by

$$h_c = h_w + h_{ow} + h_g/2, \tag{12.25}$$

where h_{ow} is the liquid height in clear head over outlet weir (Figure 12.11) and h_g is the hydraulic gradient, which the liquid has to overcome along the tray flow path. Excessive gradient could cause weep at the downcomer inlet where liquid level is low. However, hydraulic gradient is negligible for most trays and it is common to omit this term from pressure drop calculations (Ludwig, 1979; Fair, 1984). Thus, equation (12.25) can be simplified as

$$h_c = h_w + h_{ow}. \tag{12.26}$$

For segmental weir, h_{ow} can be calculated via Francis weir formula (Bolles, 1963):

$$h_{ow} = 0.48\, F_w \left(\frac{\text{GPM}}{L_w}\right)^{2/3} \text{in.}, \tag{12.27}$$

where F_w is the weir constriction factor and it can be ignored in quick sizing as $F_w \approx 1$ in most cases anyway.

Weir loading $W_L = \text{GPM}/L_W$ is an important parameter. The lower limit of weir loading is 2–3 gpm/in., while the higher limit is 7–13 gpm/in. (Kister, 1992). When weir loading is less than the lower limit, weir constriction such as picket fence should be considered to increase weir loading artificially to avoid spray. On the other hand, when weir loading is larger than the upper limit, multiple flow passes must be employed to reduce weir loading to avoid both tray and downcomer flood.

h_{ow} calculated above could be used to determine the outlet weir height h_w. For normal pressure applications, liquid level h_c is within 2–4 in. and thus h_w can be determined via

$$2 - h_{ow} \leq h_w \leq 4 - h_{ow}. \tag{12.28}$$

For vacuum conditions, h_c could be $1 \sim 1.2$ in. Thus, we have

$$1.0 - h_{ow} \leq h_w \leq 1.2 - h_{ow}. \tag{12.29}$$

With relatively high outlet weir height h_w, the clear liquid height h_c increases and vapor and liquid contact time increases. This improves distillation efficiency. However, a too high outlet weir height could affect the downcomer backup and tray capacity.

However, Bernard and Sargent (1966) showed that the estimate of clear liquid height from pressure drop calculations could be unsatisfactory for some applications. Instead, h_c could be calculated from froth height (h_f) based on froth density φ_f as

$$h_c = \phi_f\, h_f. \tag{12.30}$$

φ_f can be calculated by Colwell's correlation (1981) and the detailed discussions can be seen in Kister (1992).

12.5.3.2 Head Loss Under Downcomer Apron h$_{da}$ is calculated based on Bolles (1963):

$$h_{da} = 0.03 \left(\frac{\mathrm{GPM}}{100\,A_{da}} \right)^2. \tag{12.31}$$

A_{da} is the flow area at downcomer clearance and is also called downcomer apron. A_{da} should take the most restrictive area in the downcomer.

12.5.3.3 Tray Pressure Drop h$_t$ includes dry and wet pressure drop as

$$h_t = h_d + h_l. \tag{12.32}$$

First, let us look at the dry pressure, h_d, which is the pressure drop that the vapor experiences when it flows through holes or valves. By definition, the expression for h_d takes the format of the orifice correlation.

For the sieve tray, the dry pressure drop h_d can be expressed as

$$h_d = K \frac{\rho_v}{\rho_L} u_h^2, \tag{12.33}$$

where

$$u_h = \frac{\mathrm{CFS}}{A_h}.$$

u_h is the hole vapor velocity in ft^3/s and A_h is the hole area. K is calculated based on orifice coefficient C_v as $K = 0.186/C_v$. C_v can be found in Lockett (1986).

For valve trays, dry pressure h_d varies with valve positions, namely, partly open or fully open. At low and moderate vapor loading when the valve partly opens, dry pressure is more dependent on the valve weight than the vapor rate. In contrast, at high vapor rate when valve fully opens, the dry pressure follows the orifice correlation. Glitsch (1974) provides the following correlations of dry pressure for the valve tray

$$\text{when valve part opens,} \quad h_d = 1.35 t_m \frac{\rho_m}{\rho_L} + K_1 \frac{\rho_v}{\rho_L} u_h^2, \tag{12.34a}$$

$$\text{when valve fully opens,} \quad h_d = K_2 \frac{\rho_v}{\rho_L} u_h^2, \tag{12.34b}$$

where t_m is the valve thickness (inches) and ρ_m is the valve metal density (lb/ft^3). Pressure drop coefficients K_1 and K_2 are dependent on valve type and weight. For Koch-Glitsch A(V-1) type as an example, $K_1 = 0.20$ and $K_2 = 0.86$ for $t_m = 0.134$ in., respectively.

Now we turn our attention to wet pressure drop. As h_l represents the pressure drop that vapor goes through the aerated liquid on the tray, it can be modeled in proportion to the tray clear liquid height as

$$h_l = \beta h_c. \tag{12.35}$$

β is the tray aeration factor and has different correlations for sieve and valve trays. h_l can also be calculated from froth height by

$$h_l = \phi_t h_f, \tag{12.36}$$

where ϕ_t is the relative froth density. The most known β and ϕ_t correlations are the Smith correlation (1963) for sieve tray and Klein correlation (1982) for valve tray.

After calculating dry and wet pressure, the tray pressure drop can be determined as follows:.

For sieve tray,

$$h_t = h_d + h_l = K \frac{\rho_v}{\rho_L} u_h^2 + \beta h_c. \tag{12.37}$$

For valve tray,

$$\text{when valve part opens,} \quad h_t = \left(1.35 t_m \frac{\rho_m}{\rho_L} + K_1 \frac{\rho_v}{\rho_L} u_h^2\right) + \beta h_c, \tag{12.38a}$$

$$\text{when valve fully opens,} \quad h_t = K_2 \frac{\rho_v}{\rho_L} u_h^2 + \beta h_c. \tag{12.38b}$$

12.5.3.4 Total Hydraulic Head H_d So far, we are ready to derive the overall expressions for $H_d = h_t + h_c + h_{da}$, based on the earlier discussions for these three components.

(i) **H_d for Sieve Tray.** Combining equations (12.27), (12.31), and (12.37) yields

$$H_d = \left(K \frac{\rho_v}{\rho_L} u_h^2 + \beta h_c\right) + \left[h_w + 0.48 F_w \left(\frac{\text{GPM}}{L_w}\right)^{2/3}\right] + 0.03 \left(\frac{\text{GPM}}{100 A_a}\right)^2.$$

$$\tag{12.39}$$

Rearranging equation (12.39) based on equations (12.20), (12.26) and (12.27) as well as $A_{da} = L_w\, h_{cl}$ gives a simplified H_d expression for the sieve tray:

$$H_d = K \frac{\rho_v}{\rho_L} u_h^2 + (1 + \beta)\, h_w + (1 + \beta)\, 0.48\, F_w (W_L)^{2/3} + 0.03 \left(\frac{W_L}{100\, h_{cl}} \right)^2.$$

(12.40)

(ii) **H_d for Valve Partial Open.** Combining equations (12.27), (12.31), and (12.38a) yields

$$H_d = \left(1.35 t_m \frac{\rho_m}{\rho_L} + K_1 \frac{\rho_v}{\rho_L} u_h^2 + \beta h_c \right) + \left[h_w + 0.48\, F_w \left(\frac{GPM}{L_w} \right)^{2/3} \right] + 0.03 \left(\frac{W_L}{100\, h_{cl}} \right)^2.$$

(12.41)

Rearranging equation (12.41) gives a simplified H_d expression for partly open valve tray:

$$H_d = \left(1.35 t_m \frac{\rho_m}{\rho_L} + K_1 \frac{\rho_v}{\rho_L} u_h^2 \right) + (1 + \beta) h_w + (1 + \beta) 0.48\, F_w (W_L)^{2/3} + 0.03 \left(\frac{W_L}{100\, h_{cl}} \right)^2.$$

(12.42)

(iii) **H_d for Valve Full Open.** Combining equations (12.27), (12.31), and (12.38b) yields

$$H_d = \left(K_2 \frac{\rho_v}{\rho_L} u_h^2 + \beta h_c \right) + \left[h_w + 0.48\, F_w \left(\frac{GPM}{L_w} \right)^{2/3} \right] + 0.03 \left(\frac{GPM}{100\, A_a} \right)^2.$$

(12.43)

Rearranging equation (12.43) gives a simplified H_d expression for a fully open valve tray:

$$H_d = K_2 \frac{\rho_v}{\rho_L} u_h^2 + (1 + \beta) h_w + (1 + \beta) 0.48\, F_w (W_L)^{2/3} + 0.03 \left(\frac{W_L}{100\, h_{cl}} \right)^2.$$

(12.44)

In column design, increasing tray spacing, downcomer clearance, and the number of tray flow passes can effectively avoid downcomer backup. Increasing tray spacing is more economical because column tangent length is more expensive. In operation, reducing feed rate and reflux rate can avoid downcomer flooding.

12.5.4 Downcomer Choke

Downcomer choke occurs when large crests block the downcomer entrance and thus choke the liquid flow in the downcomer (Figure 12.9) resulting in liquid backup onto the tray deck. The root cause is too high downcomer velocity, which results in excessive frictional losses at the downcomer entrance.

Maximum downcomer velocity defines the liquid capacity limit and it can be predicted by several different correlations. According to Kister (1992), the Glitsch correlation (Glitsch, 1974) tends to predict the highest downcomer liquid load, the Koch correlation (Koch, 1982) predicts the lowest downcomer liquid load, while Nutter correlation (Nutter, 1981) provides the estimate in between. Let us focus on the Glitsch correlation next.

The Glitsch correlation (Glitsch,1974) is expressed as

$$\left.\begin{array}{l} Q_{D,\max}^1 = 250 \\ Q_{D,\max}^2 = 41\sqrt{\rho_L - \rho_V} \\ Q_{D,\max}^3 = 7.5\sqrt{H_S(\rho_L - \rho_V)} \\ Q_{D,\max} = \left[\min\left(Q_{D,\max}^1, Q_{D,\max}^2, Q_{D,\max}^3\right)\right] \times SF \end{array}\right\}, \quad (12.45)$$

where downcomer liquid load Q_D is gpm/ft^2 of downcomer area A_D. The correlation for $Q_{D,\max}^3$ in equation (12.45) is plotted for different tray spacings in Figure 4 of the Glitsch Ballast Tray Design manual (1974), which is reproduced in Figure 12.12. SF is the derating factor and SF = 1 for the case of no derating.

For quick tray sizing purposes, Summers (2011) developed a simplified version of the Glitsch correlation. First, data points on different curves in Figure 12.12 are extracted to generate a single data plot on $V_{D,\max}$ (maximum downcomer entrance velocity) and $\Delta\rho$ axis (Figure 12.13). Then a correlation is developed based on the best fit method to give

$$V_{D,\max} = 0.1747 \ln(\rho_L - \rho_V) - 0.2536 \, \text{ft/s}. \quad (12.46)$$

It must be noted that $V_{D,\max}$ and $Q_{D,\max}$ describe the same concept, namely, maximum downcomer loading, but $V_{D,\max}$ is in ft/s while $Q_{D,\max}$ is in gpm/ft^2.

The Summers' correlation expressed in equation (12.46) is independent of tray spacing, and it manifests the fact that greater density difference makes it easier for the vapor to escape from the froth in the downcomer. Summers (2011) indicated that this correlation works for nonfoaming services. For foaming services, a foaming factor must be applied. Since this correlation is independent of tray spacing and gives a slightly conservative estimate of maximum downcomer velocity, Summers (2011) demonstrated that this correlation could be effectively applied for the purpose of quick tray sizing.

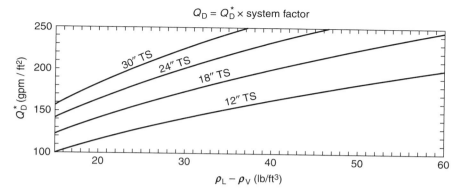

FIGURE 12.12. Downcomer design velocity curves in Figure 4 of the Glitsch Design Manual (1974).

The Koch correlation (Koch, 1982) and Nutter correlation (Nutter, 1981) are based on the minimum residence time criteria, which is converted to maximum downcomer liquid load criteria as

$$Q_{D,max} = 448.8 \frac{H_s}{12 \tau_R} \text{ gpm/ft}^2. \tag{12.47}$$

12.5.5 Weir Loading Limits

Weir loading plays a critical role in stable operation. This is because spray could occur at low weir loading, while downcomer flood could take place at high weir

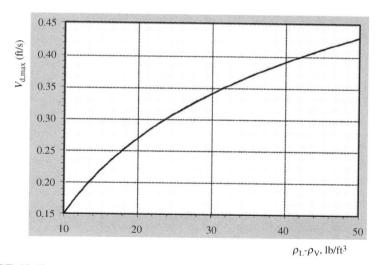

FIGURE 12.13. Maximum downcomer velocity correlation based on Glitsch's Figure 4 (1974). (From Summers (2011), reprinted with permission by AICHE.)

loading. For very low liquid loading, packing may be preferred than tray. The general guideline is: Consider packing when liquid loading is less than 4 gpm/ft^2 of column active area and use tray when liquid loading is larger than 14 gpm/ft^2 column active area. It is a judgment call for selecting either tray or packing for liquid loading is in the range of 4–14 gpm/ft^2 column active area. The discussions for weir loading next are given with tray in mind.

12.5.5.1 Minimum Weir Loading

A very low weir loading is manifested when the liquid height over outlet weir, h_{ow}, is less than the limit of 0.25–0.5 in. (Chase, 1967; Davis and Gordon, 1961; Kister, 1992). A sufficient liquid height over the weir above the limit provides a stable liquid distribution. We can apply the Francis correlation in equation (12.27) to determine the minimum weir loading:

$$h_{ow,min} = 0.48 F_w W_L^{2/3} = 0.25 \sim 0.5 \text{ in.}$$

Solving W_L yields

$$W_{L,min} = \left(\frac{h_{ow,min}}{0.48 F_w} \right)^{3/2} = 0.4 \sim 1.0 \text{ gpm/in.} \qquad (12.48)$$

The above $W_{L,min}$ is recommended by Glitsch (1974), Koch (1982), and Lockett (1986). When weir loading is close to $W_{L,min}$, spray could occur on the tray deck. According to Summers and Sloley (2007), when actual liquid loading is small, the most effective way to avoid spray regime is using picket fence outlet weirs to artificially increase weir loading.

12.5.5.2 Maximum Weir Loading

When weir loading is too high beyond the maximum limit, multiple passes are required. Kister (1990) suggested using 7–13 gpm/in. per single pass as the weir loading limit.

Resetarits and Ogundeji (2009) proposed that maximum weir loading should be determined according to tray spacing. In other words, different weir loading limits should be used for different tray spacings. For example, higher tray spacing can accommodate higher weir loading and vice versa. The intension of Resetarits and Ogundeji's argument is to minimize the number of tray flow passes as multiple passes feature lower distillation efficiency and higher tray cost than a single pass tray. This argument is consistent with Nutter's data shown in Table 12.2.

12.5.5.3 Minimum Downcomer Residence Time

Equation (12.47) can be converted to minimum residence time as

$$\tau_{R,min} = 448.8 \frac{A_d H_S}{12 \, GPM_{max}}. \qquad (12.49)$$

TABLE 12.2. Relation of Weir Loading Limits and Tray Spacing (Nutter Engineering, 1981)[a]

Tray Spacing (in.)	W_{Lmax} for Single Pass (gpm/in.)
12	3
15	5
18	8
21	10
24	13

[a]From Resetarits and Ogundeji (2009), reprinted with permission by AIChE.

As can be observed from equation (12.49), downcomer area A_d should be large enough to allow liquid to have sufficient residence time so that vapor can disengage from the descending liquid in the downcomer. Liquid can be relatively vapor free when it enters the tray below. Otherwise, too short residence time could lead to inadequate disengaging of vapor from the liquid and then choke the downcomer. For low foaming tendency, $\tau_{R,min} = 3.0$ s; for high foaming tendency, $\tau_{R,min} = 6$ to 7 s; for average foaming tendency, $\tau_{R,min} = 4$ to 5 s.

12.5.6 Excessive Weeping

Excessive weeping usually occurs at turndown operation when vapor loading is very low and liquid short-circuits the flow path and falls through tray holes directly. A small amount of weeping can be tolerated. However, when the fraction of liquid falling through holes increases and exceeds a certain percent, dumping occurs and distillation efficiency falls. As a guideline, a weeping rate of 25% corresponds to a 10% loss in distillation efficiency. If vapor rate reduces until all liquid flows through the holes and does not reach the downcomer, this condition is called dumping. The weep point can be set in the range of 25–30%. Although there are no reliable methods available to predict the weep rate accurately, trays can be designed to operate above the weep point with confidence. There is a gray area between weep point and dumping point, and the mechanism for this area is not well understood.

The mechanism for weeping is the force balance on the tray between the vapor momentum upward and the liquid gravitational force downward. Under normal operation, vapor keeps liquid on the tray when the vapor momentum is higher than the gravitational force. If the vapor momentum is too high, vapor carries liquid to the tray above and causes either spray flow and tray flood. In the opposite extreme where vapor momentum is too weak, liquid leaks through tray holes and weeping occurs. The force balance for weeping is established when vapor momentum is equal to liquid gravity as

$$h_d + h_\sigma = h_w + h_{ow}. \tag{12.50}$$

In the right-hand side, liquid force is formed by clear liquid height $h_c = h_w + h_{ow}$. On the left-hand side (LHS), the vapor force consists of h_d, the dry pressure drop required for vapor to go through the tray holes, as well as h_σ, the surface intension associated with bubble formation. h_σ is given by Fair et al. (1984) as

$$h_\sigma = \frac{0.04}{\rho_L d_h} \sigma, \tag{12.51}$$

where d_h is the hole diameter (in.) and σ is surface tension (dyne/in.).

The weep point check is important if the flexibility of large turn down operation is desired. The vapor and liquid rates under turn down are used to calculate h_d (using equation (12.33) for sieve tray and equation (12.34a) for valve tray) and h_{ow} based on Francis' equation (12.27).

The weep point corresponding to $(h_d + h_\sigma)_{min}$ can be obtained using Fair's correlation (Fair, 1963) via

$$\%\text{weep} = \frac{h_d + h_\sigma}{(h_d + h_\sigma)_{min}}. \tag{12.52}$$

The above Fair's weep point method is applied to sieve trays. Bolles (1976) extended it to valve trays and details can be found in Bolles' article. The weep rate is affected by minimum vapor loading in relation to weir loading, the number of holes for sieve tray, and the number of valves and weight of valves for valve tray. The experience shows that a well designed valve tray must maintain the weep point below the vapor loads at which the valve opens. To do this, a valve tray design should avoid use of too many valves.

To achieve an effective turndown operation, a manufacturer usually specifies two valve weights. When light valves open, the heavy ones still close, which reduces the active bubble area and thus avoids weeping. Furthermore, weeping usually occurs at the exit of the downcomer apron area. Thus, it is critical to maintain the level of the tray.

For safety, 80% weep point from equation (12.52) could be used as the weep point limit. If the %weep is larger than 80%, weep rate evaluation is warranted, which is discussed below.

Lockett and Banik (1986) developed the correlation for weep rate W as

$$\frac{W}{A_h} = \frac{29.45}{\sqrt{Fr_h}} - 44.18, \tag{12.53}$$

where Fr_h is the Froude number and it is expressed as

$$Fr_h = 0.373 \frac{u_h^2}{h_L} \frac{\rho_V}{\rho_L - \rho_V}. \tag{12.54}$$

Colwell and O'Bara (1989) suggested an alternative weep rate correlation as

$$\frac{W}{A_h} = \frac{1,841}{Fr_h^{1.533}}. \tag{12.55}$$

Colwell and O'Bara recommended applying equitation (12.53) for cases with Froude number less than 0.2 and equation (12.55) for large Froude numbers. For high pressure towers (>165 psia), the Hsieh and McNulty (1986) correlation is recommended and details are provided there.

The weep ratio is then defined as

$$w = \frac{W}{GPM}. \tag{12.56}$$

Above the weep point, $w = 0$ while $w = 1$ at the dumping point. Weeping occurs between the weep point and dumping point. Turn down operation could be still acceptable even if it is below the weep point but the weep ratio w is less than 0.1. This is because tray efficiency is not affected too severely when w is less than 0.1. Increasing vapor load and reducing the clear liquid height could help to avoid weeping.

12.5.7 Constant L/V Operating Line

For a simple tower that does not have any pump-arounds or side draws, there are two sections, namely, rectification section, which is above the feed tray, and stripping section below the feed tray. The rectification section has a vapor rate higher than the liquid rate, whereas it is reversed for the stripping section. The L/V ratio is the indication of distillation that could take place. For most towers, the L/V ratio is 0.3–3.0. L/V ratios outside this range may give sloppy or too easy distillation. Determination of L/V ratio is a major part of energy optimization, which will be discussed in detail.

When a distillation column operates under a constant L/V ratio, the operating region is reduced to a straight line as shown in Figure 12.5 with L/V as the slope and A and B as two limits. When the same column is operated under a different L/V ratio, the slope of the operating line varies. There are process variations such as changes to feed rate and compositions, and product rates and specifications as well as tower operating conditions such as pressure, temperature, and reflux rate. These variations will cause changes to the operation points within the operating window. The effects of changes will be discussed below.

12.6 TYPICAL CAPACITY LIMITS

Typical capacity limits are given here as general guidelines. Specific limits need to be developed based on process conditions and tower designs. There are three capacity limits related to vapor loading, which are spray, jet flood, and weeping. The spray limit is set as 2.78 and the flow regime is spray when spray factor S_p is less than 2.78 as calculated by equation (12.6). The tray flood limit is 75–85% where the lower limit is used for vacuum and near-atmospheric towers, while higher limit should be adopted for moderate- and high-pressure towers (>165 psia). For low-pressure towers, valve trays should be considered as they are better in suppressing spray than sieve trays. A tower should be designed with a high % flooding to avoid too much spare capacity used with unnecessarily large diameter and height. Thus, to get these two parameters right is the paramount of tower design. Weeping limit is set at 90% of Fair weep point or at 10% weeping rate.

There are two capacity limits related to liquid loading, which are the downcomer backup limit and downcomer velocity limit. The downcomer backup limit is set at 80% of liquid settling height based on the froth level. The downcomer velocity limit is 75% of maximum velocity allowed to avoid downcomer choke. The number of tray passes is the most important parameter affecting the downcomer loading and thus these two downcomer limits.

Lastly, there are two capacity limits based on weir loading. For stable operation, minimum weir loading of 2 gpm/in. is required. If weir loading is less than this limit, a picket fence weir should be considered to increase the weir loading artificially. On the other hand, when weir loading is larger than the maximum limit per pass corresponding to tray spacing (Table 12.2), multiple passes should be considered.

A typical capacity diagram or operating window is illustrated in Figure 12.14 based on the capacity guidelines, although specific applications warrant different capacity limits.

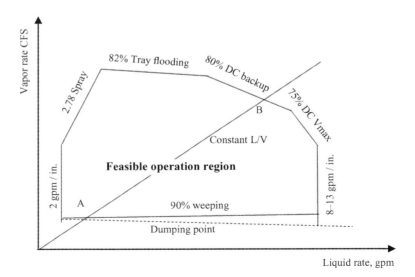

FIGURE 12.14. Typical capacity diagram.

12.7 EFFECTS OF DESIGN PARAMETERS

There are three common flow regimes for industrial towers, namely, spray, froth, and emulsion, which are the results of vapor and liquid loadings (Figure 12.15).

Froth regime is favored among these flow regimes as it enhances distillation efficiency and promotes stable operation. The fundamental reason why froth regime features higher distillation efficiency is that liquid penetrates into continuous vapor phase in the form of fine droplets, while vapor penetrates into continuous liquid phase in the form of bubbles. The feature of the highest vapor/liquid interface area from the froth regime enhances mass transfer between vapor and liquid. In contrast, spray regime features too high upward vapor momentum resulting in excessive entrainment of liquid. On the other hand, the emulsion regime has too low vapor upward momentum resulting in no penetration of liquid into the vapor phase and thus poor mass transfer. Thus, avoiding spray, reducing emulsion, and enhancing froth flow is the focus of design considerations.

The effects of different parameters can be better explained based on one simple concept, which is momentum exchange, which is captured the best by the equation below:

$$S_P = K \frac{h_c}{u_h d_h} \left(\frac{\rho_L}{\rho_V} \right)^{0.5}, \tag{12.6}$$

where h_c represents liquid gravitational force and $u_h d_h$ vapor momentum multiplied by density. The magic number for S_p is 2.78 below which spray flow occurs. Thus, the design effort strives to increase S_p above 2.78 at all cost.

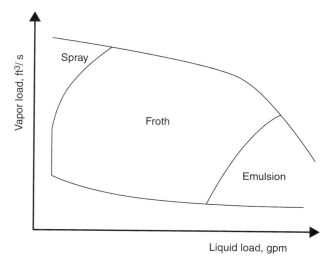

FIGURE 12.15. The flow regimes in distillation.

12.7.1   Effects of Pressure

Vacuum and low-pressure conditions enhance vapor loading at the expense of liquid loading resulting in a low S_p and thus promote spray flow. Valve trays have greater tendency to operate under froth regimes than sieve trays which is indicated by a higher K value. Thus, it is a common practice that valve trays are used for vacuum and low-pressure applications. At the same time, small valves (for small d_h) with increased number of valves (for small u_h) are used to increase S_p.

In contrast to low-pressure towers, high-pressure towers are characterized with low vapor loading but high liquid loading, and thus have a high tendency of downcomer flood. Multipass trays or counterflow trays such as UOP MD/ECMD or Shell HiFi or Sulzer high capacity or Koch-Glitsch high-performance trays can be applied to increase weir length and reduce weir loading.

12.7.2   Column Diameter and Tray Spacing

Larger diameter (D) and lower tray spacing (H_S) will increase the capability of liquid loading more than vapor loading resulting in a higher S_p. In other words, a fat and short tower (low H_S/D ratio) suppresses spray and shifts the operating point toward froth and emulsion. On the other hand, a tall and thin tower (high H_S/D ratio) can prohibit emulsion and shift the operating point to the left toward spray and froth. In addition, multipass trays also shift the operating point to the left with reduced liquid loading in comparison with single-pass trays. It is common to see fat and short vacuum towers while long and thin towers for high-pressure applications.

12.7.3   *L/V* Ratio

The rectifying section of a tower has a greater tendency to operate under spray, while the stripping section has a greater tendency to operate toward emulsion due to the characteristics of liquid and vapor loadings in these sections. That is, the rectifying section has a relatively higher vapor loading and lower liquid loading than in the stripping section.

12.7.4   Hole Diameter and Fractional Hole Area

Large hole diameter promotes spray. In contrast, small hole diameter reduces d_h and thus increases S_p and suppresses spray. Therefore, it is beneficial to use the smallest practical hole diameter possible considering fouling, tray deck material, thickness, and standard size. In contrast, a small fractional hole area promotes spray as it increases u_h and thus reduces S_p. Therefore, for a flow regime leaning toward spray, one resolution is to design tray decks with greater number of holes and smaller hole diameter. For conditions promoting emulsion regime, a lower fractional hole area can increase vapor and liquid contact via increased vapor velocity. A lower fractional area can extend the operating window for sieve tray.

12.7.5 Weir Height

There is an optimal weir height. High outlet weir promotes froth regime as it increases h_c and thus S_p. However, it is at the expense of higher tray pressure drop and low tray capacity. The pressure drop increases at a higher rate than efficiency. Thus, it is not beneficial to have too high outlet weir. On the other hand, at very low weir height, liquid inventory is low resulting in low mass transfer and hurting tray efficiency.

12.7.6 Tray Type

Sieve trays have wide applications due to lower capital cost and less maintenance required. However, valve trays can promote froth regime and thus achieve distillation efficiency up to 20% higher than sieve trays (Anderdson et al., 1976). Furthermore, valve trays can achieve 3:1 turndown compared to sieve trays in 2:1. However, the disadvantages of valve trays are higher pressure drop and cost.

12.7.7 Summary

From the discussions above, the main causes for the spray regime are: (i) low operating pressure; (ii) low weir loading; (iii) small column diameter; (iv) high tray spacing; (v) large hole diameter; and (vi) low fractional area. In short, a tall and thin sieve tray column under low pressure with large hole diameter has every tendency of spray. Another observation is that a valve tray provides a higher distillation efficiency (Anderson et al., 1976) and greater turndown ratio than the sieve tray although a valve tray is more expensive.

12.8 DESIGN CHECKLIST

A design checklist could act as a safeguard for the review of a tower design to avoid any unexpected issues occurring in the late stage. Engineering companies usually have a design checklist. Table 12.3 gives an example although this example checklist is not extensive. Extra items can be added depending on applications. As mentioned, a column is divided into sections based on feed and draw locations. Thus, the check should be performed for each section.

12.9 EXAMPLE CALCULATIONS FOR DEVELOPING OPERATING WINDOW

The following example is designed with the purpose of enhancing the understanding of column tray design, operation window and the capacity limits.

The overall tray design procedural guideline is to get the tray diameter and spacing right first with consideration of spray and tray flooding, and then size the

TABLE 12.3. Example Tower Design Checklist (Kister, 1992; Lockett, 1986; Glitsch Inc., 1974; Nutter Eng., 1981)

Check Items	Typical Limits	Calculation Basis	Comments	Recommended Values	Potential Cures
Internal L/V ratio; %	Lower limit = 0.3; Upper limit = 3	Mole percent	Slopy or easy fractionation if outside of this range. If unexpected L/V is obtained, review the vapor and liquid loading	Within the range	
Spray factor Sp; s/ft	2.78	equation (12.6) (Lockett, 1986; Summers and Sloley, 2007)	Valve tray is better in supressing spray than sieve tray	>2.78	Increase tower diameter and open area; reduce hole diameter; consider valave tray
Tray flood; %	75–85%	equation (12.18) (Glitsch, 1974)	75–77% for vacuum towers; 80–85% for normal towers; 65–77% for small tower	>50% for new design; >85% for revamp	Increase tower diameter and tray spacing
Downcomer froth backup; %	80%	equation (12.24)	It is the height of froth instead of clear liquid height	<80%	Increase tray spacing; consider multiple tray passes; increase downcomer clearance
Downcomer velocity limit; %	75	equation (12.45), (Glitsch, 1974)		>65%	Consider multiple passes and sloped downcomer if larger than the limit

Weir loading; gpm/in.	Lower limit = 2; Upper limit = 8–13	equation (12.48) for min limit (Glitsch, 1974; Koch, 1982; Lockett, 1986). Table 12.2 for max limit (Nutter, 1981)	The limits for single pass and straight weir	>4 and < max limit in Table 12.2	Use multiple passes when large than upper limit. Consider fewer passes if less than 4 gpm/in. Use blocked weir or baffled weir when less than 2 gpm/in.
Weeping	90% Fair weep point or 10% weep rate	equation (12.52) (Fair, 1963)		>10% weep rate	Reduce fractional hole area and outlet weir height. Choose right hole diameter based on liquid loading
Fractional hole area; %:	5%		Reduce hole diameter and fractional hole area if possible to enhance fractionation efficiency	7-15%	Use of fewer passes and short FPL if less than 5%
Flow path length (FPL); in.	18		For easy access in inspection and maintainance	>18	
Tray spacing; in.	15–24		For easy access in inspection and maintainance	>15	
Weir length; in.	5		For hand access when installing downcomer	>5	

TABLE 12.4. Column Overhead Conditions

Vapor				Liquid				
°F	lb/h	CFS, ft^3/s	ρ_v, lb/ft^3	lb/h	GPM, gal/m	ρ_L, lb/ft^3	σ, mN/m	μ, cP
331	19,659	28.00	0.195	17,346	42.0	51.51	32.0	0.727

downcomer adequately afterward taking into account of the downcomer backup and choke. The guideline for an "optimal" design is that the least expensive design is usually the best one when comparing alternative designs.

The calculation method adopted here is based on Summers (2011) in which tower diameter is determined based on both vapor and liquid loadings and weir loading is used as an iteration parameter to determine the number of tray passes. The method is straightforward to use and effective for a quick assessment.

Example 12.1 Tower tray design and hydraulic assessment

A column is designed for laboratory use to separate cyclohexanol and phenol mixture. The task at hand here is to generate a column tray design based on the most constrained tray i.e. column overhead conditions, as given in Table 12.4.

Working guide for diameter calculations:

Step 1: Assume H_S and W_L. Adjust them based on flood and weir loading criteria.
Step 2: Calculated maximum downcomer velocity $V_{D,max}$ based on equation (12.46), which determines the downcomer area A_d.
Step 3: Calculate tower active area A_a using equation (12.57) below.
Step 4: Total tower area $A_t = A_a + 2A_d$ and then tower diameter is calculated.
Step 5: Calculate downcomer width w_d and weir length L_w.

The calculation procedure can be summarized in Figure 12.16 while calculation details are provided below.

12.9.1 Determination of Tower Diameter

(i) First trial: First let us assume *tray spacing and weir loading*

$$H_S = 18 \text{ in.}; \quad W_L = 5 \text{ gpm/in.}$$

Downcomer Area: Applying equation (12.46) yields

$$V_{D,max} = 0.1747 \ln (\rho_L - \rho_V) - 0.2536 = 0.43 \text{ ft/s.}$$

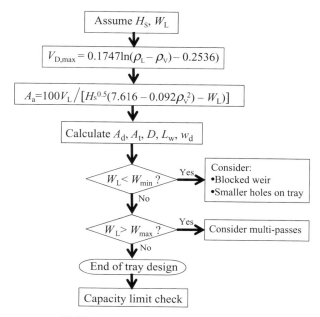

FIGURE 12.16. Tray design procedure.

Because liquid rate is 42 gpm ($0.094 \, \text{ft}^3/\text{s}$), the downcomer top area becomes

$$A_d = \frac{L}{V_{D,\text{max}}} = \frac{0.094}{0.43} = 0.22 \, \text{ft}^2.$$

Make the downcomer bottom area $0.13 \, \text{ft}^2$, which is equal to 60% of the top area.

Tray Areas: Vapor loading can be calculated based on equation (12.7):

$$V_L = \text{CFS} \sqrt{\frac{\rho_V}{\rho_L - \rho_V}} = 28 \sqrt{\frac{0.195}{51.5 - 0.195}} = 1.73 \, \text{ft}^3/\text{s}.$$

For quick tray sizing purposes, Summers (2011) derived a correlation mainly based on equation (12.18) (Glitsch flood correlation 13; Glitsch, 1974), which is expressed as

$$A_a = \frac{100 \, V_L}{H_S^{0.5}(7.616 - 0.092\rho_V^2) - 1.1 W_L}. \tag{12.57}$$

Applying equation (12.57) yields

$$A_a = \frac{100 \times 1.73}{18^{0.5}(7.616 - 0.092 \times 0.195^2) - 1.1 \times 5} = 6.44 \, \text{ft}^2.$$

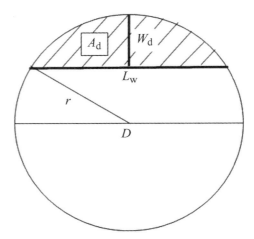

FIGURE 12.17. Derive A_d from w_d and r.

Thus, total tray area and tray diameter are calculated by

$$A_t = A_a + A_d^{top} + A_d^{bttm} = 6.44 + 0.22 + 0.13 = 6.79 \text{ ft}^2.$$

Then the tower diameter can be calculated as

$$D = \sqrt{\frac{4A_t}{\pi}} = \sqrt{\frac{4 \times 6.79}{3.14}} = 2.94 \text{ ft.}$$

Rounding up tower diameter D to 3 ft. Note that the diameter is small for trays as the tower is used for laboratory testing. Weir length L_w is the cord length in a circle, while weir width w_d is the rise as shown in Figure 12.17. From geometry, there is a relationship between arc area (A_d), rise (the same as downcomer width w_d) and radius ($r = D/2$):

$$A_d = r^2 \cos^{-1} \frac{r - w_d}{r} - (r - w_d)\sqrt{2rw_d - w_d^2}. \qquad (12.58a)$$

As $A_d = 0.22 \text{ ft}^2$ and $r = D/2 = 1.5 \text{ ft}$, w_d can be calculated as 2.5 in. based on equation (12.58a). However, the minimum downcomer width should be 5 in. and the reason for this minimum is to allow easy access by hand in installing a downcomer.

(ii) Second trial: With $w_d = 5$ in. and $D = 3$ ft, A_d can be recalculated based on equation (12.58a) to give $A_d = 0.59 \text{ ft}^2$. L_w can be calculated as

$$L_W = 2\sqrt{r^2 - (r - w_d)^2} = 2\sqrt{\left(\frac{3}{2}\right)^2 - \left(\frac{3}{2} - \frac{5}{12}\right)^2} = 2 \text{ ft.} \qquad (12.58b)$$

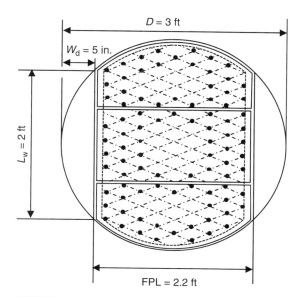

FIGURE 12.18. Tray layout for the example problem.

Flow path length (FPL) is

$$FPL = D - 2w_d = 3 - 2 \times \frac{5}{12} = 2.2 \text{ ft}.$$

The tray deck layout is shown in Figure 12.18.

Weir loading can then be calculated as

$$W_L = \frac{GPM}{L_W} = \frac{42}{2 \times 12} = 1.75 \text{ gpm/in}.$$

Obviously, the weir loading is too small and less than the lower limit of 2 gpm/in. This could cause fluidization resulting in poor distillation efficiency. To prevent this from happening, picket fence weir is used to artificially increase the weir loading. Let 60% of the weir flow length be blocked by teeth and thus the effective weir length is 0.8 ft. In this case, the weir loading becomes

$$W_L = \frac{GPM}{L_W} = \frac{42}{0.8 \times 12} = 4.37 \text{ gpm/in}.$$

(iii) Further trials: This weir loading is acceptable and calculations continue with the third trial and eventually tray layout is determined at the fourth trial. The trial results are listed in Table 12.5.

TABLE 12.5. Trials of Calculating Tray Diameter

Tray Layout	1st Trial	2nd Trial		3rd Trial	4th Trial
DC top area A_d; ft^2	0.22	0.59		0.59	0.59
DC bottom area $= 60\%A_d$; ft^2	0.13	0.36		0.36	0.36
Weir loading W_L; gpm/in.	5.00	1.75	Picket fence weir →	4.37	4.37
Active area A_a; ft^2	6.44	6.12		6.28	6.12
Total area $A_T = A_A + 2A_d$; ft^2	6.79	7.07		7.23	7.07
Net area $A_N = A_T - A_d$; ft^2	6.57	6.47		6.64	6.47
Diameter $D = (4A_T/\pi)^{1/2}$; ft	2.94	3.00		3.03	3.00
Weir length L_W; ft	1.51	2.00		2.00	2.00
Effective weir length L_{we}; ft	1.51	2.00		0.80	0.80
Weir width w_d; in.	2.52	5.00		5.00	5.00
Flow path length (FPL); ft	2.52	2.17		2.20	2.17

12.9.2 Tray Layout

Number of Passes: Single flow path is acceptable as it is less than the weir loading limit of 8 gpm/in. for tray spacing 18 in. based on Table 12.2.

Downcomer and Weir Type: Straight downcomer and chordal weir are selected for simplicity. Picket fence weir (blocked weir type) is employed to increase the weir loading artificially to avoid fluidization.

Selection of Valve or Sieve Tray: Valve tray is selected as the tower is near atmospheric pressure. Under low pressure, spray regime could become dominant and a valve tray is more effective than a sieve tray in producing froth. For this application, Koch-Glitsch A (V-1), which is a float valve, is selected.

Number of Valves: The standard valve density for KG A(V-1) type is 12/ft^2. Thus, the number of valves can be calculated by

$$N = 12A_a = 12 \times 6.12 = 73.$$

Hole Diameter: The hole diameter is calculated based on hole F-factor as

$$F_{hc} = u_{hc}\sqrt{\rho_V}. \tag{12.59}$$

Assume the selected valve will open when $F_{hc} = 9.5$. Thus, $u_{hc} = F_{hc}/(\rho_v)^{1/2} = 21.5$ ft/s. According to the relation of u_h and d_h as

$$u_h = \frac{CFS}{N(\pi/4)d_h^2}. \tag{12.60}$$

Thus, for a given $u_{hc} = 21.5$ ft/s, d_h becomes

$$d_h = \sqrt{\frac{CFS}{N(\pi/4)u_h}} = \sqrt{\frac{28}{73 \times (3.14/4) \times 21.5}} = 0.15 \text{ ft} = 1.8 \text{ in.}$$

As this design case has small liquid loading, from the concern of spray, it is more desirable to reduce vapor momentum so that the relative momentum $h_c/u_h d_h$ can increase to suppress spray flow. With this in mind, $d_h = 1.5$ in. is selected. Then the actual vapor hole velocity is

$$u_h = \frac{28}{73(3.14/4)(1.5/12)^2} = 31 \text{ ft/s} > u_{hc} = 21.5 \text{ ft/s}.$$

The valve will open since actual hole velocity u_h is large than the critical hole velocity u_{hc}. At the same time, we can derive the valve open percentage over the column diameter:

$$\psi_h = N\left(\frac{d_h}{D}\right)^2 \times 100\% = 73\left(\frac{1.5/12}{3}\right)^2 \times 100\% = 13\%.$$

Liquid Height Over Weir $\mathbf{h_{ow}}$***:*** For the purpose of quick calculation, assume weir correction factor $F_w = 1$. In most cases, F_w is close to one. Applying the Francis correlation (equation (12.27)) to the weir loading of 4.37 gpm/in. yields

$$h_{ow} = 0.48 \, F_w \, W_L^{2/3} = 0.48 \times 1 \times 4.37^{2/3} = 1.28 \text{ in.}$$

Weir Height $\mathbf{h_w}$***:*** By ignoring the liquid height gradient on the tray and based on $h_{ow} = 1.28$, applying equation (12.28) yields

$$0.72 \leq h_w \leq 2.72 \text{ in.}$$

Set $h_w = 2$ in. as an average in the range of (0.72, 2.72). Then clear liquid height becomes

$$h_c = h_w + h_{ow} = 2 + 1.28 = 3.28 \text{ in.}$$

Downcomer Clearness $\mathbf{h_{cl}}$***:*** Assume liquid seal $h_s = 0.5$ in., which implies that weir height h_w exceeds downcomer clearness h_{cl} by 0.5 in. Thus,

$$h_{cl} = h_w - 0.5 = 2.0 - 0.5 = 1.5 \text{ in.}$$

TABLE 12.6. Downcomer Layout for Example Problem

Downcomer Layout	
Weir length L_w; ft	2.0
Weir height h_w; in.	2.0
Liquid height over weir h_{ow}; in.	1.28
Clear liquid height $h_c = h_{ow} + h_w$; in.	3.28
DC seal h_s; in.	0.5
DC clearance $h_{cl} = h_w - h_s$; in.	1.5
DC clearance area $A_{da} = L_w h_{cl}/12$; ft^2	0.25
Flow passes	1
Downcomer type	Sloped
Weir type	Picket fence and Chordal
Tray type	Valve
Total number of valves N; 12 valves/ft^2	73
Valve type	KG Valve Type A(V-1)

Selection of downcomer clearness h_{cl} needs to consider a good downcomer seal; at the same time also needs to avoid too high pressure drop at the downcomer exit. Too small downcomer clearness could cause downcomer backup.

The downcomer layout is listed in Table 12.6. Until now, we have conducted a preliminary column design, which provides the column geometry and tray layout.

12.9.3 Hydraulic Performance Evaluations

It is only half of the job done after conducting tray sizing and layout design. The remaining questions are: What is the tower capacity or operating window, and can the tower give satisfactory performance? The questions can only be answered by performing hydraulic calculations for spray regime, tray flooding, downcomer backup, downcomer choke, and weeping.

Spray Flow Check: The Summers and Sloley (2007) spray factor correlation in equation (12.6) is employed:

$$S_p = K \frac{h_c}{u_h d_h} \left(\frac{\rho_L}{\rho_V}\right)^{0.5} = 2.5 \frac{3.3}{31 \times 1.5} \left(\frac{51.51}{0.195}\right)^{0.5} = 2.88$$

$K = 2.5$ is used for the valve tray in this case. The tray design point will be away from the spray regime as the spray factor is 2.88, which is larger than 2.78.

Tray Flood Check: The Fair C_F-factor correlation generalized in equation (12.15) by Summers (2011) is employed:

$$C_F = \left(\frac{18}{24}\right)^{0.5} \times \left(0.455 - 0.0055 \times 0.195^2\right) = 0.394.$$

Then we apply equation (12.14) to calculate the vapor velocity at flood based on a tray spacing of 18 in. and assuming $SF = 0.9$. This gives

$$u_F = SF \times C_F \sqrt{\frac{\rho_L - \rho_V}{\rho_V}} = 0.9 \times 0.394 \times \sqrt{\frac{51.51 - 0.195}{0.195}} = 5.75 \text{ ft/s.}$$

Applying equation (12.9) for actual vapor velocity yields

$$u_N = \frac{CFS}{A_N} = \frac{28}{6.47} = 4.33 \text{ ft/s.}$$

Fair flooding% based on equation (12.17) becomes

$$\text{flood\%} = \frac{u_N}{u_F} = \frac{4.33}{5.75} = 75.3\%.$$

As an alternative calculation, the Glitsch flood correlation 13 (Glitsch, 1974) expressed in equation (12.18) is applied. For vapor load $V_L = 1.73 \text{ ft}^3/\text{s}$, $GPM = 42$, $FPL = 2.2 \text{ ft}$ (26.4 in.) and $C_F = 0.394$; the Glitsch flooding level can be evaluated using equation (12.18):

$$\text{Food\%} = \frac{V_L + (GPM \times FPL/13{,}000)}{C_F A_a} = \frac{1.73 + (42 \times 26.4/13{,}000)}{0.394 \times 6.12} = 75.2\%.$$

Both flooding calculations give similar flood% for this example. Thus, it could be expected that tray design has flooding% less than 80% in normal operation, and thus liquid entrainment will be less than 0.1 lb liquid/lb vapor. It should be noted that the Glitsch flood correlation 13 (Glitsch, 1974) is the most commonly used flood correlation for industrial distillation.

Downcomer Backup%: According to equation (12.22), the total clear liquid height in downcomer is

$$H_d = h_c + h_t + h_{da}.$$

The height of froth in the downcomer is H_d/φ where φ is froth density. Assume the target value of 80% for the froth height in relation to the liquid setting height. For a stable operation, the liquid capacity must satisfy the condition defined in equation (12.23):

$$\frac{H_d/\phi}{H_S + h_w} \leq 80\%.$$

In the following, we will calculate individual hydraulic head losses.

Clear liquid height $h_c = h_w + h_{ow} = 2 + 1.28 = 3.28$ in.

Dry pressure drop h_d: For KG valve type A(V-1) with thickness of 0.134 in., $K = 0.86$. Based on $u_h = 31$ ft/s and $d_h = 1.5$ in., applying equation (12.34b) yields

$$h_d = K_2 \frac{\rho_V}{\rho_L} u_h^2 = 0.86 \times \frac{0.195}{51.51} 31^2 = 3.13 \text{ in.}$$

Wet pressure drop h_l is the pressure drop that the vapor goes through the aerated liquid on the tray. h_l can be calculated by using equation (12.35):

$$h_l = \beta h_c = 0.58 \times 3.28 = 1.90 \text{ in.}$$

β is the tray aeration factor, which is found to be 0.58 based on the correlation for valve tray (Klein, 1982).

Thus, *Tray pressure drop* can be calculated as

$$h_t = h_d + h_l = 3.13 + 1.90 = 5.03 \text{ in.}$$

Downcomer apron pressure drop based on equation (12.31) is

$$h_{da} = 0.03 \left(\frac{GPM}{100 A_{da}} \right)^2 = 0.03 \left(\frac{42}{100 \times 0.25} \right)^2 = 0.085 \text{ in.}$$

Total hydraulic height in the downcomer or downcomer backup is

$$H_d = h_c + h_t + h_{da} = 3.28 + 5.03 + 0.085 = 8.4 \text{ in.}$$

Froth density $\varphi = 0.6$ based on Glitsch's correlation (Glitsch, 1974). Thus,

$$\frac{H_d/\phi}{H_S + h_w} = \frac{8.4/0.6}{18 + 2} = 70\% < 80\%.$$

Thus, there will be no downcomer backup flood.

Downcomer Choke: The downcomer liquid capacity calculations are conducted based on both equation (12.47) (Koch, 1982) and equation (12.46) (Summers, 2011). Let us apply equation (12.47) first. Since a sloped downcomer is used, downcomer area A_d takes the average of the top and bottom downcomer areas:

$$A_d = (0.59 + 0.36)/2 = 0.48 \text{ ft}^2.$$

Thus, the downcomer liquid load is

$$Q_d = \frac{GPM}{A_d} = \frac{42}{0.48} = 87.5 \text{ gpm/ft}^2.$$

By assuming minimum residence $\tau_{d,min} = 5$ s, applying equation (12.47) to give the maximum downcomer liquid load as

$$Q_{d,max} = 448.8 \frac{H_S}{12\tau_{d,min}} = 448.8 \frac{18}{12 \times 5} = 134.6 \text{ gpm/ft}^2,$$

$$Q_{d,max}\% = \frac{Q_d}{SF \times Q_{d,max}} = \frac{87.5}{0.9 \times 134.6} = 72\% < 75\%.$$

Alternatively, applying equation (12.46) yields

$$V_{D,max} = 0.1747 \ln(51.51 - 0.195) - 0.2536 = 0.434 \text{ ft/s}.$$

The actual downcomer velocity can be converted based on Q_d as calculated previously

$$Q_d = 87.5 \text{ gpm/ft}^2 \Rightarrow V_d = 0.2 \text{ ft/s},$$

then

$$Q_{d,max}\% = \frac{V_d}{SF \times V_{d,max}} = \frac{0.2}{0.9 \times 0.434} = 51\% < 75\%.$$

We can double check with the liquid residence time in the downcomer using equation (12.49):

$$\tau_R = 448.8 \frac{A_d H_S}{12 \text{ GPM}} = 448.8 \frac{0.48 \times 18}{12 \times 42} = 7.7 \text{ s} > \tau_{min} = 5 \text{ s}.$$

Thus, there will be no downcomer choke in normal operation.

12.9.4 Tray Design Summary

So far, we have obtained the tray layout and conducted hydraulic evaluations. The results can be summarized in Tables 12.7–12.9.

12.9.5 Feasible Operation Window

It seems we have completed the task of tower layout design at hand. You may still not be fully satisfied with the question in mind: How good is this design in the context of the tower capacity limits? Indeed, it would be very helpful if one can visualize the tower capability diagram or operating window. This could help to know where the current tower design stands and what flexibility the tower can offer in operation. The following discussions are used to show the procedure of generating the operating window for this example problem.

TABLE 12.7. Tray Design Overall Summary

Overall Summary	
Tower diameter D, ft	3.0
Tray spacing H_s, in.	18
Number of passes	1
Type of tray	Valve
Tray thickness, in.	0.134
Total number of valves; 12/ft^2	73

TABLE 12.8. Tray Layout Summary

Layout Summary	
Column total area A_T; ft^2	7.07
Column active flow area A_a; ft^2	6.12
Downcomer type	Sloped
Weir type	Chordal
Weir shape	Picket fence
Downcomer area A_d (top); ft^2	0.59
Downcomer area A_d (bottom); ft^2	0.36
Outlet weir length L_w; ft	2.00
Effective weir length L_{we}; ft	0.80
Downcomer width W_d; in.	5.00
Liquid height over outlet weir h_{ow}; in.	1.28
Outlet weir height h_w; in.	2.00
Downcomer clearance area A_{da}; ft^2	0.25
Downcomer clearance h_{cl}; in.	1.50
Flow path length FPL ft.	2.2

TABLE 12.9. Hydraulic Performance Summary

Hydraulic Performance	
1. Spray factor (Limit = 2.78)	2.88
2. Tray flood (Limit = 82%)	
Based on Fair's correlation	75.3%
Based on Glitsch correlation	75.2%
3. Downcomer capacity limits	
Downcomer backup based on Glitsch (1974)	70%
Downcomer velocity based on Summers (2011)	51%
Downcomer velocity based on Koch (1982)	72%
Residence time, s	7.7
4. Tray pressure drop	
Dry pressure drop h_d; in.	3.13
Wet pressure drop h_l; in.	1.90
Total pressure drop h_t; in.	5.03
5. Downcomer hydraulics	
Downcomer backup, in clear liquid; in.	8.42
Downcomer froth density; φ	0.60
Downcomer apron head loss h_{da}; in.	0.08

Working guide for obtaining the operating window:

Step 1: Determine capacity limits for spray, tray flooding, downcomer backup, downcomer velocity, weeping, liquid rates, and so on.

Step 2: Define vapor and liquid rates as variables.

Step 3: Derive relationships between vapor and liquid rates based on capacity equations under these limits.

Step 4: Plot these relationships for capacity limits on a diagram of vapor and liquid rates.

12.9.5.1 Calculations of capacity limits

Spray Limit: Use spray factor $= 2.78$ and applying equation (12.6) yields

$$S_p = K \frac{h_c}{u_h \, d_h} \left(\frac{\rho_L}{\rho_v} \right)^{0.5} = 2.5 \times \frac{2 + h_{ow}}{u_h \times 1.5} \left(\frac{51.51}{0.195} \right)^{0.5} = 2.78. \qquad (12.61)$$

Applying Francis correlation (equation (12.27)) for h_{ow}

$$h_{ow} = 0.48 \left(\frac{GPM}{L_w} \right)^{2/3} = 0.48 \left(\frac{GPM}{0.8 \times 12} \right)^{2/3} = 0.106 \times GPM^{2/3}, \qquad (12.62)$$

and the hole velocity expression for u_h

$$u_h = \frac{CFS}{A_h} = \frac{CFS}{N_h(\pi/4)d_h^2} = \frac{CFS}{73 \times (3.141/4) \times (1.5/12)^2} = 1.11 \times CFS. \quad (12.63)$$

For simplicity of mathematical expressions, let $G_V = CFS$ and $G_L = GPM$. Replacing h_{ow} and u_h in equation (12.61) with the corresponding equations (12.62) and (12.63) yields

$$2.5 \times \frac{2 + 0.106 \, G_L^{2/3}}{1.11 G_V \times 1.5} \left(\frac{51.51}{0.195} \right)^{0.5} = 2.78.$$

Solving G_V gives

$$G_V = 17.56 + 0.93 G_L^{2/3}. \qquad (12.64)$$

Equation (12.64) defines the relationship between G_v and G_L for the spray limit at 2.78, which will be shown on a G_v–G_L plot.

Vapor Flooding Limit: Assume the flooding limit at 82% vapor capacity. Application of equation (12.18) with V_L replaced via equation (12.7) gives

$$Flood\% = \frac{CFS\sqrt{\rho_V/(\rho_L - \rho_V)} + GPM \times FPL/13,000}{C_F \times A_a} = 82\%,$$

where C_F is calculated by equation (12.15) as

$$C_F = \left(\frac{H_S}{24}\right)^{0.5} (0.455 - 0.0055\rho_V^2) = \left(\frac{18}{24}\right)^{0.5} \left(0.455 - 0.0055 \times (0.195)^2\right) = 0.394.$$

Thus, the above flood% can be expressed as

$$\frac{G_V\sqrt{0.195/(51.51 - 0.195)} + G_L(2.17 \times 12)/13{,}000}{0.394 \times 6.12} = 82\%.$$

Solving for G_V leads to the linear flooding limit as

$$G_V = 32.05 - 0.032G_L. \tag{12.65}$$

Equation (12.65) defines the G_V and G_L relationship which forms the vapor capacity limit at 82% flooding.

Downcomer Backup Limit: The maximum downcomer liquid height allowable is set by the tray spacing and outlet weir height. Downcomer backup flooding occurs when the liquid froth in the downcomer reaches the tray above. Assume the maximum backup limit as 80% and the froth density $\varphi = 0.6$ (Kister, 1992). Applying equation (12.23) yields

$$H_d = 0.8\,\phi(H_S + h_w). \tag{12.66a}$$

The RHS of equation (12.66a) becomes

$$0.8\,\phi\,(H_S + h_w) = 0.8 \times 0.6 \times (18 + 2) = 9.6\,\text{in.} \tag{12.66b}$$

The LHS of equation (12.66a) is

$$H_d = h_t + h_c + h_{da} = (h_d + h_l) + (h_w + h_{ow}) + h_{da}.$$

By letting $h_l = \beta h_c = \beta(h_w + h_{ow})$ and setting RHS = LHS, equation (12.66b) becomes

$$h_d + \beta(h_w + h_{ow}) + h_w + h_{ow} + h_{da} = 9.6. \tag{12.67}$$

Assume $\beta = 0.58$ (Kister, 1992). For given $h_w = 2$ in., rearranging equation (12.67) yields

$$h_d + 1.58h_{ow} + h_{da} = 6.44. \tag{12.68}$$

h_d is expressed as a function of vapor flow rate G_V (CFS) according to equation (12.34b) and u_h by equation (12.60) while h_{ow}, and h_{da} are functions of liquid flow rate G_L (GPM) based on equations (12.27) and (12.31):

$$\left. \begin{array}{l} h_\mathrm{d} = K_2 \dfrac{\rho_\mathrm{V}}{\rho_\mathrm{L}} u_\mathrm{h}^2 = 0.86 \times \dfrac{0.195}{51.51} \times \left(\dfrac{G_\mathrm{V}}{73 \times \pi/4 \times (1.5/12)^2} \right)^2 = 0.004\, G_\mathrm{V}^2 \\[3mm] h_\mathrm{ow} = 0.48 F_\mathrm{W} \left(\dfrac{G_\mathrm{L}}{L_\mathrm{w}} \right)^{2/3} = 0.48 \left(\dfrac{G_\mathrm{L}}{0.8 \times 12} \right)^{2/3} = 0.106(G_\mathrm{L})^{2/3} \\[3mm] h_\mathrm{da} = 0.03 \left(\dfrac{G_\mathrm{L}}{100 A_\mathrm{da}} \right)^2 = 0.03 \left(\dfrac{G_\mathrm{L}}{100 \times 0.25} \right)^2 = 4.8 \times 10^{-5} G_\mathrm{L}^2 \end{array} \right\} . \tag{12.69}$$

Replacing h_d, h_ow, and h_da in equation (12.68) with the corresponding expressions in equation (12.69) and then solving for G_V gives

$$G_\mathrm{V}^2 = 1610 - 41.9\, G_\mathrm{L}^{2/3} - 0.012\, G_\mathrm{L}^2. \tag{12.70}$$

Equation (12.70) defines the G_V and G_L relationship which forms the downcomer liquid capacity limit at 80% downcomer flood.

Maximum Liquid Loading Limit: Based on Table 12.2, the maximum weir loading allowed for a single pass is 8 gpm/in. Keep in mind that the effective weir length $L_\mathrm{we} = 0.8$ ft because a picket fence weir is used, although the physical weir length L_w is 2 ft. Thus, the maximum liquid loading can be calculated as

$$G_\mathrm{L,max} = 8 \times L_\mathrm{we} = 8 \times (0.8 \times 12) = 76.8\ \mathrm{gpm}. \tag{12.71}$$

The minimum residence time can be calculated from equation (12.49) as

$$\tau_\mathrm{min} = \frac{448.8\, A_\mathrm{d}(H_\mathrm{S}/12)}{G_\mathrm{L,max}} = \frac{448.8 \times (0.59 + 0.36)/2 \times (18/12)}{76.8} = 4.2\ \mathrm{s}.$$

This minimum residence time is consistent with the guideline of $\tau_\mathrm{min} = 3\text{–}5\,\mathrm{s}$ suggested by Kister (1992). When liquid flows faster than τ_min, there is no sufficient time to separate vapor from liquid in the downcomer, and large froth crest forms and blocks the flow path at the downcomer entrance resulting in downcomer choke.

Minimum Liquid Loading Limit: According to Kister (1990), liquid height over outlet weir, h_ow, must be maintained larger than 0.25–0.5 in. to keep a steady flow over the outlet weir and cover it completely during operation. By applying equation (12.27) for this case, we have

$$h_\mathrm{ow,min} = 0.48 \left(\frac{G_\mathrm{L,min}}{L_\mathrm{we}} \right)^{2/3}$$

Assume $h_\mathrm{ow,min} = 0.5$ in. for conservative consideration. Rearranging the above equation yields

$$G_{\text{L,min}} = L_{\text{we}} \left(\frac{h_{\text{ow,min}}}{0.48} \right)^{3/2} = (0.8 \times 12) \left(\frac{0.5}{0.48} \right)^{3/2} = 10.2 \text{ gpm}. \tag{12.72}$$

Minimum weir loading limit becomes

$$W_{\text{L,min}} = \frac{G_{\text{L,min}}}{L_{\text{we}}} = \frac{10.2}{0.8 \times 12} = 1.06 \text{ gpm/in}.$$

Minimum Vapor Loading Limit: Assume 60% turndown flexibility for this tower operation. For valve trays, the minimum vapor momentum to open the valve should be considered for turndown operation. Since this minimum momentum corresponds to F-Factor value of 4 at 60% turndown, applying equation (12.59) for F-Factor gives

$$F_{\text{min}} = u_{\text{h,min}} \sqrt{\rho_{\text{V}}} = 4.$$

Thus,

$$u_{\text{h,min}} = \frac{F_{\text{min}}}{\sqrt{\rho_{\text{V}}}} = \frac{4}{\sqrt{0.195}} = 9.06 \text{ ft/s},$$

and

$$G_{\text{V,min}} = u_{\text{h,min}} N \frac{\pi}{4} d_{\text{h}}^2 = 9.06 \times 73 \times \frac{3.14}{4} \times \left(\frac{1.5}{12} \right)^2 = 8.1 \text{ ft}^3/\text{s}. \tag{12.73}$$

The Design Point and Constant $G_{\text{V}}/G_{\text{L}}$ Line: The basis used for the tower layout design is 28 CFS and 42 GPM, which gives

$$G_{\text{L}}/G_{\text{V}} = 42/28 = 1.5. \tag{12.74}$$

This $G_{\text{L}}/G_{\text{V}}$ ratio is near the middle of the $G_{\text{L}}/G_{\text{V}}$ range of 0.3–3.0 (Table 12.3), which indicates a good distillation. Operation outside of this range may imply sloppy or too easy distillation.

12.9.5.2 Put All Capacity Limits Together to Generate a Capacity Diagram.
Based on the capacity limits derived above, we are ready to generate a capacity diagram based on the G_{V}–G_{L} relationships:

2.78 Spray factor :	$G_{\text{V}} = 17.56 + 0.93 \, G_{\text{L}}^{2/3}$	(12.64)
82% tray flood :	$G_{\text{V}} = 32.05 - 0.032 G_{\text{L}}$	(12.65)
80% DC backup :	$G_{\text{V}}^2 = 1610 - 41.9 G_{\text{L}}^{2/3} - 0.012 \, G_{\text{L}}^2$	(12.70)
Maximum weir loading at 8 gpm/in. :	$G_{\text{L,max}} = 76.8$	(12.71)
Minimum weir loading at 1.1 gpm/in. :	$G_{\text{L,min}} = 10.2$	(12.72)
Vapor loading when valve close :	$G_{\text{V,min}} = 8.1$	(12.73)
Design point :	$G_{\text{L}}/G_{\text{V}} = 1.5$	(12.74)

$$\tag{12.75}$$

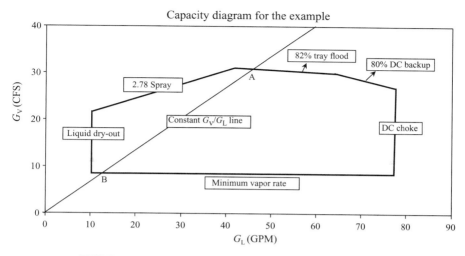

FIGURE 12.19. Operating window for the example problem.

By plotting the set of equations (12.75) onto G_V (CFS) – G_L (GPM) axis, we derive the capacity diagram or operating window for this example problem as shown in Figure 12.19. This diagram defines the feasible operating region for the example problem at hand. It must be emphasized that *a capability diagram depends on the physical conditions of specific distillation.*

12.9.5.3 Observations from the capacity diagram   Several observations can be obtained from the capacity diagram or the operating window as shown in Figure 12.19 and summarized as follows:

- The upper vapor capacity boundary of the operating region is controlled by both spray and tray flood. Spray is dominant under low liquid loadings, while tray flood becomes dominant at moderate and large liquid loadings.
- Downcomer flood imposes a constraint in large liquid loading. Downcomer width of 5 in. is used compared to 2.5 in. This is for considering the needs for installation and maintenance of downcomer.
- Based on the G_L/G_V ratio of 1.5, the operating line (*AB*) can be determined. The upper limit is the *A* point ($G_L = 45.87$ GPM and $G_V = 30.6$ CFS), which is the intersection between the constant G_V/G_L line and the tray flood line. The lower limit is the *B* point ($G_L = 12.15$ GPM and $G_V = 8.1$ CFS), which is the intersection between the constant G_V/G_L line and the minimum vapor limit. The operating flexibility is $G_{V(A)}/G_{V(B)} = 30.6/8.1 = 3.77$.
- The design point is in the middle of the operation region or *AB* line with G_L/G_V ratio of 1.5. Thus, the column can have high operating flexibility in dealing with significant process variations such as changes in feed rate and operating conditions.

Does it mean that this tower design is the optimal? The answer is "probably not" as there are other alternative tray designs available, which are yet to be determined. The same design procedure in Figure 12.16 can be employed for generating alternative designs. As mentioned, the first thing to be determined in tray design is tower diameter and tray spacing. The general trend of changes between them is that the tower diameter reduces (increases) when tray spacing is increased (reduced), thus resulting in a tall and thin (short and fat) tower. On the other hand, different tray designs provide different capacity limits resulting in different operating windows. Consequently, distillation efficiency will be different, energy for reboiling and condensing will be different, and capital cost will be different.

In general, when comparing alternative design options, the least cost, including capital and utility costs usually, is the choice by engineers who constantly seek economic solutions unless special circumstances justify a more expensive design, such as in revamp for capacity expansion. In operation, you should seek the minimum energy cost operation while satisfying throughput and product quality. In some cases, column energy use in terms of feed heating, stripping steam, and reboiling are not adjusted in proportion to process variations such as feed rate, product rates, and quality. This could result in wasted energy.

Thus, it is challenging to design a tower with the lowest overall cost (capital and utility) and simultaneously provide sufficient operation flexibility to deal with process variations. In contrast, the challenge for operation is to determine the optimal operating conditions within the feasible region under which the column is operated in the most efficient manner. Discussions and guidelines for tower operation optimization will be given in Chapter 13.

12.10   CONCLUDING REMARKS

What have we learned from the operating window discussions above? It might be helpful for us to pause a little and generalize some of the key points and stock them in our memory.

- Momentum exchange and balance between vapor and liquid is the key concept in understanding the mechanism of stable operation. A tower comes with a capacity diagram, which defines the operating window. Too high vapor load relative to liquid loading could cause tray flood. Conversely, too high liquid loading is the main cause for downcomer flood. On the other hand, too low vapor loading in relation to liquid could lead to weeping, while too low liquid loading and high vapor loading could result in spray flow.

- The spray mechanism can be best remembered by $h_c/u_h d_h$, the momentum ratio of liquid and vapor. Spray factor is defined in equation (12.6) based on Lockett (1986) with the value of 2.78 as the spray limit. To avoid spray, one needs to increase weir loading, reducing vapor loading and/or hole diameter.

- Tray flood is best described by equation (12.18) (Glitsch correlation 13; Glitsch, 1974). Both vapor load and weir loading contribute to tray flood in relation to tower diameter defining vapor open area, and tray spacing defining settling space. Thus, tower diameter and tray spacing are the two major design parameters, while feed rate, column pressure and reboiling duty are major operating parameters for avoiding tray flood.

- Downcomer flood is controlled by a combination of tray pressure drop, weir loading, and downcomer apron. Remember that it is the liquid froth level instead of clear liquid height that determines the downcomer flood. This fact is clearly stated in equation (12.24). To prevent downcomer flood in the design, downcomer size must be provided generously. From equations (12.40), (12.42), and (12.44), feed rate and reboiling duty are the major operating parameters to avoid downcomer flood.

- Weeping capacity is more related to turndown operation. Typically, 80% weep defined by equation (12.52) is used as the weep limit. Valve trays provide better turndown capacity in 3:1 versus sieve trays in 2:1. In other words, valve trays could operate in 30% of normal load with weep rate less than 10%.

- For towers operating at near atmospheric pressures or under vacuum, it is recommended to use a valve tray as it is more resistant to spray than sieve trays. For low liquid loading applications, a tray with a picket fence weir could be used to increase the liquid loading artificially. However, for much lower liquid loading where a picket fence could fail, a packing tower should be used. As a rule of thumb, use packing when liquid loading is less than 4 gpm/ft^2 column active area, and use a tray for liquid loading higher than 14 gpm/ft^2 column active area. It is a judgment call as to what tray to use for liquid loading in between.

NOMENCLATURE

English Letters

A_a active or bubble area; ft^2

A_d downcomer top area; ft^2

A_{da} area under downcomer apron; ft^2

A_h tray hole area; ft^2

A_N column net area (=Column cross section area less downcomer top area); ft^2

A_T total column cross section area; ft^2

C C-factor (describing vapor load); ft/s

CFS   vapor flow; ft^3/s

C_F C-factor at flood; ft/s

d_h tray hole diameter; in.

D column diameter; ft

f friction coefficient between vapor and liquid; dimensionless

F_{hc} hole factor

FPL	flow path length (distance from the inlet downcomer edge to outlet weir); in.
F_{hc}	hole F-factor
Fr	Froude number; dimensionless
F_w	weir correction factor; dimensionless
g	acceleration due to gravity; $32.2\,\text{ft/s}^2$
GPM	gallons per minute; gal./min
G_L	GPM
G_V	CFS
h_c	liquid height or liquid hold up; in. of liquid
h_{cl}	clearance under downcomer; in.
h_d	dry pressure drop; in. of liquid
h_{da}	clear liquid height for pressure drop through downcomer apron; in. of liquid
h_f	froth height on tray; in. of froth
h_g	hydraulic gradient on tray (high minus low clear liquid height); in. of liquid
h_l	pressure drop through aerate liquid on tray; in. of liquid
h_{ow}	liquid head over the outlet weir; in. of liquid
h_s	downcomer seal; in.
h_t	tray pressure drop; in. of liquid
h_v	height of two phase (froth and vapor) layer; in. of froth plus vapor
h_w	outlet weir height; in.
h_σ	head loss due to surface intension associated with bubble formation; in. of liquid
C_v	orifice coefficient (in. dry pressure h_d calculation); dimensionless
H_s	tray spacing; in.
H_d	downcomer liquid height; in. of liquid
H_d'	downcomer forth head; in. of forth
L	Liquid rate; gpm
L_w	outlet weir length; in.
N	number of valve per square feet; number/ft^2
p	pressure; atm
ΔP	pressure drop, in. of liquid
Q_D	downcomer liquid load; gpm/ft^2
$Q_{D,max}$	maximum downcomer liquid load; gpm/ft^2
SF	derating factor; dimensionless
S_p	spray factor; s/ft
t_m	valve thickness; in.
u_h	vapor hole velocity; ft/s
u	velocity; ft/s
u_F	vapor velocity at flood; ft/s
v	volume; ft^3
V	vapor flow rate; $\text{ft}^3\text{/s}$
$V_{D,max}$	maximum downcomer entrainment velocity; ft/s
V_L	vapor load; lb/ft^3
w	weep ratio; dimensionless

W weep rate; gpm
w_d downcomer width; in.
W_L weir liquid loading; gpm/in.

Greek Letters

α relative volatility; dimensionless
β tray aeration factor; dimensionless
λ ratio of the slop of the equilibrium curve to the slop of the component balance line
ρ density; lb/ft^3
ρ_m valve metal density; lb/ft^3
σ surface tension; dyne/in.
μ liquid viscosity; cP
τ_R downcomer residence time; s (second)
φ_d downcomer froth density; dimensionless
φ_f froth density; dimensionless
φ_t relative froth density; dimensionless
ψ valve open percentage; %
η_M Murphree tray efficiency; dimensionless
η_o column overall efficiency; dimensionless

Subscripts

l (L)   liquid
v (V)   vapor

REFERENCES

Anderson RH, Garnett G, Winkle MV (1976) Efficiency comparison of valve and sieve trays in distillation columns, *Ind. Eng. Chem., Process Des. Dev.*, **15**(1), 96–100.

Bennett BL, Kovak KW (2000) Optimize distillation columns, *Chemical Engineering Progress*, **96**(5), 19–34.

Bernard JDT, Sargent RWH (1966) Hydrodynamic performance of a sieve-plate distillation column, *Trans. Inst. Chem. Eng.*, **44**, 314.

Bolles WL (1963) in Smith BD (ed), *Design of Equilibrium Stage Process*, McGraw-Hill.

Bolles WL (1976) *Chemical Engineering Progress*, **72**(9), 43.

Chase JD (1967) Sieve tray design, *Chemical Engineering*, July, 105.

Colwell CJ (1981) Clear liquid height and froth density on sieve trays, *Ind. Eng. Chem. Process Des. Dev.*, **20**, 298–307.

Colwell CJ, O'Bara JT (1989) Paper presented at AIChE Spring Meeting, Houston, April.

Davis JA, Gordon KF (1961) What to consider in your tray design, *Petro/Chem Engineering*, Oct., 230.

Fair JR (1961) Design of direct contact gas coolers, *Petro/Chem Engineering*, **33**(10), 45.

Fair JR (1963) in Smith BD (ed), *Design of Equilibrium Stage Processes*, McGraw-Hill.

Fair JR, Steinmeyer DE, Penney WR, Crocker BB (1984) Chapter 18, in Perry RH, Green DW (eds), *Chemical Engineering's Handbook*, 6th edition, McGraw-Hill.

Glitsch (1974) *Ballast Tray Design Manual*, 3rd edition, Bulletin No. 4900, Texas, Glitsch Inc.

Hower TC, Kister HZ (1991) Solve process column problems, Parts 1 and 2, *Hydrocarbon Processing*, May and June.

Hsieh CL, McNulty KJ (1986) Paper presented at the AIChE Annual Meeting, Miami Beach, Florida, November.

Kister HZ (1990) *Distillation Operation*, McGraw-Hill.

Kister HZ (1992) *Distillation Design*, McGraw-Hill.

Klein GF (1982) Simplified model calculates valve-tray pressure drop, *Chemical Engineering*, **89**(9), 81–85.

Koch Engineering Company Inc. (1982) *Design Manual – Flexibility*, Bulletin 960-1, Kansas.

Lewis WK (1936) *Industrial and Engineering Chemistry*, **28**, pp. 399.

Lockett MJ (1986) *Distillation Tray Fundamentals*, Cambridge University Press, Cambridge, England, pp. 35.

Lockett MJ, Banik S (1986) *Industrial and Engineering Chemistry Process Design and Development*, **25**, 561.

Ludwig FE (1979) *Applied Process Design for Chemical and Petrochemical Plants*, 2nd edition, Vol. 2, Gulf Publishing.

McCabe WL, Thiele EW (1925) Graphic design of fractionation columns, *Industrial and Engineering Chemistry*, **17**, 605–611.

Murphree EV (1925) Graphical rectifying column calculations, *Industrial and Engineering Chemistry*, **17**, 960.

Nutter Engineering (1981) *Float Valve Design Manual*, Rev. 1, pp. 10, Tulsa, Oklahoma.

O'Connell HE (1946) *Transactions of AIChE*, **42**, 741.

Resetarits MR, Ogundeji, AY (2009) On distillation tray weir loadings, AIChE Spring Meeting, Florida, April.

Sakata M, Yanagi T (1979) Performance of a commercial-scale sieve tray, I. Chem. E Symposium Series, No. 56.

Smith BD (1963) *Design of Equilibrium Stage Processes*, McGraw-Hill, Inc.

Summers DR (2004) Performance diagrams: all your tray hydraulics in one place, AIChE Annual Meeting—Distillation Symposium, Austin, Paper 228f.

Summers DR (2011) A novel approach to quick sizing trayed towers, AIChE Spring Meeting, Chicago, March.

Summers DR, Sloley A (2007) How to handle low-liquid loadings, *Hydrocarbon Processing*, January, p. 67.

Weiland RH, Resetarits MR (2002) New uses for old distillation equations, AIChE Spring Meeting, New Orleans, Louisiana, March, 11–14.

13

DISTILLATION SYSTEM ASSESSMENT

Obtaining a good evaluation of separation systems can provide insights into the complex interactions in the system and help us understand how well the system operates. Although it can be a challenging task, doing it properly could yield huge benefits in enhancing operating margins as well as generating a wealth of process knowledge that can be invaluable for the operation of the system.

13.1   INTRODUCTION

Distillation is the core of a process unit for converting multicomponent streams into desirable products and accounts for the majority of energy consumptions. Improving energy utilization, reducing capital costs, and enhancing operational flexibility are spurring increasing attention on distillation column optimization during design and operation. A good understanding of distillation fundamentals, feasible operation, and equipment constraints will enable process engineers to gain insights into the distillation performance.

13.2   DEFINE A BASE CASE

The first step for tower performance evaluation is to simulate the original tower design, because it is uncommon that the original tower datasheets are unavailable or

Energy and Process Optimization for the Process Industries, First Edition. Frank (Xin X.) Zhu.
© 2014 by the American Institute of Chemical Engineers, Inc. Published 2014 by John Wiley & Sons, Inc.

inaccurate. To do this, selection of a proper VLE (vapor and liquid equilibrium) calculation package is critical. For hydrocarbon separation, the Peng–Robinson Equations of State model is a common choice. By providing process data including feed and product flows and compositions, together with tower data including temperature and pressure, feed tray, the number of theoretical stages, and reflux rate, the simulation will generate mass and composition balances as well as heat balances indicating reboiling and condensing duties.

Once the process simulation is developed, it is desirable to verify the simulation fidelity using different process conditions. The predicted product rates and purity and compositions as well as key operating parameters such as reflux rate and reboiling/condensing duties can then be compared with measurement. In some cases, performance tests are required to gather key data to compare with simulation for the accuracy and reliability of the simulation. To do this, performance tests must be conducted under steady and smooth conditions to mimic steady state operations.

If the simulation fidelity is proven to be sufficient enough, it is ready to move to the next task, which is evaluation of the tower performance, because the purpose of reproducing the original design data is to understand the tower hydraulic and thermal performances of the base case and use the well defined base case to conduct what if analysis in order to identify improvements.

An important aspect of defining the base case is gathering all the important data for the material and heat balances in one single sheet for a tower of interest. It would be very informative to have important mass flows, temperature, pressure, and composition data in one table so that a snap shot of the tower performance can be seen at a glance. Such an example is a heat pumped C_3 splitter shown in Figure 13.1 and Table 13.1. In building such a table, it is a good practice to include the tag number of the instrument for each parameter so that the data can be retrieved readily from the historian to produce the table with snapshots of different

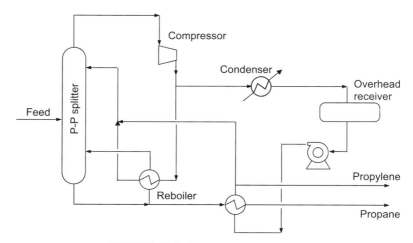

FIGURE 13.1. Heat pumped C_3 splitter.

TABLE 13.1. Major Data Set for a Heat Pumped C$_3$ Splitter

Data	Units	Tag No.	Value	Accuracy
Feed rate	bbl/d	FE-8854	4,975	−50
Feed temp.	°F	Pyrometer	87	
Top pressure	psig	PI-8831	100	
ΔP	psi	PDI-8827	9.2	
Top temp.	°F	TI-8774	53	
Bottom temp.	°F	Pyrometer	73	
Comp. suction temp.	°F	Pyrometer	55	
Comp. discharge press	psig	PC-8832	230	
Comp. discharge temp.	°F	TI-8776	119	
Comp. discharge temp.	°F	Pyrometer	135	
Main reflux rate	MSCFD	FT-8858	34,550	Too low
Main reflux temp.	°F	Pyrometer	74	
Trim reflux rate	bbl/d	FT-8857	600	
Trim reflux temp.	°F	Pyrometer	99.5	
Bottoms flow	bbl/d	FT-8864	1,060	−50
Propylene product temp.	°F	Pyrometer	110	
Propylene flow rate	bbl/d	FT-8860	3,840	−100
Overhead composition	vol% C3 =	AR 869-3	92.1	−0.5
Bottoms composition	vol% C3-	AR 869-2	97.1	−0.1

Source: From Summers (2009), reprinted with permission by AIChE.

times for evaluation of tower performance in the future. The last column in the table shows the high and low values of the corresponding parameters. The accuracy is determined by recording operation during the steady state period and noting the average high and low values of the various instruments during this period of time. This information could be very helpful when establishing heat and mass balances with indication of closure percentage. Typically, smaller flows can have a higher inaccuracy than larger streams and yet not severely affect the material balance. Therefore, it is good to know which streams have the highest reliability when determining the material or heat balance. It is also important to record the date and the time period that the data were taken for future reference. A ready reference of what data are needed in a typical tower evaluation can be seen here.

Defining a base case is to determine the base case operation of the tower of interest. This requires extracting two kinds of data. One kind is process data in terms of feed and product conditions, such as flows and compositions, while the other is tower operating data including temperature, pressure, and reflux rate. The former defines the mass and composition balances and the latter sets the heat balance around the tower with Table 13.2 giving such an example of the C$_2$ splitter column.

Due to the importance of developing a reliable base case as the basis for evaluation, Summers (2009) gave excellent discussions for this topic. For understanding the difference between simulation and measurement, readers can refer to Kister (2006).

TABLE 13.2. Heat and Mass Balances for a C_2 Splitter

Composition, wt%	Feed	Vent	Ethylene Product	Dilute Ethylene Product	Ethane Bottoms
Hydrogen	0.0016%	0.26%	0.21 ppm	0	0
CO_2	0.0001%	0.0006%	0.0002%	0.61 ppm	0
Methane	0.091%	14.45%	0.007%	0.007%	0
Ethylene	77.77%	85.28%	99.98%	80.44%	1.55%
Ethane	21.66%	0.0002%	0.0109%	19.56%	96.17%
Propylene	0.291%	0	0	0.002%	1.37%
Propane	0.0071%	0	0	0	0.033%
IsoButane and heavier	0.187%	0	0	0	0.88%
Total	100,000[a]	588	71,853	6,292	21,265
Phase	Vapor	Vapor	Liquid	Liquid	Liquid
Temperature °C	−13.0	−43.4	−29.8	−26.1	−7.0
Pressure, psig	340	250	270.5	276.2	279.7

DA-2410 condenser pressure	250	psig
DA-2410 top pressure	251	psig
DA-2404 condenser pressure	269.9	psig
DA-2404 top pressure	269.9	psig
Vent condenser duty[b]	0.73	MMBtu/h
Condenser duty[b]	49.87	MMBtu/h
Reboiler duty[b]	23.18	MMBtu/h
Side reboiler duty[b]	13.17	MMBtu/h
Reflux rate of DA 2410[a]	4,730	lb/h
DA-2410 reflux temperature	−43.4	°C
DA-2410 top temperature	−36.1	°C
Vapor rate of DA-2410[a]	5,318	lb/h
DA-2404 reflux rate[a]	349,370	lb/h
DA-2404 reflux temperature	−33.7	°C
DA-2404 top temperature	−30.4	°C

Source: From Summers (2009), reprinted with permission by AIChE.
[a]All flow adjusted to a 100 Klb feed basis to mask the true capacity of the unit.
[b]All duties adjusted to a 100 Klb feed basis.

13.3 CALCULATIONS FOR MISSING AND INCOMPLETE DATA

Plant historian data are the best source, but they are usually incomplete. This is particularly true for old process units. To avoid wasted time and rework, you need to make sure critical meters are working properly, which the instrument engineers can help verify. In most cases, design and operating data are of interest and the key is to understand the difference and reasons. For example, the knowledge about heat exchanger fouling can help in the evaluation of current operation performance considerably. Major consumption and any critical inputs must be verified carefully. The first stage of verification is to compare design data with operating data and perform some adjustments. This first pass verification can separate the important from the trivial data so that the effort for chasing high precision and gathering miniature data and nitty-gritty details can be avoided.

Since most correlations for heat exchangers are empirically based, the heat transfer calculations for exchangers are only accurate to about 85–90% when all the necessary data are known. When some data have to be estimated, the accuracy gets worse. However, this accuracy is sufficient to tell if a heat exchanger is functioning as expected or not.

Example 13.1 Obtain Missing Data

In many cases, shortcut calculations can fill in the gaps. An example used in Kenney's book (Kenny, 1984) gives good illustration for how to do it. Consider the tower in Figure 13.2. As for many plants, cooling water rates are not measured and overhead product comes off on level control. However, since feed rate and composition and overhead product composition are known, much of the missing data can be derived by energy and mass balances.

In this problem, p-xylene is to be recovered from a stream containing heavier aromatics. Neither product rate is measured, but feed rate and reflux rates and the p-xylene content of the overhead are. No heat exchanger duties are measured. With some data from a readily available source, the energy use for the tower can be estimated.

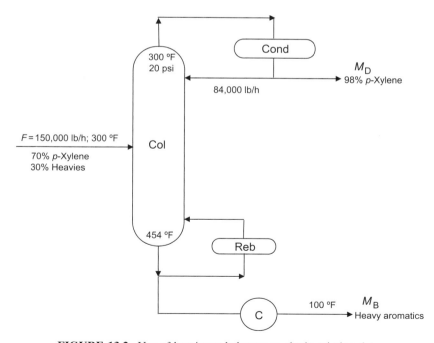

FIGURE 13.2. Use of heat/mass balances to obtain missing data.

(i) **Given Data** For p-xylene: normal boiling point $= 138.5\,°C$; latent heat of vaporization $=146.2$ Btu/lb; specific heat $= 0.38$ Btu/°C lb at $0\,°C$ and 0.43 at $41\,°C$ and 0.55 extrapolating to $140\,°C$. For heavier aromatics: specific heat $= 0.4$ Btu/°C lb for naphthalene at $87\,°C$ and 0.5 for pentadecane at $50\,°C$ and 0.8 extrapolating to $230\,°C$.

Estimate p-xylene product rate; heat duty for the overhead condenser, bottom cooler, and reboiler.

(ii) **Solution**

(a) *Calculate p-Xylene Product Rate:* Applying component balance on p-xylene and mass balance gives:

$$70\% \times F = 98\% \times m_{\mathrm{D}}; \quad m_{\mathrm{D}} = \frac{0.70 \times 150,000}{0.98} = 107,143 \text{ lb/h}$$

$$m_{\mathrm{B}} = F - m_{\mathrm{D}} = 150,000 - 107,143 = 42,857 \text{ lb/h},$$

where F is feed rate.

(b) *Calculate the Bottom Cooler Duty:* The heat rejected in the bottom cooler is

$$Q_{\text{bottom cooler}} = m_{\mathrm{B}} \times \mathrm{Cp} \times \Delta T = 42,857 \times 0.8 \times (454 - 150)/10^6$$
$$= 10.4 \text{ MMBtu/h}.$$

If the heat capacity data extrapolated are in error, the calculated duty would vary ± 0.1 Btu/(lb·°F). This is within the precision of other data.

(c) *Calculate the Overhead Cooler Duty:*

$$m_{\text{overhead}} = 107,143 + 84,000 = 191,143 \text{ lb/h}$$

Case 1: No subcooling, the condenser duty is

$$Q_{\text{condenser}} = m_{\text{overhead}} \times q_{\text{latent}} = 191,143 \times 146.2/10^6 = 27.9 \text{ MMBtu/h}$$

Case 2: Assuming $30\,°F$ subcooling, the condenser duty is the summation of latent heat duty and subcooling duty, which can be calculated as

$$Q_{\text{condenser}} = m_{\text{overhead}} \times q_{\text{latent}} + m_{\text{overhead}} \times \mathrm{Cp} \times \Delta T$$
$$= 191,143 \times (146.2 + 0.55 \times 30)/10^6 = 31.1 \text{ MMBtu/h}.$$

(d) *Calculate the Reboiler Duty:* Applying the energy balance around the tower indicates that the reboiler duty is the summation of the condenser duty and the heat required to raise the bottom from 300 to $454\,°C$.

Case 1: No subcooling in the overhead, the reboiler duty is

$$Q_{reboiler} = Q_{condenser} + m_B \times Cp \times \Delta T = 27.9 + 42{,}857 \times 0.8$$
$$\times (454 - 300)/10^6 = 33.2 \, MMBtu/h.$$

Case 2: $30\,°F$ subcooling in the overhead, the reboiler duty is

$$Q_{reboiler} = Q_{condenser} + m_B \times Cp \times \Delta T = 31.1 + 42{,}857 \times 0.8$$
$$\times (454 - 300)/10^6 = 36.4 \, MMBtu/h.$$

With these estimates, the heat duties on condenser, reboiling, and cooler are established, which provides the basis for the process simulation.

13.4   BUILDING PROCESS SIMULATION

A column simulation is conducted in a process simulation software based on tray-by-tray equilibrium calculations for mass and heat balances. Given the data for feed and products in terms of flow rates and compositions, and column operating conditions in terms of pressure and temperature as well as reboiling duty, the column simulation can mimic the mass and heat balances for the current operation. Table 13.2 above shows an example of the data required for conducting simulation of a C_2 splitter column.

For some processes that involve a process stream with many components, it could be too difficult to gather all the components for simulation. In this case, the concept of pseudocomponents is applied so that a group of components is lumped together into a pseudocomponent with similar physical properties. For oil refining processes, crude oils and refining products are a mixture of many different chemical compounds. They cannot be characterized based on chemical analysis. To characterize any crude oil and refining products, the petroleum industry applies a distillation temperature based method of describing hydrocarbon compounds by the number of carbon atoms and unsaturated bonds in the molecule, and uses distillation temperatures and properties to define crude and products. For example, commercial jet fuel can be represented by a ASTM D-86 distillation temperature profile with the kerosene boiling range of $401\,°F$ at 10% and $572\,°F$ endpoint, while naphtha jet fuel, also called aviation gasoline, can be represented by a distillation range of $122\,°F$ at 10% and $338\,°F$ endpoint.

The first step is feed simulation. If detailed feed analysis is available, which includes composition and conditions, a feed can be readily defined in simulation. Otherwise, the feed can be back-calculated as the summation of all products for which flow rates and compositions are known.

The second step is to determine feed tray position. Theoretical stages should be used in simulating a column. If tray efficiency is known, the feed tray in terms of theoretical stage can be determined from the actual feed tray and tray efficiency.

However, tray efficiency is usually unknown. In this case, a sample lab test may be warranted. It is recommended to take a side sample one tray away from the feed tray. The feed point in simulation is one stage away from the theoretical stage, which matches the sample composition the best. Taking the sample from the feed tray would give compositions that are highly influenced by the feed and hence cannot truly represent the internal compositions inside the column.

The third step is to determine the number of theoretical stages required. With the feeds defined, the feed tray determined above and product conditions given in the tabulated data, a column simulation can be established. For a simple column with two products, one from the overhead and the other from the bottom, the number of theoretical stages can be determined from the measured reflux rate. For a given reflux rate, the required number of theoretical stages is the one that can match the product specifications. As reflux rate defines the reboiler duty and hence the column heat balance for a simple column, the heat balance determines the product specifications for given column conditions. For a complex column involving side draws and pump-arounds, the column should be simulated section-by-section because the column heat balance is defined by reflux rate together with the column pump-arounds. It is recommended to simulate a complex column from top to bottom. The top section has an overhead product, a side draw, and a pump-around next to the side draw. For a given reflux rate and pump-around duty, the number of theoretical stages in the top section is determined by matching the given product specifications. The section next to the top section is then simulated similarly.

The simulation can provide a sound basis for conducting other assessment tasks, which are discussed below.

13.5 HEAT AND MATERIAL BALANCE ASSESSMENT

One of the early steps of assessing the distillation system is to obtain good material and energy balances. Otherwise, it could be possible that assessment yields misleading conclusions.

The material and energy balances can be built based on the input of feeds and energy as well as outputs of products and energy in operation. The purpose of conducting a column material balance is to make sure feeds and products are measured accurately and desirable products are obtained. The energy balance is to verify if all major sources of energy input are accounted for and if efficient use of energy is achieved.

Heat input is the driving force for fractionation. For a simple fractionation column, heat input comes from feed and bottom reboiling, while heat is removed from products and the overhead condenser. For a complex fractionation column, multiple products are produced, while pump-arounds are located to remove excess heat in the column and recover this heat for process usage.

Both material and energy balances should be conducted on the basis of steady state operation as this is a stable operation away from any transient excessive flooding or weeping operation. The steady state operation can be viewed from the historian when process data remain virtually the same within a very narrow band. On the other hand, steady state operation can be obtained in operation after a minimum time from making operating adjustments to a tower. This minimum time can be expressed as

$$t_{\min} = \frac{M_{\text{hold}} R_f}{F},$$
(13.1)

where M_{hold} is the summation of material holdup in the sump and receiver drum. R_f is the reflux ratio and F is the tower feed rate.

For any fractionation column, there are overall mass balance and component mass balances, which follow the mass conservation law:

For overall mass balance,

$$\text{Total Mass Input} = \text{Total Mass Output}.$$
(13.2)

For component balance,

$$\text{Total Input of Component } j = \text{Total Output of Component } j.$$
(13.3)

Similarly, heat balance follows energy conservation law:

$$\text{Total Heat Input} = \text{Total Heat Output}.$$
(13.4)

13.5.1 Material Balance Assessment

Good understanding of a material balance and key component balances can give insights for maximizing desirable product yields while minimizing undesirable products. Material flows are measured for feed and products, which are usually available online. Component measurement can be obtained from online analyzers or lab tests for key components. Samples are taken daily on most towers and analyzed in the plant's local laboratory. However, these laboratories are typically setup to measure for certain key compounds and do not have the capability of measuring the full spectrum of multicomponents involved in feed. Therefore, unless the tower of interest has only a few components in the feed, a complete component balance will typically need special laboratory assistance, which more than likely will come from outside the local plant. In this case, be very careful with compositions and understand the units of measurements that are provided by the laboratory. However, the component balance could be difficult to obtain for hydrocarbon distillation towers due to the fact that individual components cannot be fully characterized. However, for most chemical and natural gas distillation towers, it is possible to establish individual component balances.

When the material balances achieve at least ±10% offset ([total mass input − total mass output]/total mass input), it is acceptable for tower evaluation (Summers, 2009). For hydrocarbon processes, the closure could be as high as ±5%. It is common that a poor mass balance is caused by transmission error between pressure drop and flow rate for some of the material streams. Most flow meters are pressure drop based devises and they could give wrong readings if physical properties are not used properly for converting pressure drop to flow rate. In some cases, wrong readings can be corrected by meter calibration including proper zeroing and spanning. Consult with instrument engineers and they can help resolve the meters related issues.

Example 13.2   Overall Mass Balance Assessment

This example comes from the main fractionation tower in a hydrocracking process, which is operated to make naphtha, kerosene, and diesel products. The bottom product is called unconverted oil, part of which is recycled back to reaction for further conversion and the rest sold as fuel oil to the market. Table 13.3 shows the material balance for the fractionator indicated by expected yields versus the actual yields, as well as the flow rates for feed and products measured online.

Let us look at the overall material balance for the tower. The measured product rates in barrels per day are 18,130 of naphtha, 12,200 of kerosene, 9,968 of diesel, and 9,968 of unconverted oil, which gives a total of 50,266 bpd. The difference between the measured total of 50,266 and actual feed rate of 51,548 is 3 vol%. As the light end products are not measured, it usually accounts for around 3 vol% of feed. Thus, the material balance for the tower is in a good closure at less than 1% of uncertainty.

What could we learn from this material balance? The first observation is that 1.6% extra kerosene is produced than expected. This has a simple explanation because the tower was operated to maximize kerosene production as kerosene was more valuable in the local market at that certain time. In contrast, 4.3% less of diesel was made, which was surprising. Two plausible causes were thought of by the process engineer responsible for the tower operation. One reason was that kerosene cuts 1.6% deep

TABLE 13.3. Mass Balance Around a Fractionation Tower

	Vol% of Feed (Yield Expected) Feed	Vol% of Feed (Yield Produced)		Barrel Per Day (Produced)		Barrel Per Day Feed
		Distillate	Bottom	Distillate	Bottom	
Naphtha	35.2	35.2		18,130		
Kerosene	22.1	23.7		12,200		
Diesel	23.6			9,968		
Unconverted oil	19.1	19.3	21.8		9,968	
Total	100	78.2	21.8	40,298	9,968	51,548

into diesel, while the other was that 2.7% (4.3% − 1.6%) diesel slumps into the bottom unconverted oil. If the latter was true, it could be a significant yield loss and should be resolved.

This was only a hindsight which could be wrong. The expected yield comes from yield estimates based on empirical correlations which could give inaccurate estimates sometime. Distillation temperature for the bottom product could provide the answer to this question. Thus, a tower simulation was conducted, which shows that the 5% distillation temperature of the bottom product is 645 °F. This temperature cut should belong to the diesel range. It was clear that the unconverted oil contains a good portion of diesel, which could have been sold to the market at a premium price. The diesel price was at 2.45/gal., and thus $143,000 per day was lost under this tower operation, or $50 MM/year could be lost if this problem was not resolved. This price tag rang an alarm loud enough to secure swift action for troubleshooting.

In summary, the investigation revealed two root causes. The first one was that stripping steam at the bottom of the tower was insufficient as it was put on constant control based on the original design point. Although the throughput was increased by 10% over the years and feed became heavier, the stripping steam rate did not change. After adjusting it accordingly, the diesel recovery was improved. The reason is that more stripping steam introduced into the tower reduces hydrocarbon partial pressure which helps to lift more diesel components from the bottom. The second cause was lack of trays in the bottom section as the tower was designed for dealing with lighter feed than what it is handling now. A revamp project to add a few trays in the bottom section was scheduled for the turnaround.

Comment: A simple mass balance identifies a major yield loss.

Example 13.3 Component Mass Balance Assessment

A stripper column in a naphtha hydrotreating process unit needs to remove H_2S, which is corrosive and could poison the catalyst in a downstream naphtha reforming unit. Another objective is to remove as much C_5 as possible from the stripper bottom, as C_5 does not contribute to the naphtha reforming reaction.

The stripper was operated with these two objectives in mind. However, the lab test showed the C_5 component distribution. The stripper bottom contained 2 mol% C_5, which exceeds the targeted C_5 removal from the bottom, which is undesirable. Two negative results were observed. Because C_5 is not involved in the catalytic reforming reaction, C_5 material not only occupied the space in the reactor and hence reduced reaction throughput, but also consumed extra heat in the feed heater for the reforming reactors.

Once the problem was identified, the process engineer discussed it with the control engineer who quickly changed the set point for the reboiling duty and reflux rate. With increased reboiling duty and consequently increased reflux rate, better fractionation in the top section and hence more C_5 was stripped out of the bottom.

Comment: Therefore, the component mass balance could help determine the desirable locations for key components to go from separation. Improper separation could cost not only energy but also have a negative effect on yields.

13.5.2 Heat Balance Assessment

A major part of tower heat balance is checking reflux rate and temperature, which determines both the condenser and reboiler duties. It is important to measure reflux temperature as it affects the heat balance significantly when the reflux is subcooled. Example 13.1 demonstrated the effect of reflux subcooling.

The common problem with measuring reflux rate is that reflux meters are typically set at startup and then never adjusted again. Therefore, the reflux flow rate is typically not reliable. The reflux ratio is checked and monitored as an important operating parameter, but the absolute value of the reflux rate is rarely monitored. However, to have a correct heat balance, the reflux flow meter must be checked and calibrated to achieve at least $\pm 5\%$ closure of heat balance ([total heat input − total heat output]/total heat input). Only with this accuracy of heat balance, tray efficiency can be accurately determined (Summers, 2009).

13.6 TOWER EFFICIENCY ASSESSMENT

A benchmarking efficiency for a tower should be established. By comparing the actual efficiency with the benchmark efficiency, we can obtain a trend of efficiency over time and see the sign of poor separation. A good efficiency indicates a healthy operation of the tower in general, while a poor efficiency identifies signs of unstable operation, which warrants a tower rating assessment to reveal root causes of abnormality and thus determine actions for corrections.

Calculation of distillation efficiency requires process simulation. From the number of theoretical stages (N_{eq}), column section and overall efficiency can be determined, respectively based on actual number of trays (N_{act}). For a simple tower with two products, one from the top and one from the bottom together with one condenser and one reboiler, the separation efficiency can be calculated via

$$\eta_o = \frac{N_{eq}}{N_{act}}. \tag{13.5}$$

As an example for illustration, the McCabe–Thiele diagram (McCabe and Thiele, 1925) in Figure 13.3 indicates 12 actual stages required in comparison with 8 theoretical stages. Partial condenser and partial reboiler are counted in both the theoretical stages and actual stages. Thus, the overall tower efficiency is 67%.

For a complex column, equation (13.5) cannot give the right answer as it is only applied to each section. Instead, fractionation correlation plots by O'Connell (1946) (Figure 12.3) is widely used for overall tower efficiency, which is the standard for

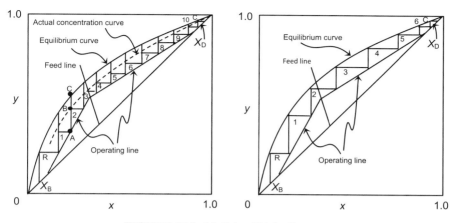

FIGURE 13.3. McCabe–Thiele diagram.

industrial tower efficiency. Lockett converted the O'Connell's plots into a generalized equation as

$$\eta_o = 0.492(\mu\alpha)^{-0.245}, \tag{13.6}$$

where μ is the viscosity of liquid and α relative volatility, which are calculated based on average temperature and pressure between column top and bottom. Thus, the O'Connell correlation states that higher viscosity leads to lower efficiency due to greater liquid phase resistance, while higher relative volatility also reduces efficiency as it increases the significance of the liquid phase resistance.

O'Connell's correlation plots (Figure 12.3) and Lockett's equation (13.6) were developed based on efficiency data points for industrial towers and do not reveal fundamental reasons for what to do, why and how, to improve efficiency.

Thus, the natural question is: What things affect tower separation efficiency? Mainly there are three kinds of parameters. The first is flow properties such as relative volatility and viscosity, which are intrinsic. The second one is tray layout design such as tray type (sieve or valve), flow path length, tower diameter, tray spacing, and weir length, which affect the liquid and vapor equilibrium and flow regime. The third one is process conditions such as tower feed rate and reboiling duty. The common effect of these parameters is in impacting the balance between vapor and liquid loadings.

Efficiency varies very little in the region of stable operation as discussed in Chapter 12, while efficiency falls off the cliff outside the feasible region. Figure 13.4 shows a typical trend of tower efficiency dependent on the balance of vapor and liquid rates. In the middle of the efficiency curve corresponding to stable operation, there is a relatively flat region although with marginal variation. Trays with good turndown features such as valve tray, compared with sieve tray, have a wider flat or stable operating region. On either side of the curve, efficiency drops off dramatically. Efficiency declines under low feed rate corresponding to turndown operation and

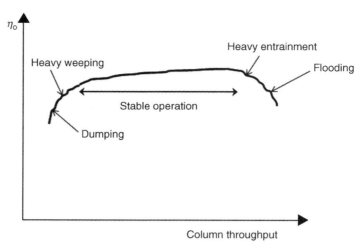

FIGURE 13.4. A typical trend of tower efficiency.

falls off the cliff when dumping occurs. On the hand, efficiency reduces at excessive entrainment and thus plummets when flooding happens. Optimization in design and operation tends to push the tower toward the boundary of stable operation. Detailed discussions on these controlling mechanisms are provided in Chapter 12, which shed insights into how to optimize tower design and operation while achieving stable operation.

Efficiency assessment can detect the section(s) with poor efficiency from which root causes can be found. A section or whole column could be flooded due to too high vapor or liquid loading. This could be caused by changes in the conditions of feed and products in terms of rates, compositions, and product specifications. It could be also caused by too high reboiler duty, high feed temperature, and low column pressure or a combination of these. For the case of changes in feed compositions, it could be traced back to processing issues upstream.

From a retrofit point of view when dealing with too high liquid loading, enhanced capacity trays could be used such as UOP MD/ECMD or Shell HiFi or Sulzer high capacity or Koch-Glitsch high-performance trays. For the case of too low liquid loading, use of blocked weir or packing could be the cure. For too high vapor loading, valve trays could be considered. Tray damage could also cause malfunction of a column operation. Whichever the case, identification of low fractionation efficiency triggers the search for the root causes and solutions.

Example 13.4 Overall Efficiency Estimate Using O'Connell's Correlation

This example comes from Wankat (1988). A sieve tray distillation column is separating a feed that is 50 mol% n-hexane and 50 mol% n-heptane. The feed is a saturated liquid. Tray spacing is 24 in. The average column pressure is 114.7 psia. Distillate composition is 99.9% mol of n-hexane and 0.1% mol of n-heptane. Feed rate is 1000 lbmol/h. Internal reflux ratio L/V is 0.8. The column has a total reboiler and total condenser. Estimate the overall efficiency.

Solution. To apply equation (13.6), we need to estimate α and μ at the average temperature and pressure of the column. The column temperature can be obtained from the modified DePriester chart (Dadyburjor, 1978) as shown here.

x_{C6}	0.000	0.341	0.398	0.500	1.000
y_{C6}	0.000	0.545	0.609	0.700	1.000
$T\,°C$	98.4	85.0	83.7	80.0	69.0

Relative volatility is $\alpha = (y/x)/[(1-y)/(1-x)]$. The average temperature can be estimated in several ways:

Average temperature $T = (98.4 + 69.0)/2 = 83.7$; x and y at $T = 83.7\,°C$ can be interpolated based on the table above. Thus, $\alpha = 2.36$ at $T = 83.7\,°C$. If average at $x = 0.5$, $T = 80$, $\alpha = 2.33$ at $T = 80\,°C$. Not much difference. Use $\alpha = 2.35$ corresponding to $T = 82.5\,°C$.

The liquid viscosity of the feed can be estimated (Reid et al., 1977) from

$$\ln \mu_{mix} = x_1 \ln \mu_1 + x_2 \ln \mu_2. \tag{13.7}$$

The pure component viscosities can be estimated from

$$\log_{10}\mu = A\left(\frac{1}{T} - \frac{1}{B}\right), \tag{13.8}$$

where μ is in cP and T in K (Reid et al., 1977).

nC_6: $A = 362.79$; $B = 207.08$; nC_7: $A = 436.73$; $B = 232.53$.

The two equations just presented for μ_{mix} and μ give $\mu_{C6} = 0.186$, $\mu_{C7} = 0.224$, and $\mu_{mix} = 0.204$. Thus, $\alpha\mu_{mix} = 0.479$. Applying equation (13.6) gives $\eta_o = 58.9\%$, which agrees well with $\eta_o = 59.0\%$ obtained from O'Connell correlation plots (Figure 12.3). The lower value should be used for conservative purposes.

13.7 OPERATING PROFILE ASSESSMENT

Another simple assessment method is based on tower profiles generated from simulation, which include flow, temperature, pressure, and composition profiles. What can we learn from these profiles? In a nut shell, tower profiles can allow us to observe what is going on inside the tower, like X-ray photos by vision.

(i) *The flow profile* shows internal liquid and vapor flows across the column, which can vary from tray to tray with sudden change at feed stage and withdraw stages. In general, in the rectifying section above the feed stage up to the condenser, the vapor flow is higher than the liquid flow, while it is opposite in the stripping section below the feed stage. As part of the flow estimates, the feed is flashed at the feed tray

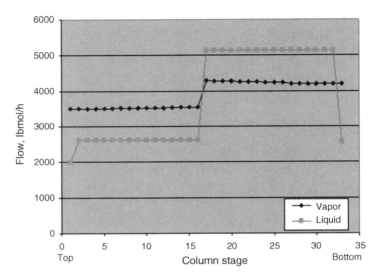

FIGURE 13.5. Example column flow profile.

conditions. The importance of flow estimates is not so much the absolute values but the ratio of L/V, which determines the internal reflux and the slope of the operating line. This behavior can be observed in Figure 13.5.

(ii) *The temperature profile* shows a general trend of monotonic reduction in temperature from the reboiler to the condenser. Figure 13.6 shows an example temperature profile where the steep parts of the curve would be where light and heavy keys are significantly separating. In some cases, temperature profiles feature plateaus in certain trays where little temperature change occurs. In these flat regions, it

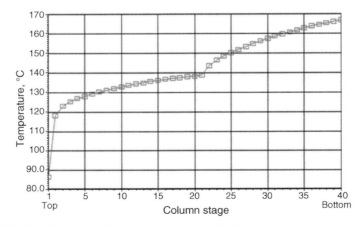

FIGURE 13.6. Example column temperature profile for a benzene–toluene separation.

indicates virtually no separation taking place although nonkey components are being distributed. When there is a large number of stages, these plateaus can be more self evident. These stages represent the pinch region where the operating line is very close to the equilibrium curve. In this pinch region, the ratio of relative volatility between key components is very small corresponding to a difficult fractionation.

Obviously, when a column or a section is flooded or dumping, a flat temperature profile can be obtained since there is no fractionation taking place.

(iii) A tower *pressure drop profile* can also indicate abnormal operation. A too low pressure drop across a tower or a section indicates potential dumping or flow channeling, while a too high pressure drop manifests flooding operation.

Lieberman (1991) recommends a simpler method for assessing flooding condition based on his operation experience. Lieberman's method indicates occurrence of flooding when

$$\frac{\Delta P}{\text{Sp}_L N_T H_S} \geq 22 - 25\%, \tag{13.9}$$

where ΔP, inches of water, is the overall column pressure drop between the column overhead and reboiler outlet or section pressure drop. Sp_L is the average specific gravity of liquid on a tray; N_T is the number of trays and H_S is tray spacing, in inches.

(iv) The *composition profile* can reveal the details of separation taking place inside the tower. For component balances, it is highly important to know the compositions of the feed and product streams around the tower. Samples are taken daily on most towers and analyzed in the plant's local laboratory. However, these laboratories are typically setup to measure for certain key compounds that can contaminate the final product and do not have the capability of measuring the full spectrum of a multicomponent column's feed. Therefore, unless the tower of interest has only a few components in the feed, a full component balance will typically need special laboratory assistance, which more than likely will come from outside the local plant. One needs to understand the units of measurements for compositions by the laboratories. Frequently, they will not provide the units such as molar, volume, or weight percentages.

Figure 13.7 shows the composition profiles for the separation of toluene (light key, LK) from ethyl benzene (heavy key, HK). The liquid mole fractions for these four components are given in the figure. Stage 21 is the feed stage.

Let us follow the toluene mole fraction curve, which is more obvious. Concentration of LK toluene increases through the tower from bottom up in a monotonous manner until it peaks at the top of the tower, but dips at the receiver. This is because benzene, the nonkey light component, is the most volatile component and peaks at the receiver.

On the other hand, the concentration of the heavy key, ethyl benzene, enriches toward the bottom of the tower monotonously through the tower and peaks a few stages above the reboiler because it is the least volatile component.

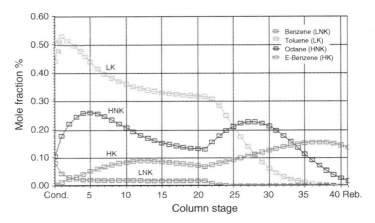

FIGURE 13.7. Example composition profile for toluene–ethyl benzene separation.

The concentration profile for the nonheavy key (HNK), octane, is the most confusing one as it goes up and down throughout the tower with two maxima because of the competition with other compounds on those trays and how they concentrate. From the feed stage 21 to stage 28, the separation takes place between the LK toluene versus HK ethyl benzene and HNK octane. As a result, the concentrations of both ethyl benzene and octane increase while toluene concentration reduces. From stage 29 to stage 40 where LK toluene concentration continues to decrease to reach a very low level, the separation takes place mainly between HK ethyl benzene and HNK octane. In this section, octane concentration reduces steeply and ethyl benzene concentration increases but drops slightly at the bottom. In summary, octane concentration increases from the feed stage until it peaks at stage 28 and then starts to decrease toward the bottom. This creates the first maxima in octane concentration.

From stage 15 to stage 5, LK toluene is separated from HK ethyl benzene. At the same time, the separation between HK ethyl benzene and HNK octane occurs where ethyl benzene concentration goes down as octane concentration steps up until octane peaks at stage 5. Above stage 5 toward the top of the tower, the separation takes place between LK toluene against both HK ethyl benzene and HNK octane where toluene concentration climbs and peaks at the top. In contrast, HNK octane concentration plummets and HK ethyl benzene concentration reduces to distinction. In summary, octane concentration increases from stage 15 until it peaks at stage 5 and then it reduces sharply. This creates the second maxima in octane concentration.

13.8 TOWER RATING ASSESSMENT

Tower simulation is only mimicking the current operation and can provide the vapor and liquid loadings as the basis for rating assessment while rating assessment will tell how the tower is operating under current conditions in relation to the feasible operating window which is discussed in details in Chapter 12.

Tower rating is applied to assess the effects for changing process conditions on tower performance. It can also be applied when an existing tower is considered to be used for a new service. Briefly, a tower rating can answer three questions:

(1) Can the tower operate with increased throughput or with changing process conditions within the feasible operating window? The calculations for the operating window were shown in details in Chapter 12. With the internal L/V from simulation, we can determine the current operating point in relation to the operating window.

(2) What is the hydraulic performance of the tower under new conditions? The operating pressure drops can be calculated based on hydraulic calculations discussed in Chapter 12.

(3) What are the limiting factors of the tower under new conditions? The limitations could come from the size of the tower, tray spacing, and down-comer geometry, and so on.

The criteria can be established as necessary and sufficient conditions for the suitability of an existing tower for changing conditions or new services:

(i) The operating point must fall within the operating window.

(ii) Operating pressure drop must be less than allowable pressure drops.

When these two conditions are fulfilled, an existing tower is suitable for different conditions for which it is rated. When the process conditions undergo significant changes, the rating assessment should be performed to make sure the tower can perform the task satisfactorily under new conditions. Otherwise, either operating conditions should be altered or modifications to the existing tower need to be determined.

Why would we want to use a tower for a service that it was not designed for? The main reason is that it is less expensive and quick to modify an existing tower than to purchase a new one. It is rare that the existing tower provides a perfect fit to a new service. However, engineers are keen to take the challenge of modifying existing equipment as it is their second nature of seeking the most economic solution for existing assets.

Tower rating assessment can be conducted using tower rating software by vendors. A tower simulation provides basic data required for tower rating. In transfering data from simulation to rating assessment, a tower is divided into sections and the stage with highest vapor loading in each section is selected to represent this section as this tray is the most constrained tray for the whole section. Thus, the data for this stage is entered into the tower rating software. The input data include: (i) vapor and liquid loadings and physical properties for both vapor and liquid, which are obtained from simulation; and (ii) tower tray layout. Execution of the rating software will give capacity limits (Figure 12.14) such as tray flooding, downcomer backup and dumping, vapor maximum capacity, liquid maximum capacity, froth/spray transition, pressure drop, dry tray pressure drop, downcomer velocity, weir loading etc.

The rating assessment software will indicate the current operating point in relation to the operating window and thus reveal what operating limits the tower is encountered. Performance diagrams, if plotted with vapor and liquid volume loadings, can represent tray performance independent of operating pressure and composition.

13.9   COLUMN HEAT INTEGRATION ASSESSMENT

The methodology for column heat integration is explained by Figure 13.8, which can be applied for new and retrofit designs. Let us explain the methodology step-by-step as below. The $T–H$ plot is the column grand composite curve (CGCC), which is discussed in Chapter 11.

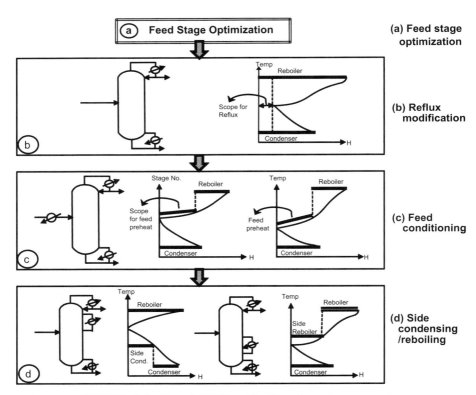

FIGURE 13.8. Use of CGCC to identify column improvements.

13.9.1 Feed Stage Optimization

The feed stage location of the column is optimized first in the simulation prior to the start of the column thermal analysis since the feed stage may strongly interact with the other options for column improvements. This can be carried out by trying alternate feed stage locations in simulation and evaluating its impact on reboiling duty.

In principle, there could be several stages that can be used as feed stages. When the feed stage is too low, there is a big jump in temperature in the region below the feed stage since too much change in composition is happening than necessary. To get the composition change needed to meet bottoms specifications, more reboiler duty is required leading to high boil-up and liquid and vapor traffic in the bottom section. Because of the higher flow rates, the bottom section could flood in operation or requires a larger diameter. Having the feed too high does not have the dramatic change as having the feed too low.

Since the objective for feed stage optimization is to minimize reboiling duty for the separation without the need for additional trays, a plot of reboiler duty versus stage number can be obtained as Figure 13.9. The optimal feed stage should be in the flat region away from the steep change.

After the feed stage optimization is accomplished, the CGCC for the column is then obtained, which is used as the basis for the next step optimization.

13.9.2 Reflux Rate Optimization

The next step is to optimize reflux rate for the column. As shown in Figure 13.8b, the horizontal gap between the vertical axis and CGCC pinch point is the scope for reflux improvement. The CGCC will move closer to the vertical axis when the reflux ratio is reduced. The reflux rate optimization must be considered before other thermal

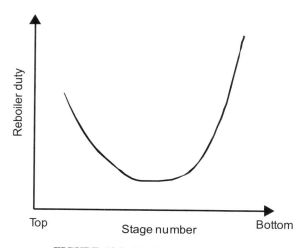

FIGURE 13.9. Feed stage optimization.

modifications since it results in direct heat load savings from the reboiler and the condenser. For a new column the reflux rate can be reduced by addition of stages. For an existing column, the reflux rate can be reduced by reducing reboiling duty when product quality is better than specifications.

13.9.3 Feed Conditioning Optimization

After reflux improvement the next step is to address feed preheating or cooling. In general, feed conditioning offers a more moderate temperature level than side condensing/reboiling. Also, feed conditioning is external to the column and is therefore easier to implement than side condensing and reboiling. Feed conditioning opportunity is identified by a "sharp change" in the stage-H (H: enthalpy) CGCC close to the feed point as shown in Figure 13.8c. The extent of the sharp change approximately indicates the scope for feed preheating. Successful feed preheating allows heat load to be shifted from reboiler temperature to the feed preheating temperature. The analogous procedure applies for feed precooling.

13.9.4 Side Condensing/Reboiling Optimization

Following the feed conditioning, side condensing/reboiling should be considered. Figure 13.8d describes CGCCs, which show potential for side condensing and reboiling. An appropriate side reboiler allows heat load to be shifted from the bottom reboiling to a side reboiling without significant reflux penalty.

13.10 GUIDELINES FOR REUSE OF AN EXISTING TOWER

The guidelines discussed next are mainly recommended by Wankat (1988), and the following things in order of increasing costs, can be explored when the existing column cannot produce the desired product purities:

- Find out whether the product specifications can be relaxed. A purity of 99.5% is much easier to obtain than 99.99%.
- Increase reflux rate and see if it can meet product specifications. Remember to check if column vapor capacity is sufficient as flooding could be an issue with an increased reflux rate. Also check if the existing reboiler and condenser are large enough. If the tower can make purer products, usually reducing reflux rate can make product back to specification, which also reduces operating cost.
- Change the feed temperature. This change may require altering of feed stage and could result in an optimal feed location.
- Will a new feed stage at the optimal stage allow meeting product specification?
- Consider replacing the existing column internals with more efficient or tighter-spaced trays or new packing. This is relatively expensive but is cheaper than a new column.

- Find out if two existing columns can be hooked together in series to achieve the desired separation. Feed can be introduced at the feed tray of either column or in between. Since vapor loading requirements are different in different sections of the column, the columns do not need to be the same diameter.
- Add a stub column to increase the total number of trays.

If the column vapor loading is more than the limit implying the existing column diameter is not large enough, engineers can consider:

- Operating at a reduced reflux ratio, which reduces vapor loading; however, this could make it difficult to meet product specifications;
- Operating at a higher pressure, which increases vapor density. Need to check if the column can operate at the increased pressure;
- Using two columns in parallel;
- Replacing the existing downcomers with large ones;
- Replacing the trays or packing with higher capacity ones.

On the other hand, if the column diameter is too large, vapor velocities will be too low. Trays will operate at too low efficiency and in severe cases they may not operate since the liquid may dump through the holes. Engineers can consider:

- Decrease column pressure to decrease vapor density and hence vapor velocity;
- Increase reflux ratio (or increase reboiling duty);
- Recycle some distillate and bottom products.

Using existing columns for new services often requires creative solutions. Thus, it can be both challenging and fun.

NOMENCLATURE

Cp	specific heat
F	feed rate
H_S	tray spacing
M	mass flow
N_T	total number of trays
N_{eq}	number of theoretical trays
N_{act}	number of actual trays
ΔP	pressure drop
q	latent heat
Q	heat content
R_f	tower reflux ratio
Sp	average specific gravity of liquid on tray

t time
T temperature
ΔT temperature difference

Greek Letters

α relative volatility
μ liquid viscosity
η_o overall tower efficiency

REFERENCES

Dodyburjor DB (1978) SI units for distribution coefficients, *Chem. Eng. Prog.*, April, p. 85, AIChE.

Kenny WF (1984) *Energy Conservation in the Process Industries*, Academic Press, Inc.

Kister HZ (2006) *Distillation Troubleshooting*, AIChE-John Wiley & Sons.

Lieberman N (1991) *Troubleshooting Process Operation*, 3rd edition, PennWell.

McCabe WL, Thiele EW (1925) Graphic design of fractionation columns, *Industrial and Engineering Chemistry*, **17**, 605–611.

O'Connell HE (1946) *Plate efficiency of fractionating columns and absorbers, Trans. AIChE*, **42**, p. 741.

Reid RC, Prausnitz JM, Sherwood TK (1977) *The Properties of Gases and Liquids*, 3rd edition, McGraw-Hill, New York.

Summers DR (2004) Performance diagrams: all your tray hydraulics in one place, *AIChE Annual Meeting—Distillation Symposium,* Austin, Paper 228f.

Summers DR (2009) How to properly evaluate and document tower performance, *AIChE Spring Meeting,* April 27, Florida.

Wankat PC (1988) *Equilibrium Staged Separations*, PTR Prentice Hall, New Jersey.

14

DISTILLATION SYSTEM OPTIMIZATION

Process operation involves strong trade-off between energy use and product recovery. Optimizing this trade-off could yield huge benefits for increased operating margins as well as energy savings. The optimization is a challenging task as there are many hidden opportunities to be uncovered and many constraints need to be explored.

14.1   INTRODUCTION

Distillation is the core of a process unit and also the major energy user. The design and operation of distillation columns involves a trade-off between energy use and product recovery. When energy usage is less than design, product recovery and quality may suffer. On the other hand, when energy is more than design, product quality is better than the specification, which is called product spec giveaway. In abnormal operations, little product recovery can be achieved regardless of how much energy is used.

However, reducing energy usage in a distillation system is not straightforward. This is because a distillation system involves many operating parameters including those within and outside the process battery limit. In particular, variations in conditions of feed and products as well as prices of feeds and products add much complexity to the economic operation of the process. This feature leads to strong dynamic behaviors of operating parameters. Furthermore, most of the parameters

Energy and Process Optimization for the Process Industries, First Edition. Frank (Xin X.) Zhu.
© 2014 by the American Institute of Chemical Engineers, Inc. Published 2014 by John Wiley & Sons, Inc.

interact in a nonlinear manner and have numerous constraints on their operation, which further complicate the task of process optimization. If some of the constraints can be relaxed, this could improve operating margins significantly.

The concept of tower feasible operating region was discussed in Chapter 12 and performance assessment in Chapter 13. This chapter focuses on economic operation within the feasible operation region.

14.2   TOWER OPTIMIZATION BASICS

Tower optimization is a difficult task as product pricing and unit constraints often change daily or weekly, but changing unit operating philosophy and addressing hardware constraints can take weeks or even months to accomplish. Even after the measures to improving optimal performance have been identified and implemented, if the desire to improve is declining, operation tends to return to the older, more comfortable routine. Thus, it is highly recommended that for a complex system, performance optimization should be implemented in advanced process control (APC), which can maintain tower operation under the most economic mode on a regular and consistent basis and in an automatic manner.

To establish sustainable tower optimization, key operating parameters must be defined and correlations must be developed to understand the relationships between key parameters. Finally, an optimization objective function must be developed to determine the optimal set points for the key parameters. The optimization can be conducted in two ways; one is semimanual based while the other is APC based. With the semimanual-based approach, some of the operating parameters are manually adjusted while optimization is done in an offline (online) manner. In contrast, with the APC approach, operating parameters are automatically adjusted while optimization is done online. However, both methods adopt the common ground of optimization: Using an objective function to derive the optimal set points for key parameters based on economic trade-offs; correlations to represent relationships between key parameters and constraints to define process and equipment limits. Noticeably, the optimization pushes operating limits in obtaining the optimal solution, and relaxation of sensitive constraints could generate significant benefits.

14.2.1   What to Watch: Key Operating Parameters

As discussed extensively in Chapter 4, it is important to define major operating parameters or key indicators as they can describe the process and energy performance. A key indicator can be simply an operation parameter like desirable product rate, column overhead reflux ratio, column overflash, column temperature, pressure, and so on. By the name of key indicator, the parameter identified is important and has a significant effect on process and energy performance.

Although primary operating parameters affecting both fractionation and energy use are tower specific, common operating parameters can be identifiedand and good understanding of them is critical in generating good correlations and optimization model.

14.2.1.1 *Reflux Ratio*

Reflux ratio is defined as the ratio of reflux rate to distillate rate (R/D) or the reflux rate to feed rate (R/F). In essence, a reflux rate is to set a tower top temperature required for making the distillate (overhead product) to meet specification. Furthermore reflux rate provides sufficient internal liqued flow to balance the vapor flow in the tower. Reflux is generated by energy either via tower reboiler or feed heater. Lower reflux rate saves energy but too low reflux rate could affect product quality. On the other hand, too high reflux rate could be wasting of energy if product quality is better than the specifications. In this case, the quality that is better than the specification is given away for free because there is no credit in pricing for the extra better quality.

Optimal reflux rate in operation depends on the operating margin, which is defined as the difference of product sales minus feed cost and energy cost. When energy cost is too high, it could drive the operation toward lower reflux rate and vice versa for the case of lower energy cost.

In tower design, the optimal reflux ratio is determined based on the trade-off between operating cost in the reboiler and capital cost for the tower. In other words, use of more separation stages requires less reflux rate and in turn less reboiling energy, but at the expense of additional capital cost. The minimum reflux ratio is calculated based on Underwood (1948). A tower requires an infinite number of stages to achieve the minimum reflux ratio. To make the tower feasible in operation and affordable in cost, a reflux ratio larger than the minimum is used. The typical reflux ratio is 1.1–1.3 of the minimum reflux ratio. With a high reflux ratio, the number of theoretical stages is lower resulting in lower capital cost for a tower, but at the expense of higher reboiler duty.

14.2.1.2 *Overflash*

Overflash is defined as the ratio of internal reflux at the feed vaporization zone and the feed rate. By definition, overflash represents the percentage of feed vaporized more than the amount of products drawn from above the feed tray. Overflash is a function of reflux rate, feed temperature, and tower pressure.

Overflash is generated from the overhead reflux rate while reflux rate is generated by reboiling. Thus, it can be said that overflash is generated by reboiler or feed heater. Overflash is an indication of reflux rate sufficiency for proper separation throughout the tower. A small overflash implies less reboiling duty and thus saves energy, but it could negatively affect the fractionation efficiency and hence product quality and vice versa. Therefore, overflash connects fractionation efficiency and energy efficiency for a tower. A tower could be making poor product quality even with high overflash when the tower is operated under abnormal operation such as flooding or dumping.

Overflash is typically controlled between 2 and 3%. An operation policy focusing on throughput would operate a tower at very low overflash; it is not uncommon to observe that a tower is operated at close to 1% overflash. This low reflux operation could be beneficial if the tower produces intermediate products, which will be processed further downstream. In this case, this operation could lead to energy efficiency as well as high economic margin.

14.2.1.3 Pressure   Lower pressure typically saves energy. This is because the lower the tower pressure the less heat required for liquid to vaporize and thus less energy required. This results in better fractionation as it is easier for vapor to penetrate into liquid on the tray deck.

The condenser pressure controls the tower pressure and thus the feed tray pressure. There is a pressure valve in the overhead, which can be used to control tower pressure. The lower limit of the tower pressure is defined by the column overhead condensing duty, net gas compressor capacity, and column flood condition. During extended turndown periods, reducing pressure up against an equipment limit can avoid dumping. Many of the new APC systems have pressure control implemented.

Heat exchanger fouling in overhead condensers could cause higher pressure drop and thus result in high tower pressure. On the other hand, higher reflux rate could lead to high pressure drop in the overhead loop causing high tower pressure.

14.2.1.4 Feed Temperature   A hotter feed can increase feed vaporization and thus reduce reboiling duty. However, higher temperature feed could cause too much vapor resulting in rectification section flooding. For an existing tower, the optimal feed temperature corresponds to the lowest reboiling duty, while the tower can meet product specifications.

14.2.1.5 Stripping Steam   Some towers may have stripping steam in the feed zone. Stripping steam reduces flash zone pressure and thus helps to increase the lift of light components from the bottom product. Stripping steam for a fractionation tower is controlled based on the lift while stripping steam for a stripper is controlled based on stripper product specification. Be aware of too much stripping steam as it could lead to high energy cost and also cause vapor loading limitations in the overhead system.

14.2.1.6 Pump-Around   Many fractionation towers have pump-arounds to remove excess heat in the key sections of the tower. The effect of increasing pump-around rate is reduced internal reflux rate in the trays above the pump-around, but increased internal reflux rate below the pump-around. Thus, change in pump-around duty affects fractionation. On the other hand, pump-around rates and return temperature have effects on heat recovery via the heat exchanger network. It is not straightforward in optimizing pump-around duties and temperatures since the effects on both fractionation and heat recovery can only be assessed in a simulation model. An APC application incorporated with process simulation should be able to handle this optimization.

14.2.1.7 Overhead Temperature   In hot weather, the tower overhead fin fan condenser could be limited and thus the tower top temperature can go up. As a result, valuable components could be vaporized into overhead vapor leading to yield loss. There are a number of ways to reduce the overhead temperature such as increasing cooling water rate, turning on spare overhead fan for air cooler, and increasing reflux rate. On the other hand, when overhead temperature is too cold, sulfur or salt condensation in the condenser could occur and cause corrosion.

14.2.2 What Effects to Know: Parameter Relationship

How do operating parameters relate to each other? Which parameters are more sensitive to fractionation and energy use? What is the impact of changing one parameter? Understanding these could provide insights and guidelines for operational improvements. The objective of developing key indicators is to understand the strong interactions between process throughput, yields, and energy use so that the trade-off among them can be optimized with the objective of maximizing operating margins. In the traditional view, energy use is regarded as a supporting role. Any amount of energy use requested from processes is supposed to be satisfied without question and challenging. This philosophy loses sight of synergetic opportunities available for optimizing energy use for more throughput and better yields.

In developing correlations, one needs to connect energy with product yields and quality. One such example is discussed in detail in Chapter 4. The correlations can be applied for operation optimization. For automatic control, the correlations can be implemented into an APC system, which determines the set points for independent operating parameters. For manual control, operating targets for independent parameters can be obtained based on the correlations.

A process simulation could be a very good vehicle in developing correlations of primary parameters. To do this, a simulation model for the tower can be developed based on the feed conditions (rate and compositions) and tower conditions (temperature, pressure, theoretical trays). With product specifications established as set points in simulation, operating parameters such as reflux rate and reboiling duty can be adjusted to meet product specifications. The simulation model is verified and revised against performance test data based on clean conditions. Different operating cases can be generated and simulation results can be transferred to a spreadsheet with relationship between dependent and independent variables. Then the regression method is applied to derive the correlations.

When dealing with correlations involving multiple variables, an economic sensitivity analysis is essential to determine the most influential parameters on process economics. For example, in a debutanizer, reflux rate and reflux drum pressure are very sensitive to product quality and reboiling duty more than any other operating parameters. Getting the most sensitive parameters right in operation can get the greatest bang.

14.2.3 What to Change: Parameter Optimization

A tower is built to make separation of products. Therefore, tower optimization is to maximize operating margins and minimize energy usage. This processing goal can be described mathematically in an objective function with the parameters in the objective function connected to other processing parameters. All these parameters are defined as constraints in two forms: inequality equations (larger and smaller than), which are used for describing operating minimum and maximum operating limits; and equality equations representing relationship between these papameters. Therefore, these constraints form a feasible operating region in which the objective function is constrained during optimization. The objective function plus the set of constraints form an optimization model. The results of solving this model yield the set points for operating parameters, which can be adjusted in operation to achieve the maximal operating margin defined in the objective function.

A generic form of a process optimization model is provided as

$$\text{Objective function}: \text{Maximize } Z = \sum_i c_i P_i - \sum_j c_j P_j - \sum_k c_k Q_k$$

$$\text{Subject to}: f_n(X_c, X_m, P, F, Q) = 0; \quad n = 1, 2, \ldots, N_1$$

$$X_{c,min} \le f_l(X_c) \le X_{c,max}; \quad l = 1, 2, \ldots, N_2$$

$$X_{m,min} \le f_p(X_m) \le X_{m,max}; \quad p = 1, 2, \ldots, N_3,$$

where c_i are the unit prices of products, c_j are the unit prices of feeds, while c_k are the unit costs of energy including steam, fuel, and power. F's and P's are the mass flows of feed and products while Q's are the amount of energy. Essentially, the objective function Z represents the upgraded value from feed to products at the expense of energy. X's are the operating parameters, while X_c's are the control variables and X_m's are the manipulated variables for the tower. $f(X_c, X_m, P, F, Q)$ are the relationship constraints.

Maximizing the objective function Z under these constraints with proper limits of the operating parameters will change the related parameters to the values under which economical value Z will be maximal. In this case, the operating parameters achieve optimal values, which can be used either as set points for the closed-loop or open-loop control.

In building this optimization model, the most important thing is to include all the major operating parameters that affect operating margins. Then, correlations are developed to describe the relationship among these major parameters and between these major parameters and other operating parameters. When defining operating limits, it is very important to distinguish soft and hard constraints. Hard constraints refer to mechanical performance limits, for example, the tower tray flood limit, or the compressor flow rate limit, or the furnace heat flux limit, and so on. While making sure that hard constraint must be satisfied, relaxing soft constraints could play a significant role in improving operating margin.

14.2.4 Relax Soft Constraints to Improve Margin

Finding ways to relax plant limitations is one of the most important tasks in improving operating margins. Mathematically, relaxing constraints will lead to a large operating region and push the objective function toward the edge of the enlarged region. But the bottom line is: What can be done in reality for relaxing plant limitations? Equipment rating analysis could be a very effective way to identify equipment spare capacity and limitations. Utilization of spare capacity can allow process capacity expansion up to 15–20% and accommodate improvement projects without or with little capital cost. The important part of a feasibility study is to find ways to overcome soft constraints, at some times, hard constraints if they can be overcome using cheap options. Details for equipment rating analysis are discussed in Chapter 22.

Three general limitations for equipment are pressure, temperature, and metallurgy as each equipment is designed with the limits for these parameters. If it is identified that equipment will operate at a higher pressure than the design limit, one option is to replace the equipment, which comes with a cost and it is usually expensive. This is similar to the case when operating temperature will exceed the design limit. Another option is to overcome these constraints if process conditions could be changed. However, if metallurgy is found to be less than required, this could be a major hard constraint and it needs to be flagged out for metallurgist's attention as early as possible. In some cases, the equipment could still be usable if the metallurgist agrees, and the plant takes actions for routing inspection.

For fired heaters, the limitations could be heat flux or tube wall temperature. The former is applied to heaters with pressure drop larger than 20 psi, which is the most common. The latter is for low-pressure drop heaters. When the heater duty must be increased to handle duty much larger than the design, the existing heater may be insufficient in meeting the limits. Installing a new heater could be very expensive. The most effective way to avoid this is to increase feed preheating via process heat recovery, adding a feed steam preheater or adding tube surface area into the heater.

For separation/fractionation columns, the major limitation is the tray loadings. Typically, the column is design with 80–85% tray flooding. In operation, the column can run harder to reach 90% or even 95% of the flood limit. It is possible to push throughput to reach the flooding limit if product specifications can be relaxed. If higher throughput is desired in revamp, higher capacity trays could be used to replace the old ones partially or completely. Although it is costly, it is cheaper than installing a new column.

For the compressor, when it reaches the capacity limit in the flow or head, the direct solution is reducing the gas flow. The alternative solutions include increasing speed by gear change or adding wheels to the rotor or adding a booster compressor. Similarly for pumps, the major constraint is insufficient capacity in the flow or head. The low-cost solution is replacing the impeller with a larger size. The alternative solution is to add a booster pump.

For heat exchangers, the constraint is insufficient surface area. In resolving this constraint, adding surface area via more tube counts or tube enhancement to the existing exchanger could help; but area addition could only be up to typically

20%. Beyond this, adding a new shell in parallel may be required, which could reduce ΔP.

The above constraints occur in the ISBL (internal system battery limit). However, it is important to identify outside system battery limits (OSBL) as well, such as offsite utility, layout, piping, substations, and so on. For example, a revamp project requires additional high-pressure (HP) steam for process use, but the existing steam system may reach the boiler capacity limit of HP steam generation. Installing a new boiler could kill the economics of a revamp project. Another revamp case could require installing a new exchanger, which is well justified from ISBL conditions; however, the piping to connect two process streams in the exchanger is too long in distance, which becomes cost prohibitive. A recent revamp project determined the great benefit of installing a new motor of 5 MW as a replacement of the steam turbine to run the recycle gas compressor in a reforming unit. The reasons for the moterization is that the turbine has a low power generation efficiency and the electricity is very cheap. However, the project was deemed infeasible because the electricity requirement of the new motor is beyond the capacity of the nearby substation. Installing a new substation could cost several millions of dollars. Numerous ISBL and OSBL limitations could occur, which are not listed here in detail.

14.3   ENERGY OPTIMIZATION FOR DISTILLATION SYSTEM

Operation of distillation columns involves a trade-off between energy use and product recovery. The optimal trade-off determines targets for operating parameters. The operating parameters include reflux ratio, feed temperature, column temperature, reboiling duty, column pressure, and so on. The question is: How to achieve the most economic operation of the column?

Let us look at an example that consists of a debutanizer column (Figure 14.1) for which White (2012) gave excellent explanations. This example gives perspective and guidelines in principles for optimizing a separation column and this example is reproduced here with permission from AIChE.

The feed and product specifications and prices for this example are listed in Table 14.1. Both products have tiered prices: on-specification product is priced much higher than the one that does not achieve specification. If the top product, butane, achieves the specification, that is, less than 3% C_5, it is sent to the downstream unit for further processing leading to eventual sales. Off-spec butane will be used as fuel, which has a low value than selling as a product. Similarly, the bottom product, pentane, if achieving specification, is used for making high value product. Otherwise, it will be sent to tank for reprocessing. In operation, the operator changes column temperature and reflux to feed ratio to achieve the product specification.

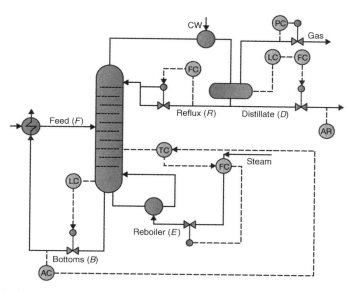

FIGURE 14.1. Debutanizer example: energy optimization based on reflux ratio. (From White (2012), reprinted with permission by AIChE.)

TABLE 14.1. Product Specifications and Prices

Stream	Composition/Specification	Value
Feed, 20,000 bbl/d	25% C_3	$60/bbl
	25% nC_4	
	25% nC_5	
	25% nC_6	
Bottoms product = C_5	$\leq$5% C_4	$80/bbl
	>5% C_4	$60/bbl
Top product = C_4	$\leq$3% C_5	$60/bbl
	>3% C_5	$40/bbl
Steam		$15/MBtu

Source: From White (2012), reprinted with permission by AIChE.

14.3.1 Develop Economic Value Function

The objective function (economic value) representing the economic operating margin for the column is shown in Figure 14.2, which is defined as the difference between the value of products (top butane product and pentane bottom pentane product) and costs of feed and energy. The value function features two discontinuities. The first, which occurs when the composition of the bottom product is about 1% butane, corresponds to a change in the top product from off-spec to on-spec. The second discontinuity occurs when the bottom product becomes off-spec at 5% butane.

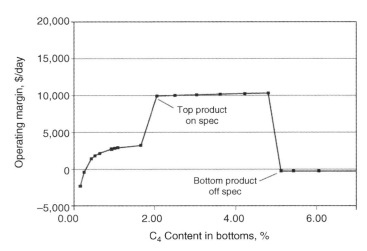

FIGURE 14.2. Operating margin as a function of the bottom composition. (From White (2012), reprinted with permission by AIChE.)

14.3.2 Setting Operating Targets with Column Bottom Temperature

To choose the bottom temperature target, first assume that the reflux rate is fixed, and that the bottom product is on-spec, but the top product off-spec because of its high pentane content. This would correspond to a very hot bottom temperature. When the bottom temperature is slowly reduced, the amount of bottom product increases, but the percentage of butane in the bottom also increases simultaneously. As the amount of pentane product increases, the total product value improves. The middle line in Figure 14.2 represents this operation.

Normally, one would select a temperature target such that the bottom composition is as close to the specification limit as possible. There will always be some variability in the control performance due to external disturbances and limitations on loop control action. If composition control is poor and has a high variance, the observed composition probability distribution could look like a normal distribution in relation to operating margin as shown in Figure 14.3. More detailed explanation for using normal distribution to represent operating data can be found in Chapter 4.

The product composition target is the mean value of observed composition distribution as shown in Figure 14.3. The mean value of the operating margin in Figure 14.3 is calculated based on the weighted average composition of the observed distribution—that is, the percentage at each composition is multiplied by the margin value at that composition to determine the mean value.

Figure 14.3 shows that part of the column operation is the bottom product being off-spec. The mean product value does not correspond to the value at the mean of product composition, which is also the operating target. This is because of the nonsymmetrical nature of the operating margin and low value of off-spec products.

After reducing the variability through improved control valve performance and reduced measurement error, the new mean value of the operating margin increases at

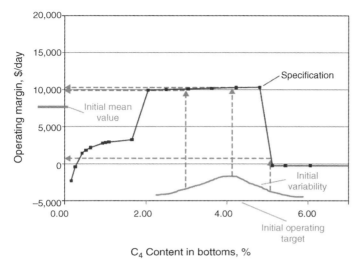

FIGURE 14.3. Observed composition normal distribution versus operating margin. (From White (2012), reprinted with permission by AIChE.)

the same operating target or bottom composition target (Figure 14.4). It can be seen that reduced variability results in increase in the mean value of operating margins.

14.3.3 Setting Operating Targets with Column Reflux Ratio

The earlier discussions involved constant reflux ratio. Next consider the situation where the reflux rate is varied and bottom temperature is constant. Fundamentally,

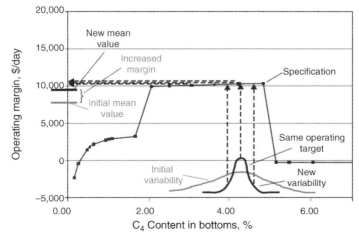

FIGURE 14.4. Improved composition normal distribution versus operating margin. (From White (2012), reprinted with permission by AIChE.)

reflux rate provides internal reflux needed for separation in the tower and it is generated by either feed heater or reboiler. Thus, lower reflux rate saves energy, but too low reflux rate could affect product quality. On the other hand, high reflux rate could improve production of more valued product.

In tower design, the reflux ratio is determined based on the trade-off between operating cost in the reboiler and capital cost for the tower. In other words, use of more separation stages requires less reflux rate and thus less reboiling energy but at the expense of additional capital cost.

However in operation, optimal reflux ratio is determined based on the trade-off product value and energy cost. When the reflux ratio increases, the separation improves at the expense of increased reboiling duty (Figure 14.5). As a result, top product rate decreases while the bottom product rate increases. As shown in Figure 14.5, the cost of reboiling duty presents a linear relationship with reflux ratio but the product rate is nonlinear and presents a different trend as reboiling duty.

Figure 14.6 shows the operating margin for different energy prices, assuming constant product prices. The optimum reflux rate depends on the price of energy. At a high energy price, the optimum reflux rate is at the minimum value, which allows the column to maintain the top product in specification. At the lower energy prices, the optimum reflux rate increases.

The conclusion is that operating targets should be a function of energy costs rather than a fixed number even with fixed composition limits. It is common to observe separation columns operating at reflux rate that are 50% higher than the optimum. For the debutanizer column operation discussed here, such an operation could cost operating margins in excess of $500,000 per year.

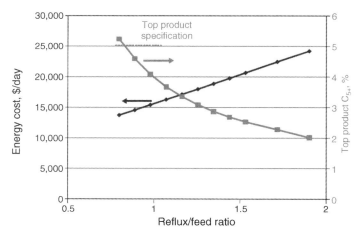

FIGURE 14.5. Energy separation trade-off: Energy cost increases linearly as reflux rate, while the top product quality improves. (From White (2012), reprinted with permission by AIChE.)

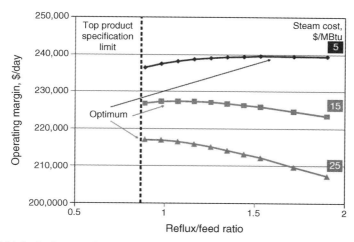

FIGURE 14.6. Optimum reflux rate depends on energy price. (From White (2012), reprinted with permission by AIChE.)

14.3.4 Setting Operating Pressure

It is generally known that reducing the operating pressure of separation columns reduces energy consumption. This is because the lower the tower pressure, the less heat required for liquid to vaporize (thus less energy required) and the easier for vapor to penetrate into liquid on the tray deck (thus better separation). Yet many columns are operated well above their potential minimum pressure. One may ask: If benefit of reducing pressure is well known, why is it not widely implemented? There appears to be three primary reasons for this.

First, changing column pressure requires simultaneously changing the bottom temperature set point to hold the product composition at their targets. This is difficult to do manually—advanced composition control is required.

Second, changes in column pressure have other impacts such as changes in the off-gas rate, the amount of reboiler duty, and hydraulic profile of the plant. In the case of partial condensation, pressure control can interact with the overhead receiver level. While these effects are real, their magnitude is sometimes exaggerated and cited as reasons for not making any changes.

Finally, plant personnel frequently do not agree on the amount of operating margins required to handle major disturbances. For instance, questions often arise about the dynamic response of an air-cooled condenser to a rainstorm and the ability of the overall control system to handle such conditions. A well-designed overall control system for the column can compensate for such disturbances.

The condenser pressure controls the tower pressure and thus the feed tray pressure. There is a pressure valve in the overhead, which can be used to control tower pressure. The lower limit of the tower pressure is set by column overhead condensing duty, net gas compressor capacity, and column flood condition. Many of the new APC systems

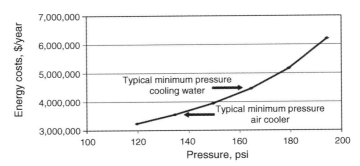

FIGURE 14.7. Pressure has significant effect on energy cost. (From White (2012), reprinted with permission by AIChE.)

are using pressure control to save energy. During extended turndown periods, reducing pressure up against an equipment limit can improve efficiency (Figure 14.7).

14.4 OVERALL PROCESS OPTIMIZATION

This example comes from Loe and Pults (2001) and is reproduced, with permission from AIChE. This example explains the operational improvements for a single deisopentanizer fractionation tower. To do so, the current operation is simulated and assessed. Then improvement opportunities are identified and the limiting factors are determined. Optimal solutions are obtained by optimizing tower ISBL conditions and OSBL conditions. The improvements on this single tower have generated over $500,000 as compared with historical operation over the first six months.

14.4.1 Basis

The deisopentanizer (DIP) tower, shown in Figure 14.8, processes light straight-run naphtha from two sources. The debutanizer bottoms consists of primarily iso and normal pentane (iC_5 and nC_5), with small fractions of butane and C_6+ components. The overhead from the naphtha fractionator tower contains mostly C_5 and C_6 paraffin compounds, with some benzene and C_6 naphthenes and a small amount of butane. The combined feed to the DIP is typically in the range of 9,000–15,000 bpd.

The DIP overhead product is normally rich in isopentane, and is routed to gasoline blending along with other high octane, low Reid vapor pressure (RVP) gasoline components. The DIP bottom, which is rich in nC_5 and C_6 paraffins, is routed to the light naphtha isomerization unit, along with light raffinate from the aromatics extraction unit.

The DIP tower has 50 trays in comparison with 70+ trays used in a typical deisopentanizer. As a consequence, the DIP tower often has a difficult time making a good split between isopentane and normal pentane components.

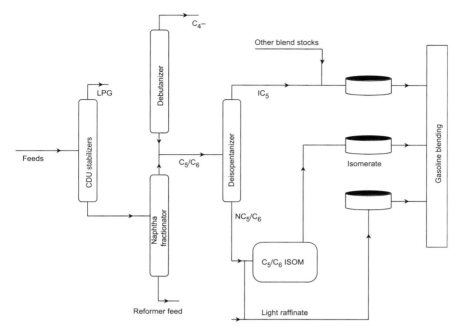

FIGURE 14.8. Deisopentanizer flow scheme. (From Loe and Pults (2001), reprinted with permission by AIChE.)

14.4.2 Current Operation Assessment

Historically, the DIP tower had been operated with a target of 10% nC_5 in the overhead, and 20–30% iC_5 in the bottom product. The tower was reported to be limited by reboiler or condenser duty. One of the two steam reboilers had been out of service for some time, and 20–30% of the condenser fin fan motors were not operating and in need of repair. The DIP equipment had not been a maintenance priority, in part because no economic penalty had been calculated for having a reboiler or condenser out of service. There was also a concern that the tower could flood if both reboilers were placed in service.

The DIP process control was accomplished with a Distributed Control Stystem (DCS) system equipped with an APC algorithm. The controller was set to target 10% nC_5 in the overhead and 10% iC_5 in the product, and would increase reboiler steam and reflux rate until reaching the maximum limits for these flows. The tower pressure was also controlled within a specified range by the APC, and this could indirectly limit the reboiler duty as well, if the tower pressure increased beyond its maximum limit. Inferential estimates for product iC_5 and nC_5 qualities were calculated based on tower temperatures and pressures, and a bias for these values was continually updated based on daily laboratory (lab) data.

To establish the historic performance, the operating and laboratory data were collected from the past year, eliminating periods of known equipment failure or

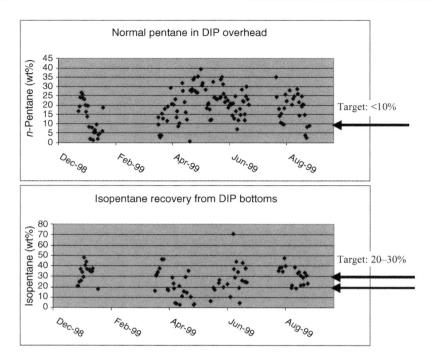

FIGURE 14.9. Variation of DIP performance. (From Loe and Pults (2001), reprinted with permission by AIChE.)

poor unit volume balance. The data showed an average of 19% nC_5 in the DIP overhead, and 27% iC_5 in the bottoms product, with a wide variation in the product quality, as shown in Figure 14.9. An average of 18% of normal butane was observed in the overhead product, indicating poor debutanization in the upstream fractionation towers.

The process design for the tower showed an available reboiler duty of 51,000 lb/h of steam, but the average for the data collected showed only 27,000 lb/h. The data show an erratic variation in reboiler duty.

14.4.3 Simulation

Using the averaged process and lab data, a simulation model for the DIP was developed. The feed rate and compositions to the tower were fixed, as well as the reflux rate, tower pressures, and overhead rate. The model results for reboiler duty, tower temperatures, and compositions compared favorably with the unit operating data, as shown in Table 14.2.

The calibrated model was used to evaluate DIP performance at the design reboiler duty, to determine if available condenser duty and tower tray capacity would be adequate for this operation. As shown in Table 14.3, the predicted condenser duty for

TABLE 14.2. Simulation Results Versus DIP Operating Data

Result	Data	Model
Reboiler duty (MMBtu/h)	23.6	26.7
Top temperature (°F)	143	153
Bottom temperature (°F)	179	183
nC_5 in overhead (wt%)	18%	16%
iC_5 in bottoms (wt%)	31%	30%

Source: From Loe and Pults (2001), reprinted with permission by AIChE.

TABLE 14.3. Simulation Results at Designed Reboiler Duty

Result	Design	Model
Reboiler duty (MMBtu/h)	48.8	48.8
Condenser duty (MMBtu/h)	45.0	48.1
Max tray flood	85%	74%
Max downcomer backup	50%	37%
nC_5 in overhead (wt%)	NA	5%
iC_5 in bottoms (wt%)	NA	18%

Source: From Loe and Pults (2001), reprinted with permission by AIChE.

this operation was only slightly above the design value, and tower tray parameters indicated that flooding was unlikely. Also, the separation of iC_5 and nC_5 improved dramatically versus the historical operation as would be expected with the increased tower traffic.

14.4.4 Define the Objective Function

The profitability of the DIP column was determined based on the value of separating iC_5 for direct blending to gasoline and nC_5 to be used as feed for the C_5/C_6 isomerization unit, less the utility and downstream isomerization unit opportunity costs incurred to do so. Lighter feed components, such as n-butane, were assumed to always be fractionated into the DIP overhead, and components heavier than nC_5 were assumed to always be found in the DIP bottoms stream. Thus, only the disposition of iC_5 and nC_5 components were considered in the profitability calculation.

Therefore, the objective function for optimizing the DIP tower is defined as

DIP Upgrade Value = Overhead Value + Bottoms Value − Feed Value

− Reboiler Steam Cost − Isom Operating Cost − Isom Capacity Penalty.

The DIP feed and overhead values are calculated as the gasoline blending value of the iC_5 and nC_5 in this stream, with corrections for road octane and RVP of these components versus those of conventional regular gasoline. The DIP bottoms stream is normally processed at the isomerization unit, where 75% of the exiting C_5's are assumed to be iC_5. After this equilibrium conversion, the value of the resulting iC_5/nC_5 stream is calculated at gasoline blending value as described for feed and overhead above. The reboiler steam cost is calculated assuming a 70% generation efficiency from refinery fuel gas, and the isomerization unit operating cost (for fuel, power, and catalyst) was taken to be the same value per barrel as used in the refinery planning model.

During some periods, the isomerization unit has more feed available than can be processed. If additional DIP bottom is produced, less capacity is available to process light raffinate from the aromatics extraction unit. The Isom capacity penalty, or the opportunity cost for processing additional DIP tower bottoms at this unit was therefore estimated by evaluating the octane upgrade of light raffinate.

14.4.5   Offline Optimization Results

Once the economic evaluation criterion was determined, it was implemented in the simulation model. Numerous case studies were conducted via process simulation to determine the optimum operating point for the DIP tower under different scenarios. It quickly became apparent that in nearly all economic and operating situations, maximizing the DIP reboiler duty up to the maximum limit gave the highest profitability.

For subsequent case studies, the simulation was completed with maximum reboiler duty, and tower pressure and nC_5 content of the overhead product were also fixed. These constraints completely specified the tower operating conditions.

The DIP profitability was first examined for scenarios where the isomerization unit has available capacity. The DIP feed rate and overhead nC_5 content were varied, and profitability calculated, as shown in Figure 14.10. This analysis showed that the optimum target was around 5% of nC_5 in the overhead, regardless of the tower feed rate.

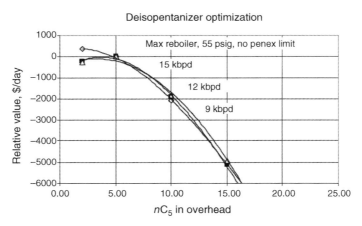

FIGURE 14.10. Optimization without Isom capacity constraint. (From Loe and Pults (2001), reprinted with permission by AIChE.)

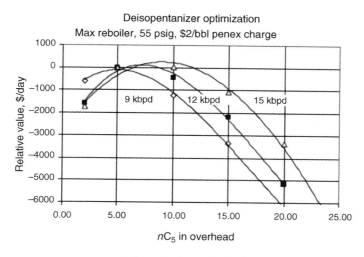

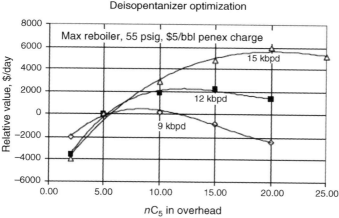

FIGURE 14.11. Optimization with Isom capacity constraint. (From Loe and Pults (2001), reprinted with permission by AIChE.)

Profitability was then examined assuming the isomerization unit was at its maximum charge rate, and additional production of DIP bottoms would result in bypassing of light raffinate around the Isom, direct to gasoline blending. The cost of losing the light raffinate octane upgrade can vary between $2 and $5 per barrel, and so simulation cases were completed for both of these scenarios as shown in Figure 14.11. In these cases, the optimum nC_5 in the DIP overhead target is dependent on the charge rate to the tower. At low charge rates, the available reboiler duty is sufficient to obtain good separation between iC_5 and nC_5 components, so that minimal iC_5 is lost into the DIP bottoms when targeting 5% nC_5 in the overhead. At higher charge rates, more iC_5 is lost to the bottoms stream, and it is more profitable to increase the overhead nC_5 target, reducing the DIP bottoms rate to the Isom unit and allowing additional raffinate upgrading. Thus, the optimal nC_5 in overhead target varies between 5 and 20%, depending on DIP charge rate and the value of light raffinate upgrading.

Based on a comparison of the optimal tower operation as determined above and the historical performance, an incentive of around $1.5 million per year was identified to improve DIP fractionation.

14.4.6 Optimization Implementation

In order to realize the benefit indicated by the optimization results above, several unit hardware, process control, and operating philosophy changes were needed. First, it was clear that both reboilers would be required, so the spare bundle was leak-tested and returned to service by operations. Several fin fan motors were also quickly repaired to ensure that design condenser duty was available and overpressurization of the tower would not limit the reboiler duty that could be applied.

Second, the optimum target for the DIP overhead nC_5 was implemented into the APC system. The APC controller on the DIP DCS system was reconfigured to operate at this nC_5 target, while maximizing the reboiler duty as limited by the high limit on tower pressure. This APC system allowed the DIP operation to be maintained at an economic optimum, accounting for the isomerization unit capacity and the economics of the day.

Third, communication of the new operating philosophy was also critical in improving DIP performance. Operators and unit supervisors were trained on the importance of maximizing the reboiler duty and setting the nC_5 in the overhead target. The iC_5 content of the bottoms stream was still measured by lab and inferred analysis, but this was no longer a tower control variable. The reboiler duty and overhead nC_5 were tracked on a daily basis, and performance for these key performance indicators were discussed at weekly operations and planning meetings.

14.4.7 Online Optimization Results

The economic benefit of improving DIP fractionation was tracked on a monthly average basis shortly after implementation of the new optimization strategy. Economic performance versus the baseline operation is shown in Figure 14.12, using actual monthly averaged economics and unit operating and lab data. Monthly benefits of over $100,000 were achieved in several cases during the summer months, when octane values were at their highest level. As octane values dipped during spring and fall months, benefits from the improved DIP fractionation dropped off as well.

A significant drop in benefits can be seen for July. This was due to poor operation of the DIP tower, caused by a high butane content in the tower feed from the crude unit stabilizers. The C_4's caused the DIP tower pressure to increase up to its safe operating limit, and the reboiler duty was cut back to avoid overpressuring the tower. This resulted in a reduction in fractionation efficiency and profitability during part of July, and represented one of the challenges encountered in sustaining the improved DIP performance.

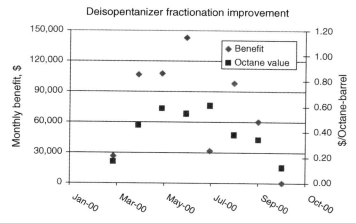

FIGURE 14.12. DIP economic improvements. (From Loe and Pults (2001), reprinted with permission by AIChE.)

14.4.8 Sustaining Benefits

Without ongoing attention, optimization improvements from initiatives such as that on the Deisopentanizer tower tend to fade over time, for a multitude of reasons. Challenges to the new level of performance must be tackled as they arise, whether they result from hardware or control problems, misunderstandings, or operating changes. Tracking the economic benefits of the initiative is a critical element of sustaining the change, since knowing the lost profits associated with a loss in performance helps to set work priorities within the refinery. Several problems occurred during the first six months of improved DIP operation, which reduced the profit derived from this tower, and these issues were quickly addressed to sustain the improvement.

During the summer months, a change in routing of a portion of the refinery condensate resulted in a hydraulic constraint on the amount of condensate that could be removed from the DIP reboilers into the condensate header. This caused reboiler duty reduction leading to fractionation efficiency drop. The economic calculations clearly showed that the benefit derived from the additional reboiler duty was much higher than the value of recovering the condensate. For this reason, condensate was safely spilled into the sewer until normal condensate header operation was restored. After the reboiler duty was back to maximal the DIP operating margin went up again.

A second challenge occurred when it was noticed that the DIP reflux drum temperature had increased above 140 °F, which was higher than the recommended run-down limit to tankage. Initially, reboiler duty was decreased to ensure safe operation. However, in meeting with the tank farm operators about the problem, it was found that the DIP overhead mixed with several other much larger streams before entering a gasoline blending tank. Calculations showed that the effect of the higher DIP run-down temperature on the tank temperature was minimal, and DIP reboiler duty was again increased.

Another problem in maintaining reboiler duty was identified when the DIP tower pressure increased due to butanes in the feed, as mentioned above. It was found, however, that the maximum tower pressure was set well below the vessel design pressure, and so the safe operating limit of the tower could be increased after an appropriate management of change review. Simulation modeling showed that operation at the higher pressure did not significantly impact the tower profitability if the reboiler duty could be maintained near the maximum. The pressure limit was increased and performance of the tower again improved.

A fourth reduction in DIP profitability was noticed when the APC controller began cutting reboiler duty for no apparent reason. Further investigation showed that the inferred nC_5 content of the overhead stream had been deviating from the lab value for a few days, because of a computer glitch. This was quickly corrected and DIP operation returned to normal.

Although all of these obstacles reduced profitability for a short period, timely identification and resolution of the constraints averted potentially long periods of underperformance.

14.5 CONCLUDING REMARKS

The profit improvement process consists of several phases. First, current performance must be assessed, evaluating upstream and downstream constraints and unit equipment limitations. An understanding of how the unit operation can be optimized can then be gained by use of an appropriate process simulation, and a new operating strategy is then developed to improve profitability. Implementation of this strategy requires good communication of its benefits throughout the plant. Upgrading or repair of process equipment may be needed to allow operation under desired conditions. Tracking of key operating parameters as well as economic performance of the unit and distributing these results within the plant is essential to sustaining the improvement, as this process flags deterioration of profitability. The profit improvement process requires input from technical, operations, safety, environmental, and economics groups within the plant.

REFERENCES

Loe B, Pults J (2001) Implementing and sustaining process optimization improvements on a deisopentanizer tower, *AIChE Spring Meeting*, April, Houston.

Underwood AJV (1948) Fractional distillation of multi-component mixtures, *Chemical Engineering Progress*, **44**, 603.

White DC (2012) Optimizing energy use in distillation, *Chemical Engineering Process (CEP), AIChE*, March, 35–41.

PART 4

UTILITY SYSTEM ASSESSMENT AND OPTIMIZATION

15

MODELING OF STEAM AND POWER SYSTEM

Modeling of key equipment in the steam and power system is essential for building steam and power balances, which can represent process steam and power demands and respond to variation in production.

15.1 INTRODUCTION

Equipment in a steam and power system usually contains boilers, steam turbines, letdown valves, steam flash recovery, desuperheaters, temperature control, steam traps, a condensate collection system, deaerator, and so on. Steam generated from boilers reaches end users via a piping distribution system, which includes steam mains or steam headers (large pipes) and take-off pipes connecting to the individual users. Steam reduces pressure via letdown valve or turbine for users rated at lower pressure. Steam traps are placed in the back end of the steam heating equipment to keep steam inside of the equipment until it condenses so that latent heat of steam is effectively transferred to the process. Steam traps also are installed in the lower points of steam pipes to collect condensate to avoid water slugs to form water hammers. High-pressure condensate goes to flash tanks to recover steam, while low-pressure condensate goes to central condensate tanks to be pumped to a deaerator where it mixes with makeup water. A water conditioning system may be used to provide various levels/types of water treatment depending on the source of makeup water. Low-pressure steam is injected to the deaerator to remove noncondensable

Energy and Process Optimization for the Process Industries, First Edition. Frank (Xin X.) Zhu.
© 2014 by the American Institute of Chemical Engineers, Inc. Published 2014 by John Wiley & Sons, Inc.

gases, in particular oxygen, because dissolved oxygen could cause serious corrosions to steam pipes and process equipment. Most steam heaters are designed for saturated conditions while steam is generated in superheated conditions. Thus, desuperheaters are used to convert superheated steam to saturated conditions by mixing water with steam. For applications where process temperature requires controlling, a temperature controller is used to control the steam flow rate.

Although general and detailed discussions for equipment of steam and power system are widely available in public sources, this chapter focuses on modeling of key equipment to fulfill the needs of the modeling steam and power system.

15.2 BOILER

There are many types of boilers, which mainly depend on steam pressure, the amount of steam generation, and fuel type. The heat transfer pattern is what gives each boiler type its name. The two main patterns are "fire-tube" and "water-tube." If the combustion gases are on the inside of the tubes and the water on the outside, the boiler is called "fire-tube" boiler. On the other hand, if the water is inside of the tubes and the flame is on the outside, the boiler is called a "water-tube" boiler. Both types are older technology, which have been used since the beginning of the industrial age, and each continues to dominate its own size and pressure range. Other types of boilers are much less significant and in narrower ranges of applications.

The pressure of the HP (high-pressure) steam is limited around 42 barg (610 psig) as the maximum due to the metallurgy limit of carbon steel. Alloy must be considered beyond this pressure, which can be very expensive in comparison with carbon steel. The pressure of VHP (very high pressure) steam can be as high as 1500 psig and the advantage of VHP steam is higher power generation efficiency than HP steam. Thus, it is an economic decision for the selection of VHP steam in a process plant. It is very common to observe that VHP steam is used in steam turbines for electricity generation, which are located in the boiler house. In this way, the run length of the VHP steam header is minimized.

15.2.1 Fire-Tube Boiler

The fire-tube boiler (Figure 15.1) features a large cylindrical shell, which imposes mechanical limitations on steam pressure and quantity. The economic limit on steam pressure is around 20 bar (290 psia). The fire-tube boiler is the most economical and dominant in small and medium steam production and at lower pressures.

15.2.2 Water-Tube Boiler

The water-tube boiler (Figure 15.2) features high steam pressure and output. Water-tube boilers come in many configurations. The most common configuration is where a large number of individual tubes are located in the convection section, which connects with the steam drum at the upper of the tubes and the water drums in the

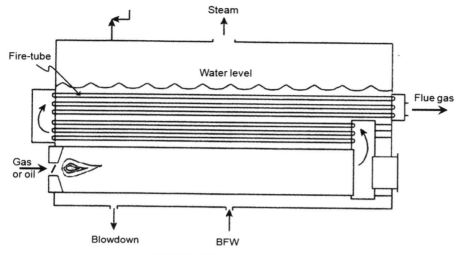

FIGURE 15.1. Fire-tube boiler.

bottom. The dirt and the residue of water treatment settle in the water drum, which is removed by blowdown on a regular basis. For this type of boilers, steam pressure could be at 1500 psig, which is commonly used in a power plant for making electricity or at around 610 psig for process use.

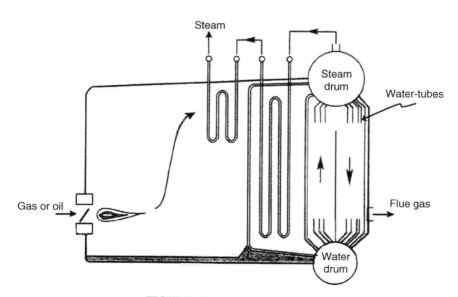

FIGURE 15.2. Water-tube boiler.

15.2.3 Packaged Boiler

The packaged boiler is so called because it comes as a complete package with burner, level controls, feed pump, and all necessary boiler fittings and mountings. Once delivered to site it requires only the steam, water, blowdown pipe work, fuel supply, and electrical connections to be made for it to become operational. The main feature of the packaged boiler is small in size and readiness for use due to its modular design. Consequently, this feature allows quick installation and will reduce heat loss and plot space. It is very important to understand the steam requirements when selecting a package boiler. There are many variables that could affect the successful implementation of a package boiler.

15.2.4 Blowdown

A boiler requires blowdown to remove concentrated dissolved solids and control the water quality. The lack of blowdown could result in a higher pH of boiler feed water (BFW) in the boiler, which could potentially lead to corrosion. Insufficient blowdown could also cause impurities to carryover to steam. On the other hand, excessive blowdown wastes energy, water, and chemicals. The optimum blowdown rate is determined by various factors including the boiler type and capacity, operating pressure, water treatment, and makeup water quality. Blowdown rate is 2–4% for relatively large boilers and 4–8% for small boilers. It can be up to 10% if makeup water contains high concentrations of solids. Industrial standards for blowdown are available and can be referenced that indicate the amount of blowdown depending on the type and pressure of the boiler.

15.2.5 Boiler Efficiency

Three operating parameters affect boiler efficiency, namely, excess air, temperatures of air, and BFW temperature. Excess air is controlled to fulfill complete combustion. Too much excess air costs extra fuel to bring cold air to combustion temperature. The hotter the air and BFW enter the boiler, the less fuel is required. That is why air and BFW preheat is commonly adopted.

15.2.6 Boiler Modeling

Both types of boilers can be described by Figure 15.3 and equations (15.1) and (15.2).

$$Q_{Steam} = M_{Steam} \times (h_{Steam} - h_{BFW}) \tag{15.1}$$

$$\eta_{Boiler} = \frac{Q_{Steam}}{Q_{Fuel}}, \tag{15.2}$$

where Q is the heat duty in Btu/h; M_{Steam} is the net steam flow in lb/h; h is the enthalpy in Btu/lb; and η is the boiler efficiency. Some of the steam generated from a boiler is used internally for spinning steam turbines to run water circulation pumps and air fans.

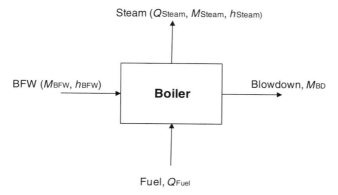

FIGURE 15.3. Modeling of boiler.

15.3 DEAERATOR

A deaerator (Figure 15.4) is a device that is widely used for the stripping of oxygen and other dissolved gases from boiler feed water. In particular, dissolved oxygen in boiler feed water can cause serious corrosion damage to steam system by attaching to the walls of metal piping and other metallic equipment. BFW must reach the saturation temperature of the steam being used to deaerate the water. If the water is cold, it is an indication that water is not completely treated.

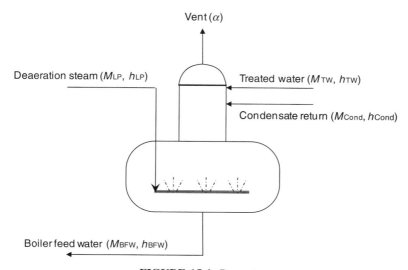

FIGURE 15.4. Deaerator.

The heat and mass balances around the deaerator can be described as

$$M_{LP} \times (h_{LP} - h_{TW}) \times (1 - \alpha) = M_{BFW} \times (h_{BFW} - h_{TW}) - M_{Cond} \times (h_{Cond} - h_{TW})$$
$$(15.3)$$

$$M_{BFW} = M_{LP}(1 - \alpha) + M_{Cond} + M_{TW}, \qquad (15.4)$$

where M is mass flow in klb/h and h is enthalpy in Btu/lb. α is the fraction of deaeration steam vented, typically 3%.

15.4 STEAM TURBINE

It is common that steam turbines are used in the process industry to drive pumps and compressors as well as generate electricity. In many cases, a condensing turbine generates electricity, which uses VHP steam (1500 psig) for higher power generation efficiency, while back pressure turbines are used to run process drivers and exhausted steam at lower pressure is used for process heating. This kind of application is called cogeneration of heat and power. Although power generation efficiency is lower than a condensing turbine, a back pressure turbine has a high cogeneration efficiency that is close to boiler efficiency.

For large steam turbines, isentropic efficiency can be between 50 and 70% while efficiency for small turbines is much lower with a wide variation depending on speed, horse power (hp), and pressure conditions. Efficiency varies with turbine loading, which can be described by an efficiency curve. This curve can be used to determine actual steam rates based on specific turbine loading.

15.4.1 Modeling

Two parameters, namely, theoretical steam rate and turbine isentropic efficiency, can be used to determine actual steam rate required by the steam turbine to make a certain amount of power. The theoretical steam rate is described by ideal expansion. The assumption behind ideal expansion is that there is no thermal and hydraulic losses in the expansion process resulting in zero change in entropy (s) (see Figure 15.5). Thus, the theoretical steam rate is the minimum required for making a certain amount of power. In reality, losses occur in steam expansion and isentropic efficiency describes the irreversible losses in the real expansion causing deviation from the ideal expansion:

$$\eta_{is} = \frac{h_1 - h_2'}{h_1 - h_2}. \qquad (15.5)$$

If η_{is} is known, the actual power generated from the steam turbine can be calculated as

$$W = (h_1 - h_2')/3413 = (h_1 - h_2)\eta_{is}/3413 \text{ kWh/lb}, \qquad (15.6)$$

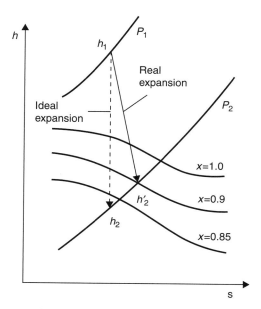

FIGURE 15.5. Ideal and real expansions.

where 3413 Btu/kWh is the conversion factor and enthalpy of steam (h) is in Btu/lb. Enthalpy change for isentropic expansion can be found from the Mollier steam diagram (Mollier, 1923).

Example 15.1 Estimate Steam Rate for Steam Turbine

Consider an HP-MP turbine (Figure 15.6) with isentropic efficiency of 50% and steam conditions given as: HP inlet at 750 °F and 600 psig; MP exhaust at 150 psig. The turbine is used to drive a pump with power demand of 300 kW. What is the actual steam rate required?

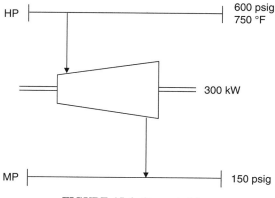

FIGURE 15.6. Steam turbine.

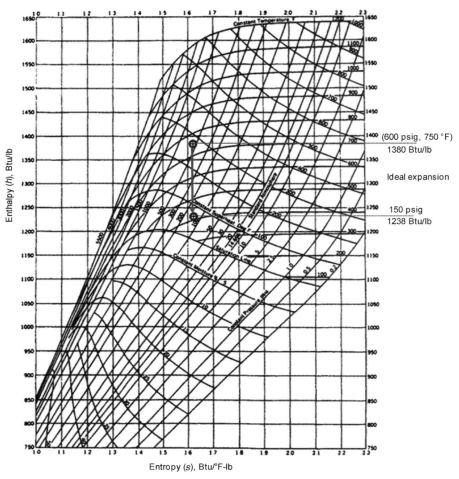

FIGURE 15.7. The Mollier (h–s) steam diagram. (From Kutz (1998), reproduced with permission of Wiley.)

Answer
The Mollier h–s diagram (Figure 15.7) (Mollier, 1923) is used to determine steam enthalpy based on steam conditions. The inlet steam enthalpy is found to be 1380 Btu/lb. Following the vertical line (zero change in entropy), the inlet steam expands and reduces its pressure to exhaust at 150 psig with enthalpy of 1238 Btu/lb. Therefore, the enthalpy change ($\Delta h_{\text{theoretic}}$) is 142 Btu/lb or 330 kJ/kg to give the theoretic steam rate as

$$m_{\text{theoretic}} = \frac{3600 \text{ kJ/kWh}}{\Delta h_{\text{theoretic}} \text{ kJ/kg}} = \frac{3600}{330} = 10.9 \text{ kg/kWh} = 24 \text{ lb/kWh.} \qquad (15.7)$$

The specific steam rate is

$$m_{actual} = \frac{m_{theoretic}}{\eta_{is}} = \frac{24}{0.5} = 48 \text{ lb/kWh}. \qquad (15.8a)$$

Thus, actual steam rate is

$$\begin{aligned} M_{actual} &= m_{actual} \times W \\ &= 48 \text{ lb/kWh} \times 300 \text{ kW} = 14{,}400 \text{ lb/h}. \end{aligned} \qquad (15.8b)$$

So far, we defined the isentropic efficiency, which describes the deviation of actual expansion from ideal expansion. However, what can be used to describe the thermodynamic efficiency for power generation? It is power generation cycle efficiency (η_{cycle}), which is defined in terms of power generated versus fuel consumed:

$$\eta_{cycle} = \frac{\text{Shaft power in kWh}}{\text{Fuel equivalent in kWh}} = \frac{W}{Q_{fuel}}. \qquad (15.9)$$

As steam is the heat input for a steam turbine, steam heat must be traced back to fuel as

$$Q_{fuel} = FE_{steam} \times M_{actual,steam}. \qquad (15.10)$$

Thus, cycle efficiency can be specifically expressed for steam turbines as

$$\eta_{cycle} = \frac{W\,(\text{kW}) \times 3414(\text{Btu/kWh})}{FE_{steam}(\text{Btu/lb}) \times M_{actual,steam}\,(\text{lb/h})} = \frac{3414}{FE_{steam} \times m_{actual,steam}}. \qquad (15.11)$$

As an indication of magnitudes, a condensing turbine with steam input of VHP at 1500 psig has η_{cycle} of 37.5% and specific steam rate of 9.09 lb/kWh. Condensing turbines with steam inlet pressure of 610 psig (HP) and 200 psig (MP), have η_{cycle} of 32.0% and 24.5%, as well as specific steam rates of 10.67 and 13.9 lb/kWh, respectively.

In applications, some steam turbines have multiple extractions. Modeling of such steam turbines becomes more complex and it requires simulation tools to help. The basis for modeling such complex turbine is the turbine capability diagram, which is provided by manufacturing. This diagram defines the operating window for any feasible turbine operation. Each operating point in the diagram indicates steam input, extraction rates and total power generation. Consider a turbine with HP steam input, one MP extraction and one condensing flow. In operation, the HP steam input is known as it is measured online while the MP extraction rate is imposed by the MP steam header balance. With these two flow rates given, the turbine generates a certain amount of power. To mimic this in simulation, this complex turbine is simplified into two simple turbines: one is HP-MP extraction turbine and the other is MP-Condensing turbine. The condensing rate is determined as the difference of HP input and MP extraction. An average isentropic efficiency is used for both turbines. For given HP steam input and the extraction rate, the simulation determines the isentropic efficiency under which the two turbines make the total power equal to the amount shown on the capability diagram. As a result of simulating a set of representative operating points in the diagram, a matrix of isentropic efficiency can be obtained. Using the regressive method, an efficiency correlation can be generated based on the matrix, which becomes the model for the complex turbine and can be used to predict power generation for any given HP steam rates and extraction rates.

15.5   GAS TURBINE

A gas turbine is a type of internal combustion engine. It has a compressor for air compression coupled to an expansion turbine for power generation and a combustion chamber in-between (Figure 15.8). Fuel enters the combustor to mix with compressed air and ignite. Flue gas, the combustion product at high temperature and pressure expands and spins the expander to make power. Part of the power is used to run the compressor and the remaining is the net power output.

In many applications in the process industry, gas turbine is mainly used in electricity generation although it can be used as process drivers in some cases. Gas turbine could be coupled with steam turbine to form a CCGT (combined cycle gas turbine). Exhaust gas has a high temperature of 450–550 °C. When it is used for steam generation, steam can spin steam turbines for more power generation. This boosts power generation efficiency greatly resulting in 60% or higher for CCGT. In contrast, the power generation efficiency on standard alone basis is in the range of 20–40% depending on the type and size of the gas turbine.

Gas turbines are available in two types, namely, industrial type and aero-derivative type. In comparison, the former is cheaper, robust with single shaft and longer period of operation without overhaul but at lower efficiency. The latter is

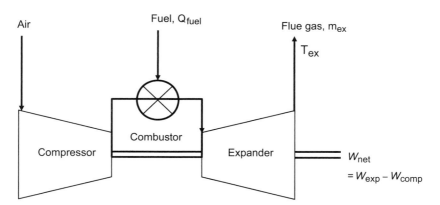

FIGURE 15.8. Gas turbine.

lighter, has a twin shaft, high efficiency, and better part-load performance. Industrial gas turbines are available in standard frame sizes ranging from 250 kW to 250 MW.

15.5.1 Modeling

In modeling a gas turbine in the context of steam and power system, the purpose is to know the fuel consumption (Q_{fuel}) for a given net power output (W_{net}) and fuel efficiency. It is also important to know the exhaust mass flow (m_{ex}) and temperature (T_{ex}), which determine the integration opportunity with steam turbine cycle. Gas turbine performance curves are provided by manufatcturer. As an example for ilustration, the key parameters are described by the following equations (Manninen and Zhu, 1999):

For industrial gas turbine,

$$Q_{fuel} = 2.84 \times W_{net} + 7.33 \text{ MW} \tag{15.12}$$

$$m_{ex} = 2.9 \times W_{net} \text{ kg/s} \tag{15.13}$$

$$T_{ex} = 0.4 \times W_{net} + 493.42 \,°C. \tag{15.14}$$

For aero-derivative gas turbine,

$$Q_{fuel} = 2.35 \times W_{net} + 7.75 \text{ MW} \tag{15.15}$$

$$m_{ex} = 3.1 \times W_{net} \text{ kg/s} \tag{15.16}$$

$$T_{ex} = -1.0 \times W_{net} + 512.24 \,°C. \tag{15.17}$$

For both types of gas turbines,

$$\eta_{cycle} = \frac{W_{net}}{Q_{fuel}}, \tag{15.18}$$

where Q_{fuel} and W_{net} are in MW.

15.6 LETDOWN VALVE

In a plant, both the steam pressure and temperature may be reduced for process steam use. This is because the process equipment is not rated for the high pressure and temperature. Also, steam latent heat at lower pressure is higher so less steam circulation is required with lower pressure steam. The device for reducing pressure is the steam letdown valve.

Reducing the steam pressure increases superheating at lower pressure but does very little to reduce the temperature. The letdown valve can be also viewed as a way of superheating steam. The major feature of the letdown process through a pressure reduction valve is that the enthalpy remains substantially the same.

TABLE 15.1. Valve Characteristics for the HP Letdown Control Valve

Position %	10	20	30	40	50	60	70	80	90	100
C_v	7.3	15.8	25.6	36.5	49.2	62.5	77.9	95.6	111.7	122.4

15.6.1 Modeling

For illustration purposes, consider the example valve characteristic (C_v) as shown in Table 15.1, which is provided by the manufacturer.

Depending on the flow characteristics in the control valve, the mass flow rate is different for the same valve opening. The mass flow equations for the example valve in Table 15.1 are provided by the manufacturer and are listed for illustration as

$$M = 1.31\, C_v\, P_i \text{ for choke flow when} \frac{P_o}{P_i} \le 58\% \qquad (15.19)$$

and

$$M = 1.72\, C_v \sqrt{(P_i + P_o)(P_i - P_o)} \quad \text{for nonchoke flow,} \qquad (15.20)$$

where M is the steam flow, lb/h; P_i and P_o are inlet and outlet pressure of the control valve, psia.

The choke flow occurs when outlet pressure is much lower than inlet pressure with the ratio of outlet and inlet pressure less than 58%. As expected, steam flow in the choke mode is much less than that under nonchoked mode. Thus, it is critical in determining if a valve is operating under choke or nonchoke flow. Inappropriate choice will lead to either underestimate or overestimate of letdown flow rates.

In the field, valve opening is usually measured. The modeling task thus becomes developing a correlation for steam flow based on valve position.

To do this, the first step is selection of which equation to use based on either choke flow or nonchoke flow. The second step is to apply the equation selected to calculate the steam flow according to C_v and then incorporate the flows into the valve characteristic table and expand it to Table 15.2. The last step is to develop a correlation for steam flow based on the valve position.

TABLE 15.2. Relationship Between the Valve Position and Steam Rate Equations (15.19) and (15.20)

Position %	10	20	30	40	50	60	70	80	90	100
C_v	7.3	15.8	25.6	36.5	49.2	62.5	77.9	95.6	111.7	122.4
M_1, klb/h	7.3	16.0	25.8	36.8	49.6	63.0	78.5	96.4	112.6	123.4
M_2, klb/h	6.0	13.0	21.0	29.9	40.4	51.3	63.9	78.5	91.6	100.4

M_1: nonchoke flow; M_2: choke flow.

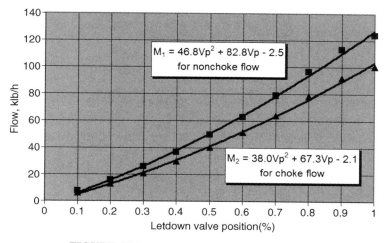

FIGURE 15.9. Modeling of steam letdown valve.

Typically, a second-order polynomial equation is sufficient to capture the nonlinearity. To do this, correlations are developed as shown in Figure 15.9 based on Table 15.2, which describes the relationship between valve position V_p and steam flow, M:

$$M_1 = 46.8V_p^2 + 82.8V_p - 2.5\text{klb/h} \quad \text{for nonchoke flow.} \tag{15.21a}$$

$$M_2 = 38.0V_p^2 + 67.3V_p - 2.1\text{klb/h} \quad \text{for choke flow.} \tag{15.21b}$$

15.7 STEAM DESUPERHEATER

To reduce the temperature of the steam, a device is utilized that adds water to the superheated steam. Typically, clean condensate is used as it is important that good quality of water is used to cool the steam. By varying the amount of water added, the temperature can be controlled. If the temperature is lowered to within 10 °C of that of saturated steam, at the actual pressure involved, this unit is called a desuperheater (Figure 15.10).

The heat and mass balances can be derived as expressed in the following equations. Steam temperature and pressure of the inlet steam are given, and the balance equations can be applied to determine the amount of inlet steam and water for a given steam temperature and amount in the downstream header.

$$M_{\text{Sat}} \times h_{\text{Sat}} = M_{\text{Sup}} \times h_{\text{Sup}} + M_{\text{BFW}} \times h_{\text{BFW}} \tag{15.22}$$

$$M_{\text{Sat}} = M_{\text{Sup}} + M_{\text{BFW}}. \tag{15.23}$$

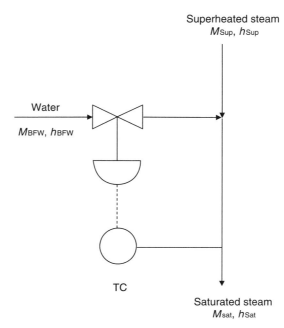

FIGURE 15.10. Desuperheater.

15.8 STEAM FLASH DRUM

The steam flash drum is a device for steam recovery. Flash steam occurs at the drum where steam condensate or boiler blowdown experiences a drop in pressure causing some of the condensate or boiler blowdown to evaporate forming steam and thus produces steam at the lower pressure (Figure 15.11). For low-pressure condensate, flash steam is negligible and thus it is not worth to recover. However, for medium- and high-pressure condensate, it is important to recover flash steam.

The heat and mass balances for the flash drum can be applied to determine the amount of flash steam based on the inlet and outlet conditions.

$$M_{FS}^{LP} \times h_{FS}^{LP} = M_{in} \times h_{in} - M_{BD} \times h_{Cond}^{LP} \tag{15.24}$$

$$M_{FS}^{LP} = M_{in} - M_{BD}. \tag{15.25}$$

15.9 STEAM TRAP

The steam trap is an automatic valve used in every steam system to remove condensate and noncondensables. For process heating, steam traps keep the steam inside the equipment until the steam condenses so that the steam latent heat is transferred to the process. When steam condenses to condensate, steam trap valves open and remove

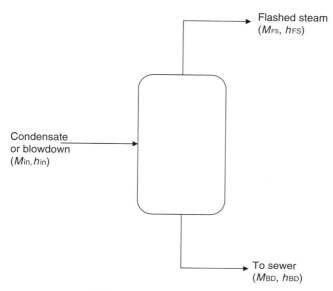

FIGURE 15.11. Steam flash tank.

the condensate and sends it to the condensate system and finally to deaerator. Another common application is that steam traps are also used in steam pipes to remove condensate from pipes to avoid water hammers, which could cause damage to the facility. The condensate is then sent to the condensate system.

If operating properly, steam traps have negligible loss of steam. However, when steam traps malfunction, the valve sticks either in an open or closed position. Failure in the open position can be analogous to having a hole in the steam system the same size of the trap's internal discharge passages where steam leaks out of the passages. If the steam leaks into the condensate piping, water hammer could occur as hot steam contacts condensate that has cooled below the temperature of the steam. Failure in the closed position causes steam equipment to cease operating because it cannot discharge condensate. Failure of drip traps in the closed position is dangerous because slugs of condensate remain in steam lines leading to water hammers, which could cause significant damage. Some traps tend to fail in an open position, while others tend to fail in a closed position. This tendency may also be influenced by the characteristics of the steam system.

There are hundreds of steam traps all over the place in a site and many of them could malfunction if there is no best practice of maintenance in place. That is the reason why the most common energy waste is steam leakage from "bad" steam traps.

The leakage rate can be easily estimated by using the general orifice equation:

$$M_{\text{leak}} = C_v \left(\frac{1}{4}\pi d^2\right)\sqrt{2gH} \qquad (15.26)$$

where m_{leak} is the leaked steam in m³/s; C_v is the orifice coefficient; d the equivalent diameter of hole (meter); g is acceleration from gravity (9.81 m/s²); and H is the hydraulic head acting on orifice (meter).

15.10 STEAM DISTRIBUTION LOSSES

Steam is distributed from the sources of generation to the locations of end users via steam headers and take-off pipes. There could be significant steam and condensate losses due to poor insulation of steam pipes, steam leaks, steam trap losses, condensate losses, venting, and letdown. The total energy loss associated with steam and condensate losses could be in the range of 10–30% of fuel fired in the boilers depending on the steam system design and operation. The losses need to be reflected in the steam balance, which will be discussed in chapter 16.

NOMENCLATURE

C_v valve orifice coefficient
d equivalent diameter of hole
FE fuel equivalent
g acceleration from gravity (9.81 m/s²)
h enthalpy
ΔH the head acting on valve orifice
M mass flow
m specific flow rate
P_i, P_o inlet and outlet pressure of the control valve
Q heat duty
V_p opening position of a control valve
W shaft work

Greek Letters

α fraction of deaeration steam vented
η efficiency
η_{cycle} power generation efficiency

REFERENCES

Kutz M (1998) *Mechanical Engineers' Handbook*, 2nd edition, John Wiley & Sons.

Manninen J, Zhu XX (1999) Optimal gas turbine integration to the process industries, *Ind. Eng. Chem. Res.*, **38**, 4317–4329.

Mollier R (1923) Ein neues diagram für dampfluftgemische, *ZVDI*, **67**(9), 869–872.

16

ESTABLISHING STEAM BALANCES

Building a steam balance is an importent step toward effective energy management because it can give an overview as to how much energy is generated versus how much is actually used. The gap can direct attention to improvement opportunities in operational and capital projects. Inaccurate steam balance could mislead evaluation of energy improvment projects.

16.1 INTRODUCTION

Although energy is used in the processes, the cost is measured and paid at the boiler house for any amount of imported natural gas, power, and steam. As an asset manager or asset supervisor, your ultimate goal in managing the steam and power system is to satisfy the process energy demand in the lowest expense, which is measured by the invoice the boiler house receives on a regular basis.

It is important to establish a steam and power balance. To do this, a steam balance must feature high accuracy. A good steam balance can tell how much is used versus generation, and thus can quickly identify opportunities of low hanging fruit.

We will learn two things from this chapter. The first is how to build a complete steam balance from scratch illustrated by a real example, while the second is to verify the balance to make it more trustworthy.

Energy and Process Optimization for the Process Industries, First Edition. Frank (Xin X.) Zhu.
© 2014 by the American Institute of Chemical Engineers, Inc. Published 2014 by John Wiley & Sons, Inc.

16.2   GUIDELINES FOR GENERATING STEAM BALANCE

The first thing to do in building the steam balance is gathering the data related to steam generation and usage for all steam pressure levels as well as steam letdown flows, condensate return rates, boiler feed water (BFW), and so on. The most common steam system has three steam headers [high pressure (HP), medium pressure (MP), and low pressure (LP)] in the process industry, although there are steam systems with four-header systems (very high pressure (VHP), HP, MP, and LP) or (HP, MP, LP, LLP). Occasionally, two-header systems (HP and LP) are used. Typically, the pressure levels for VHP, HP, MP, LP, and LLP are 1500, 600, 180, 60, and 25 psig, respectively. LLP stands for very low pressure.

Once the data are extracted, it is time to sketch a steam balance flow diagram (the so-called steam process flow diagram, PFD) and put these data together for each steam header in the PFD (Figure 16.1). The steam PFD is a very powerful visual aid as it allows an overview of a complex steam system at a glance.

For a common steam system with three steam headers, the balancing procedure usually starts from the HP header. HP steam is mainly generated from steam boilers as well as from convection sections of process furnaces. While HP steam is mainly used for steam turbines as process drivers for pumps and compressors, it can also be used for column reboiling operating at relatively high temperatures.

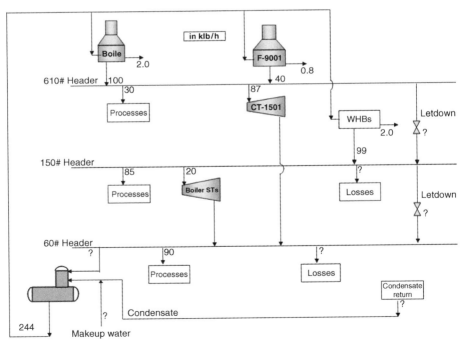

FIGURE 16.1. Step 1: Develop a steam PFD sketch with missing flows.

As the best practice, HP steam is measured at each point of generation and usage. The total HP generation is calculated by adding all the individual generation from boilers and processes, which should be balanced against all the HP usage. The total HP usage is the summation of all individual users including HP steam feeding to the letdown valves.

The same procedure is applied to the MP and LP steam headers. MP and LP steam are mainly used as reboiling and stripping for separation columns as well as ejectors for vacuum columns. In addition, LP steam is used for miscellaneous usages such as atomizing steam in furnaces, tracing, off-site tank heating, boiler feed water preheating, deaeration, and so on. MP and LP steam are generated from different sources, including waste heat steam generators from processes, exhaust steam from steam turbines, and letdown valves.

For end users that do not have meters for measurement, correlations should be developed to estimate steam flows based on process conditions. The key for developing decent correlations is to identify major operating variables that affect energy use.

As a guideline, an accuracy of at least 97, 95, and 90% should be maintained for HP, MP, and LP steam headers, respectively. The inaccuracy could be attributed by meter miscalibration, uncertainty in steam losses, and estimates for unmeasured flows. The most difficult task in making the steam balance is to balance the LP header with high accuracy as there are many unmeasured LP flows and losses all over the site, which are not measured.

However, as a starting point, one should try to make good HP and MP balances. If they are not in good balance, the reasons causing the imbalance should be identified and actions be taken to maintain high accuracy because HP and MP steam are very expensive.

16.3 A WORKING EXAMPLE FOR GENERATING STEAM BALANCE

The purpose is to develop a steam balance for operational supervision as well as for identification of improvement opportunities in the steam system. Models for boilers, turbines, deaerators (DAs), letdown valves, desuperheaters, and steam flash tanks are discussed in the previous chapter. Historian and distributed control system (DCS) data will be connected to steam balance so that the steam balance is capable of dynamically balancing the steam and power demands due to process variations, units on or off, and weather change.

As a general guideline, the focus in the early stage of developing a steam balance should be on capturing all the major users and generations and ensure the measurement and calculations are correct for these flows. High precision should not be rigorously pursued as it is more important to put things together at this early stage, while precision can be improved in the later stage. It will be difficult to perform steam balances for an old facility as there will be a lack of measurement and information for steam flows and conditions. In this case, estimates based on experience and calculations will be much required.

The examples below are used to illustrate two different methods and essential steps of developing a steam balance.

16.3.1 Top–Down Methodology for Steam Balance

When the HP steam generations are measured and the amount from each source is known, a top–down approach should be applied in which the HP header balance is conducted first and move down to MP, LP, and DA balances.

16.3.1.1 Step 1: Develop a Steam System PFD with Unknown Flows The challenge in this step is to develop a steam PFD and put together steam use and generation based on measurement and estimates as shown in Figure 16.1. Steam generation and major steam usages are usually measured online, while some usages are estimated based on design duty and experience. It is common that quite a few steam flows are not measured resulting in missing flows to be determined from the steam balance. For simplicity of illustration, let us assume that the missing flows include letdown flows, makeup water, LP for deaeration, condensate return, and steam losses. It is also assumed with 2% blowdown on average for all steam generators.

16.3.1.2 Step 2: Calculate Missing Flows

(i) HP Header Balance. The balancing starts the HP steam header. HP steam is generated from the on-purpose boiler and the convection section of a process furnace (F-9001) where the radiation section is used for process feed heating. HP steam is used in some high-temperature processes and for steam turbine CT-1501, which runs a gas compressor. All these flows are measured online, but the remaining HP flow, which is not measured for this case, goes through the letdown valve as supplementary provision for MP steam. Table 16.1 shows the balance for the HP header.

We have

$$140 = 117 + \text{HP letdown}. \tag{16.1}$$

Thus,

$$\text{HP letdown} = 23\,\text{klb/h}.$$

TABLE 16.1. HP Header Balance

Input (kpph)		Output (kpph)	
Boilers	100	HP letdown	?
F-9001	40	CT-1501	87
		Processes	30
Total	140	Total	$117 + \text{HP letdown}$

TABLE 16.2. MP Header Balance

Input (kpph)		Output (kpph)	
HP letdown	23	MP letdown	?
WHBs	99	Processes	85
		Boiler turbines	20
		MP losses	?
Total	122 + HP letdown	Total	105 + MP letdown + losses

(ii) MP Header Balance. For the MP header, the major input comes from two waste heat boilers (WHBs). On the other hand, MP steam is used for process reboiling and stripping as well as in two steam turbines to run a water circulation pump and fan in the boiler house. MP letdown is unknown and will be determined together with the MP losses. The steam losses are due to steam leaks and trap losses. With HP letdown equal to 23, Table 16.2 shows the balance for the MP header.

Obviously, we can derive

$$122 = 105 + \text{MP letdown} + \text{MP losses}. \tag{16.2}$$

We cannot calculate MP letdown and MP losses as there are two variables in one equation. We have to rely on other equations.

(iii) LP Header Balance. For the LP header, the extraction turbines provide the main supply, while the MP letdown is supplementary, which is unknown. The LP users include processes and deaeration. The closure is the LP losses, which is unknown. In some cases, various LP consumptions such as tracing, tankage heating, and atomization are not measured, and the amount for each individual usage is not known in the plant. In the steam balance, the aggregated amount can be included in the steam losses, but effort must be taken later to gain good understanding and develop reasonable estimates. The LP balance is developed as shown in Table 16.3.

We have

$$107 + \text{MP letdown} = 90 + \text{DA LP} + \text{LP losses}. \tag{16.3}$$

Solving for DA LP from equations (16.2) and (16.3) gives

$$\text{DA LP} = 34 - \text{MP losses} - \text{LP losses}. \tag{16.4}$$

TABLE 16.3. LP Header Balance

Input (kpph)		Output (kpph)	
Boiler PTs	20	Processes	90
CT-1501	87	Deaerators	?
MP letdown	?	LP losses	?
Total	107 + MP letdown	Total	90 + LP to DA + losses

TABLE 16.4. Deaerator Balance

Input (kpph)		Output (kpph)	
LP to DA	?	Boiler feed water	244
Cond return	?		
Makeup	?		
Total	DA LP + cond return + makeup	Total	244

(iv) Deaerator Balance. Deaerator balance is given in Table 16.4, which has three unknowns, namely, DA LP steam, treated water for makeup, and condensate return. Boiler feed water is calculated, which is equal to the steam generation from the boiler, furnace F-9001, and WHBs plus blowdown. A blowdown of 2% on average is assumed. A small fraction of the LP steam, typically 5%, is vented from the deaerator to remove any noncondensable. If there is no noncondensable, no venting is needed. For simplicity, no venting in the deaerator occurs.

We can derive

$$\text{DA LP} + \text{Cond return} + \text{Makeup} = 244. \qquad (16.5)$$

(v) Solving for Unknowns. Equations (16.4) and (16.5) have five unknowns, namely, DA LP rate, MP losses, LP losses, condensate return, and makeup rate. It is impossible to solve three equations with five unknowns. To do this, we have to make some assumptions.

In most cases practically, LP losses are higher than MP losses. Thus, the first assumption is MP losses are half of LP losses and this assumption reduces two unknowns to one. The second assumption is 70% condensate return for process steam use. This assumption gives the condensate return of 144 klb/h. Third, DA LP rate is assumed as 12% of BFW, which gives 29 klb/h. As a result of these three assumptions, five unknowns are reduced to two, which can be solved by two equations (16.4) and (16.5).

Assuming deaeration steam is 29 klb/h, the LP losses and MP losses can be calculated from equations (16.4) as

$$\text{LP losses} = 3.3 \text{ and MP losses} = 1.7 \text{ klb/h.}$$

Then MP letdown is determined from equation (16.2) as

$$\text{MP letdown} = 15.3 \text{ klb/h.}$$

Makeup is obtained from equations (16.5) as

$$\text{Makeup} = 72 \text{ klb/h.}$$

So far, we resolved all unknowns and thus developed a closed balance as shown in Figure 16.2.

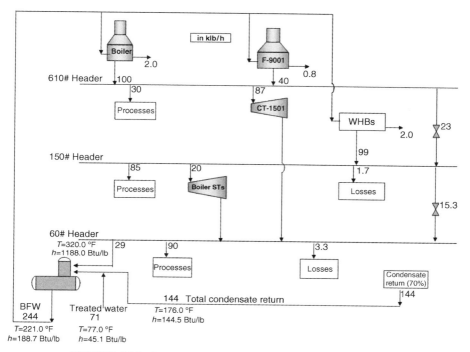

FIGURE 16.2. Step 2: Assume DA LP to resolve missing flows.

The missing flows are specific for each plant and depend on measurement. For example, if LP steam for deaeration is measured, condensate return could be the missing flow to be determined from the DA balance. In cases where both DA steam and total condensate return is measured, the LP and MP losses could be determined from the balance procedure.

16.3.1.3 Step 3: Calculate DALP and Revise the Balances

29 klb/h of DA LP steam was assumed previously based on experience and it is verified here based on the heat and mass balances around the deaerator:

$$m_{DA} \times (h_{DA} - h_{TW}) \times (1 - \alpha) = m_{BFW} \times (h_{BFW} - h_{TW}) - m_{Cond} \times (h_{Cond} - h_{TW}),$$
(16.6)

$$m_{DA}(1 - \alpha) + m_{Cond} + m_{TW} = m_{BFW},$$
(16.7)

where m is the mass flow in klb/h, h is the enthalpy in Btu/lb, and α is the fraction of DA steam vented. Typically, α is 5% if DA steam is not condensing. Otherwise, α is zero. For simplicity, assume $\alpha = 0$ for this example.

For given BFW rate of 244 klb/h and condensate return of 143 klb/h, DA LP flow can be derived based on equation (16.6):

$$m_{DALP} = \frac{m_{BFW} \times (h_{BFW} - h_{TW}) - m_{Cond} \times (h_{Cond} - h_{TW})}{(h_{DALP} - h_{TW}) \times (1 - \alpha)}, \quad (16.8)$$

that is,

$$m_{DALP} = \frac{244 \times (188.7 - 45.1) - 144 \times (144.5 - 45.1)}{(1188 - 45.1)} = 18 \text{ klb/h}.$$

With DA LP rate of 18 klb/h newly calculated, equation (16.4) becomes

$$18 = 34 - \text{MP losses} - \text{LP losses}.$$

Based on the assumption of MP losses equal to half of LP losses, thus

$$\text{LP losses} = 10.6 \text{ klb/h and MP losses} = 5.3 \text{ klb/h}.$$

The MP letdown can then be calculated from equation (16.2):

$$\text{MP letdown} = 122 - 105 - \text{MP losses} = 11.7 \text{ klb/h}.$$

Solving equation (16.7) for makeup as

$$\text{Makeup} = 244 - \text{DA LP} - \text{Cond return} = 82 \text{ klb/h}.$$

The converged steam balance is shown in Figure 16.3.

What about if make up rate is measured, but condensate return rate is unknown? Discussions for this scenario is given below.

16.3.2 Bottom-Up Methodology for Steam Balance

When some of the HP steam generation is unknown, a bottom-up approach may be the right one for setting up a steam balance. This approach is applied to the same example above, but assuming the boiler steam generation is unknown and to be determined from the steam balance.

16.3.2.1 Step 1: Develop a Steam System PFD with Unknown Flows   The steam PFD is shown in Figure 16.4, which is the same except that the boiler steam is unknown.

16.3.2.2 Step 2: Calculate Missing Flows   The balancing starts from the LP header and goes upward to MP and HP headers until the boiler. The initiator of the balancing procedure is the assumption of LP to deaerator as 25 klb/h.

(i) LP Header Balance (Table 16.5). Thus,

$$107 + \text{MP letdown} = 126 \rightarrow \text{MP letdown} = 126 - 107 = 19 \text{ klb/h}.$$

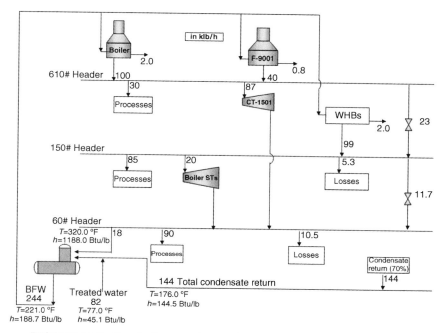

FIGURE 16.3. Step 3: Converged steam balance based on deaeration steam.

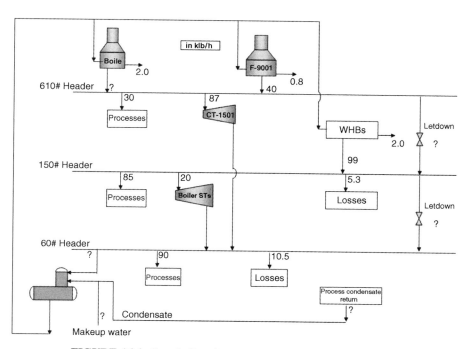

FIGURE 16.4. Step 1: Develop a steam PFD with missing flows.

TABLE 16.5. LP Header Balance

Input (kpph)		Output (kpph)	
Boiler PTs	20	Processes	90
CT-1501	87	Deaerators	25
MP letdown	?	LP losses	11
Total	107 + MP letdown	Total	126

TABLE 16.6. MP Header Balance

Input (kpph)		Output (kpph)	
HP letdown	?	MP letdown	19
WHBs	99	Processes	85
		CT-1601	20
		MP losses	5
Total	99 + HP letdown	Total	129

(ii) MP Header Balance (Table 16.6). Thus,

$$99 + \text{HP letdown} = 129 \rightarrow \text{HP letdown} = 129 - 99 = 30 \, \text{klb/h}.$$

(iii) HP Header Balance (Table 16.7). Thus,

$$40 + \text{boiler steam} = 147 \rightarrow \text{boiler steam} = 147 - 40 = 107 \, \text{klb/h}.$$

(iv) BFW Balance. Boiler feed water can then be calculated as the summation of BFW demands for the boiler, F-9001, and WHBs. Assuming 2% lowdown, we have

$$\text{BFW} = (107 + 40 + 99) \times 1.02 = 251 \, \text{klb/h}.$$

(v) Deaerator Balance (Table 16.8). Thus,

$$25 + \text{Cond return} + \text{Makeup} = 251 \rightarrow \text{Makeup} = 251 - 25 - \text{Cond return}.$$

TABLE 16.7. HP Header Balance

Input (kpph)		Output (kpph)	
Boilers	?	HP letdown	30
F-9001	40	CT-1501	87
		Processes	30
Total	40 + Boiler steam	Total	147

TABLE 16.8. Deaerator Balance

Input (kpph)		Output (kpph)	
LP to DA	25	Boiler feed water	251
Cond return	?		
Makeup	?		
Total	25 + Cond return + makeup	Total	251

By assuming 70% for condensate return, we have

$$\text{Cond return} = (30 + 85 + 90) \times 0.7 = 144 \, \text{klb/h}$$

$$\text{Makeup} = 251 - 25 - 144 = 82 \, \text{klb/h}.$$

Until now, all the missing flows are resolved and the results are shown in Figure 16.5. The question is: Is this steam balance converged? You may think about it and the answer will be given later.

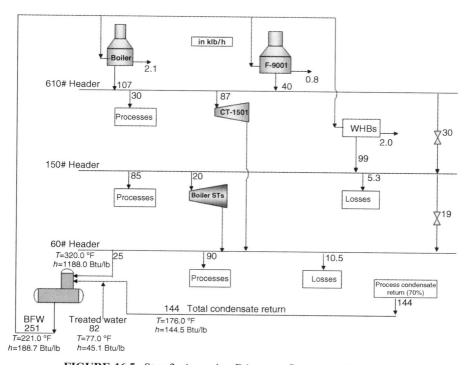

FIGURE 16.5. Step 2: Assuming DA steam flow to start iterations.

16.3.2.3 Step 3: Iterate Calculation of DALP for Convergence

To check the convergence, duration steam must be verified based on equation (12.8) as

$$m_{\text{DALP}} = \frac{m_{\text{BFW}} \times (h_{\text{BFW}} - h_{\text{TW}}) - m_{\text{Cond}} \times (h_{\text{Cond}} - h_{\text{TW}})}{(h_{\text{DALP}} - h_{\text{TW}}) \times (1 - \alpha)},$$

that is,

$$m_{\text{DALP}} = \frac{251 \times (188.7 - 45.1) - 144 \times (144.5 - 45.1)}{(1188.0 - 45.1) \times (1 - 0.05)} = 19 \text{ klb/h}.$$

With the DA steam of 19 klb/h newly calculated, the same procedure as above for calculating the missing flows is repeated to give the closed steam balance as shown in Figure 16.6.

It can be observed that 19 klb/h of DA steam is determined based on the previous BFW rate of 251 klb/h. However, the newly derived steam balance indicates the BFW rate of 245 klb/h. Thus, the DA steam must be recalculated based on the BFW rate of 245 klb/h and new iterations are required until the DA steam rates are the same between two immediately adjacent iterations.

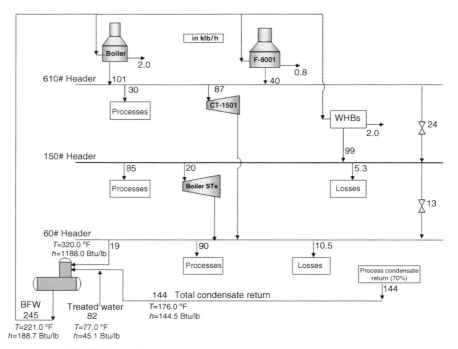

FIGURE 16.6. Step 3a: Calculate DA steam flow for iterations.

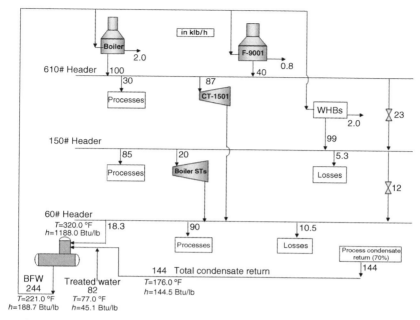

FIGURE 16.7. Step 3b: Converged steam balance based on DA steam iterations.

After several iterations of recalculating DA steam, the converged steam balance is finally obtained as shown in Figure 16.7. It can be observed that this steam balance is the same as Figure 16.3. The balance in Figure 16.3 is developed based on the top–down approach because the HP steam generation is known. In contrast, the balance in Figure 16.7 is obtained by applying the bottom-up method since one of the HP steam generation is unknown. However, different methods can reach the same converged balance for the same steam system.

16.4 A PRACTICAL EXAMPLE FOR GENERATING STEAM BALANCE

The example in Figure 16.8 is more complex than the previous one. The method for steam balancing is the top–down approach because steam generations are measured and known. Another difference is that deaeration steam is measured. It will be shown how to calculate the total condensate return based on the DA steam.

16.4.1 HP Header Balance (Table 16.9)

High-pressure steam is generated from three on-purpose boilers, together with the waste heat boiler (H-9001) in the hydrogen plant and the convection section of a

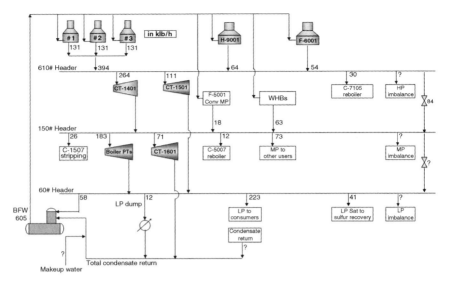

FIGURE 16.8. A practical example of the steam system with missing flows.

process furnace (F-6001). In this case, net steam generation is known. HP steam is used for reboiling of column C-7105 and for steam turbines CT-1401 and CT-1501, which run gas compressors. The remaining HP goes through the letdown valve as supplementary provision for MP steam. All these flows including the HP letdown are measured online.

Thus,

$$\text{HP imbalance} = 512 - 489 = 23 \, \text{klb/h}.$$

This imbalance is caused by metering error and corresponds to 5% of total HP steam input, which is not acceptable compared with the best practice guideline suggesting 3% or less. We will discuss this later for identifying the root causes of the imbalance and to close the balance with higher accuracy.

TABLE 16.9. HP Header Balance

Input (kpph)		Output (kpph)	
Boilers	394	HP letdown	84
H-9001	64	CT-1401	264
F-6001	54	CT-1501	111
		C-7105 reboiler	30
		HP imbalance	?
Total	512	Total	489 + HP imbalance

TABLE 16.10. MP Header Balance

Input (kpph)		Output (kpph)	
CT-1401	264	MP letdown	?
F-5001 Conv MP	18	C-1507 stripping steam	26
WHBs	63	Boiler PTs	183
HP letdown	84	CT-1601	71
		C-5007 reboiler	12
		Other users	73
		MP imbalance	?
Total	429	Total	365 + MP letdown + imbalance

16.4.2 MP Header Balance

For the MP header, steam input comes from the process-fired heater F-5001 where the radiation section is used for process feed heating, while the convection section for MP steam generation. MP steam is also made from the extraction of CT-1401 turbine, the HP letdown, and two WHBs. On the other hand, MP steam is used as column reboiling and stripping as well as in the steam turbines and MP letdown. CT-1601 turbines run a compressor and boiler PTs include four small turbines used for running a boiler water circulation pump and three AD fans. Since MP letdown is not measured, it is unknown and to be determined together with the MP imbalance. The MP imbalance includes metering error, steam losses due to steam traps, leaks, and so on. Table 16.10 shows the balance for the MP header.

16.4.3 LP Header Balance

For the LP header, several extraction turbines provide input together with the MP letdown, which is unknown. The LP users include saturated LP for sulfur recovery process, deaerator, other users, and LP dump. The closure is the LP imbalance, which is caused by metering error, unmetered flows, steam trap losses, leaks, and so on. Other users include tracing, tankage heating, and so on. LP dump is for LP condensation, which is the better way to handle LP long than venting to the atmosphere. Similarly, the LP balance is developed as shown in Table 16.11.

TABLE 16.11. LP Header Balance

Input (kpph)		Output (kpph)	
Boiler PTs	183	Other users	223
CT-1501	111	LP dump	12
MP letdown	?	Sulfur recovery unit	41
		Deaerators	58
		LP imbalance	?
Total	294 + MP letdown	Total	333 + LP imbalance

TABLE 16.12. Deaerator Balance

Input (kpph)		Output (kpph)	
Total cond return	?	Boiler feed water	605
LP to DA	58		
Makeup	?		
Total	58 + Cond return + makeup	Total	605

The best practice recommends that the letdown flows should be measured. However, it is not for this example plant where the MP and LP letdowns are not measured. This results in several unknowns. Let us see how to solve them to complete the balances. To do this, some assumptions must be made to complete the balances.

16.4.4   Deaerator Balance

The deaerator balance (Table 16.12) has two unknowns, namely, treated water as makeup and condensate return. Boiler feed water is provided for three boilers, steam generation in convection sections of furnaces H-9001, F6001 and F-5001 as well as two WHB's. Assume blowdown rate of 2% in average, the total amount of BFW can be calculated as BFW rate $= (394 + 64 + 54 + 18 + 63) \times 1.02 = 605$ klb/h.

16.4.5   Solving for Unknowns

For the MP header,

$$365 + \text{MP letdown} + \text{MP imbalance} = 429.$$

For the LP header,

$$294 + \text{MP letdown} = 333 + \text{LP imbalance}.$$

Solving for MP letdown for the above equations gives

$$71 + \text{MP imbalance} = 96 - \text{LP imbalance}$$

or

$$\text{MP imbalance} = 25 - \text{LP imbalance}.$$

In most cases practically, LP imbalance is higher than MP imbalance. If we assume MP imbalance is half of that from LP imbalance, this assumption gives 8.3 and 16.7 klb/h for MP and LP imbalances, respectively. Thus, MP letdown $= 333 - 294 + \text{LP imbalance} = 55.7$ klb/h.

The last two parameters, namely, condensate return and makeup water, are yet to be determined. Table 16.12 provides only one equation as below and we need to find another equation.

$$58 + \text{Cond return} + \text{makeup} = 605.$$

The equation (16.6) could be applied to derive the calculation for the total condensate return flow as

$$m_{\text{Cond}} = \frac{m_{\text{BFW}} \times (h_{\text{BFW}} - h_{\text{TW}}) - m_{\text{DALP}} \times (h_{\text{DALP}} - h_{\text{TW}})}{(h_{\text{Cond}} - h_{\text{TW}})}, \quad (16.9)$$

that is,

$$m_{\text{Cond}} = \frac{605 \times (188.7 - 45.1) - 58 \times (1188.0 - 45.1)}{(144.5 - 45.1)} = 207 \, \text{klb/h}.$$

Then,

$$\text{Makeup} = 605 - 58 - 207 = 340 \, \text{klb/h}.$$

$$\text{Process condensate return} = 20.7 - 12 - 71 = 124 \, \text{Klb/h}.$$

Thus, all the missing flows are resolved and the overall steam balance is completed, which is shown in Figure 16.9.

From this steam balance, some key observations can be made quickly as:

- Condensate return is only 43% compared with the best practice of 70–80%. Condensate loss is a waste of money (water purchase, pumping, and chemical

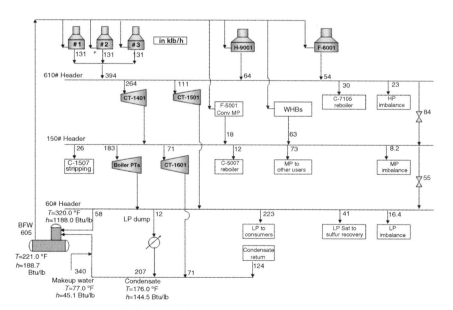

FIGURE 16.9. Completed steam balance.

treatment) and heat (hot condensate). Assume 70% condensate return will be achieved, which can bring the total condensate return of 203 klb/h versus 125 klb/h currently obtained at 43% of condensate return. Since hot condensate is at 120 °C, 13.5 MMBtu/h could be saved. In addition, about $1.3 MM/year could be saved for condensate recovery of 78 klb/h (=203 − 125) assuming condensate at $2/klb.

- There exist large letdown flows, which are at the cost of potential power. For example, 84 klb/h HP–MP letdown could make 1.9 MW power if a steam turbine is installed in place of the letdown valve, while 56 klb/h MP–LP letdown could make 1.5 MW power.

- There is a large condensing turbine and the thermal efficiency of the condensing turbine is less than 30%, which is very low compared with a back-pressure turbine or motor.

16.5 VERIFY STEAM BALANCE

It is a daunting task to put together overall steam balance as it may require weeks or months of time to develop such a comprehensive steam balance based on operating data, modeling, correlations, and experience. In the end of this endeavor when putting everything together, you may punch your fist to the sky and say: I got it, baby!

It will be very understandable if you are exuberant and happy after you build the steam balance. It gives you a sense of pride when you visualize the complex steam network of energy generation, distribution, and demand at a glance.

Your colleagues may ask you: Is it ready to be applied for operational supervision and opportunity assessment? It is a million dollar question as the stake on the table is high. Opportunities for a steam and power system are worth millions of dollars.

Using the steam balance in Figure 16.9 as an example, is it good enough to be deployed? The answer is a big "NO"! Reason number one is that the HP imbalance is 5% in comparison with the best practice of 3% or less. This imbalance cascades down to MP and LP headers. Reason number two is the use of constants for many small steam users, which are determined based on nominal process conditions. However, when the production and weather conditions vary, the measured users will change accordingly but these constants will not. Although most of the unmeasured users can be small, accumulatively they could have significant effects on the overall steam balance.

16.5.1 Control Valve Verification

The letdown flow of 84 klb/h in Figure 16.9 was obtained from online measurement. To detect any error in this measurement, the tasks involve verification of the correlation used, field inspection of the control valve for corrosion, and checking the digital transmission from the control valve to DCS.

16.5.1.1 Task 1: Verify the Correlation   For the HP letdown control valve in Figure 16.9, the nonchoke flow correlation was used in DCS to model the letdown valve. Is it correct? To answer this question, the conditions for choke flow were inspected from the manufacturing manual in which two equations were found as

$$m = 1.31\, C_v P_i \text{ for choke flow when } \frac{P_o}{P_i} \leq 58\% \qquad (16.10)$$

and

$$m = 1.72\, C_v \sqrt{(P_i + P_o)(P_i - P_o)} \text{ for nonchoke flow,} \qquad (16.11)$$

where m is the steam flow, lb/h, and P_i and P_o are inlet and outlet pressure of the control valve, psia.

For the control valve at hand, the ratio of the inlet pressure (624.7 psia) and outlet pressure (164.7 psia) is only 26%. Thus, the choke flow equation (16.10) is applied to give the letdown flow of 71.2 Klb/h based on the value position at 75% openning. The choke flow rate is smaller than the nonchoke flow. By correcting this mistake, there is still a large imbalance in the HP header. What else went wrong? The search continues.

16.5.1.2 Task 2: Inspect the Control Valve for Erosion   The field inspection indicated severe erosion occurring. This implies that the correlation in DCS could give an underestimate of actual letdown flow. To prove this suspicion, meter calibration was conducted for each HP user and then the letdown flow was back-calculated to give a letdown flow of 97 klb/h for the meter reading of 84 klb/h. A factor was added to the DCS correlation for modeling the control valve to reflect the erosion. At the same time, a maintance order was requested for fixing the problem.

16.5.2 Verify Steam Turbine

The large turbines such as CT-1401, CT-1501, and CT-1601 in Figure 16.9 are used to drive process compressors and steam flows are measured. The other four small turbines are grouped together named as "Boiler PTs" and they are used to run the boiler water circulation pump and three AD fans. These small turbines are not measured. These small turbines are modeled based on the design performance in the steam balance. However, these turbines have been running over 15 years without replacement of turbine blades, although maintenance was conducted a few years ago. Could these turbines operate at the same performance level as design? To verify this suspicion, we went out to the field in a chilly and damp morning and took temperature measurements using an infrared gun. As the measurement was taken on the outside pipe, the following correction was made to convert the skin temperature to the fluid temperature inside the pipe:

$$T_f = T_m + 0.06(T_m - T_a), \qquad (16.12)$$

where T_f, T_m, and T_a are the fluid, measured, and ambient temperature in °F (Lieberman, 1991).

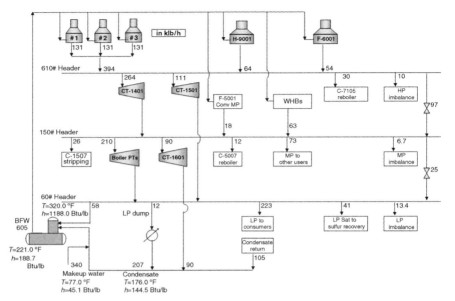

FIGURE 16.10. Adjusted steam balance.

With measured temperature after correction, the actual steam turbine perform-ance was assessed. First, the turbine design performance can be found in the datasheet, which also indicates the steam rate, adiabatic efficiency, and shaftwork output for running the pump. Then the steam rate for the turbine is calculated based on the measured temperature and given steam header pressure for inlet and outlet steam.

It was found that estimated steam rates for current conditions were 10% higher on average than the design performance. The reason was because the turbine perform-ance deteriorates over time. This was reflected by a hotter outlet temperature. Therefore, the steam rates for the boiler drivers and CT-1601 were adjusted to 210 (vs. 183) and 90 (vs. 71) klb/h, respectively.

In the end of verification, the steam balance was modified with newly calculated flows to give the revised steam balance as shown in Figure 16.10. The imbalances are well within the best practice guideline.

16.6 CONCLUDING REMARKS

This is an important step in developing site steam balance because it allows visualizing a complex steam network at a glance, provides insights for individual generation and usage, and helps to identify synergetic opportunities in the boiler house and processes.

In developing the steam balance, the focus at the early stage should be placed on ensuring all users with reasonable steam usage are included; some with measured data and the other based on educated guesses. Excessive precision should be avoided in this stage until imbalance analysis indicates the need of more precise data.

The level of precision of the steam balance depends on the applications. In the case of using the steam balance as a means of brainstorming for improvement ideas, it is not necessary to have the steam balance with high precision. For example, the steam balance in Figure 16.3 is good enough to pinpoint directional opportunities associated with large letdown flows, condensing turbines, and large LP losses potentially.

However, if the steam balance is applied for operational supervision, the steam balance with high precision and steam dynamic response to process variations is required. This is because the steam balance reflects the changing conditions of process energy demand and sets the basis for energy supply optimization to achieve the goal of minimizing the overall cost for generating steam and power via optimizing loadings of boilers and turbines, and optimizing the routes of providing energy.

For application of the steam balance in evaluating energy projects, the steam balance must have two features. One is the high precision, of course, and the other is that the steam balance must represent a good base case corresponding to the most common process operation to represent the true values of opportunities to be assessed. The above aspects will be further discussed in detail in the next chapters.

NOMENCLATURE

C_v valve orifice coefficient
m steam flow rate
P_i, P_o inlet and outlet pressure of the control valve
T temperature

Greek Letter

α fraction of deaeration steam vented

REFERENCE

Lieberman NP (1991) *Trouble Shooting Process Operations*, 3rd edition, PellWell Books.

17

DETERMINING TRUE STEAM PRICES

Inaccurate steam pricing could mislead energy optimization and business investment decisions. In most plants, reported steam costs are based on average steam prices for a typical production. Although average steam cost provides a convenient way for corporate financial budgeting and reporting, it has little value in managing the steam system to minimize true costs. In contrast, marginal steam prices can indicate true steam costs at the point of use.

17.1 INTRODUCTION

There can be confusion and misunderstanding from plant personnel on steam prices. Are we talking about steam price based on the point of use (marginal steam price) or total steam from the boiler house (average steam price)? Should the steam be priced at the point of generation or point of use? Does the steam price include both fixed and variable costs? Does the average steam price vary for different production rates? Why is the steam price obtained based on steam enthalpy instead of costs? These are the questions that will become the focus of this chapter.

Whenever possible, true steam prices should be set in the very early stages to foster plant energy efficiency. With steam priced based on its cost, engineers and operators will strive to reduce energy operating cost and identify beneficial

Energy and Process Optimization for the Process Industries, First Edition. Frank (Xin X.) Zhu.
© 2014 by the American Institute of Chemical Engineers, Inc. Published 2014 by John Wiley & Sons, Inc.

energy-saving projects. Therefore, steam should be priced to fulfill the following objectives:

- Optimize operation for the steam and power system
- Capture market spot opportunity for power export/import
- Optimize driver selection between motor and steam turbine
- Promote process heat integration
- Minimize energy losses
- Identify investment opportunity for plant modernization including cogeneration
- Provide credits to capital projects with energy efficiency improvements

It is important to understand true variable and fixed steam costs when determining the benefits of an energy-saving project. Inaccurate steam costs could lead to wasted investment for energy projects and mislead energy operation improvements.

17.2 THE COST OF STEAM GENERATION FROM BOILER

Calculations of specific costs for boiler steam is straightforward and can be clearly defined as discussed below. Ambiguity arises when pricing medium-pressure (MP) and low-pressure (LP) steam, which will be discussed in the following sections.

There are several cost components contributing to boiler steam generation:

- Variable costs per 1000 klb steam:
  - Fuel (C_F)
  - Fresh raw water supply (C_{RW})
  - Water chemical treatment (C_{CH})
  - Power requirement for water pumping and boiler air fans (C_{PW})
- Fixed cost:
  - *Depreciation ($R \cdot C_{Inv}$):* C_{Inv} is capital investment and R as the fraction of C_{Inv} depreciated annually
  - *Maintenance and Labor (C_M):* Includes the material and labor
  - *Other Costs (C_O):* Including environmental control.

Thus, the specific total cost C_{ST} for boiler steam generation can be expressed as

$$C_{var} = C_F + C_{RW} + C_{CH} + C_{PW}, \$/klb \tag{17.1}$$

$$C_{fix} = \frac{R \cdot C_{Inv} + C_M + C_O}{\sum m}, \$/klb \tag{17.2}$$

$$C_{ST} = C_{fix} + C_{var}, \$/klb \tag{17.3}$$

where C_{fix} is annualized fixed cost and C_{var} is variable cost, while Σm (klb/year) is the total steam generation from boilers.

Example 17.1   Variable and Fixed Costs for Boiler Steam

Calculate the average cost for boiler steam generation in Figure 17.1. Consider two boilers with capacity of 250 klb/h high-pressure (HP) steam at 615 psia and 700 °F with installed cost of $20 MM for the boilers. The rated capacity of the boiler is higher than the normal steam production capacity for reserve factors and process variations. Two operators at $60,000/man-year are required to operate the boilers and related equipment. The average boiler efficiency is 80% and fuel price is $6/MMBtu. Other related data are given in Table 17.1.

Solution. Assume the capital depreciation is 15% of capital and maintenance cost is 2% of the capital. Based on equation (17.2), the annual fixed cost can be calculated as

$$C_{fix} = \frac{(0.15 + 0.02) \times (20 \times 10^6) + 2 \times (60,000) \times 3}{250 \times 365 \times 24} = \$1.7/\text{klb}.$$

The factor of 3 applied to the annual salary is to include miscellaneous costs for the employee from the company. It should be noted that the fixed cost will remain the same regardless of amount of steam generation. In contrast, the variable cost reflects the amount of steam generation. Based on equation (17.1) and the data in Table 17.1, the variable cost can be calculated as

$$C_{var} = 1.56 \times 6 + 0.02 + 0.74 + 0.62 + 1.10 + 0.05 = \$11.9/\text{klb}.$$

Thus, based on equation (17.3), the specific steam cost is

$$C_{ST} = C_{fix} + C_{var} = 1.7 + 11.9 = \$13.6/\text{klb}.$$

Some explanation is required for specific fuel consumption of 1.56 MMBtu/klb steam. About 5–10% of the boiler gross steam is used internally for spinning steam turbines to run water circulation pumps and air fans. As an average, 7% internal

TABLE 17.1. Basic Data for Calculations of Fixed and Variable Costs

	per 1000 lb steam	Boiler capital; MM$	20
Average boiler fuel; MMBtu	1.56	R depraciation factor; % of capital	15
Freshwater; $	0.02	Maintenance cost; % of capital	2
Water treatment cost; $	0.74	Two employees; $/year	120,000
Water preheating and pumping; $	0.62	Employee cost factor	3
Deaeration steam; $	1.10		
FD fan; $	0.05	C_{fix} (fixed cost); $/1000 lb steam	1.7
C_{var} (variable cost); $/1000 lb steam	11.9	$C_{ST} = C_{var} + C_{fix}$; $/1000 lb steam	13.6
Fuel Price: $6/MMBtu			

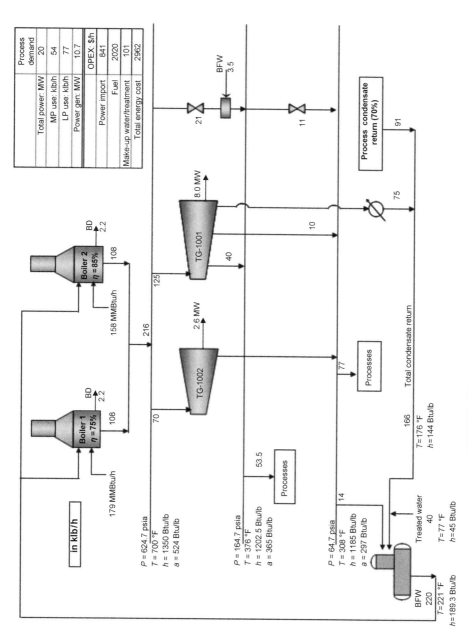

FIGURE 17.1. Example steam system configuration.

369

steam use is assumed. Boiler efficiencies are 75 and 85% for boilers 1 and 2, respectively. Thus, gross fuel consumption can be calculated as

$$Q_{B1} = \frac{(1 + 0.07) \times 108 \times (1350 - 189.3)}{0.75 \times 1000} = 178.8 \, \text{MMBtu/h},$$

$$Q_{B1} = \frac{(1 + 0.07) \times 108 \times (1350 - 189.3)}{0.85 \times 1000} = 157.8 \, \text{MMBtu/h}.$$

The specific fuel consumption is

$$q_{\text{fuel}} = \frac{178.8 + 157.8}{216} = 1.56 \, \text{MMBtu/klb}.$$

17.2.1 Cost Breakdown for Boiler Steam

There are three major cost components: fuel, boiler feed water (BFW) including freshwater, chemical treatment, pumping and heating, as well as fixed cost including capital and maintenance and environmental related costs. Table 17.2 shows the breakdown for these three costs based on fuel prices at \$6/MMBtu. In this case, fuel cost is dominant, which accounts for 69%, and all other costs for 31%.

Although the cost components should be estimated rigorously in principle, it could be a good approximation to estimate the variable steam price with the approximation based on the fuel cost:

$$C_{\text{var}} = (1 + x) \times C_F, \tag{17.4}$$

where x is the fraction of other costs and it is typically within the range of 20–40% with an average of 30%.

If applying equation (17.4) to example 17.1, C_{var} can be quickly estimated as \$12.2/klb ($=(1 + 0.3) \times 9.4$) based on C_F alone (see Table 17.2). This C_{var} is reasonably close to \$11.9/klb calculated previously based on itemized BFW related costs.

The steam costs above are the average costs for steam generation. The operating steam costs could vary for different product rates, steam system operation, and fuel and power prices. Regarding fixed cost, it is important to understand at what levels the fixed cost will change—for example, when do we need to add or reduce the number of operators? The fact that steam costs could be affected by production rates is not intuitively obvious to some people. However, for the given calculation details shown previously, readers should be able to figure out the reason why.

TABLE 17.2. Cost Breakdown for Boiler Steam Cost

Items	\$/klb	%
Fuel (C_F)	9.4	68.9
BFW (C_{BFW})	2.5	18.7
Fix cost (C_{Fix})	1.7	12.6
Total cost	13.6	100

17.3 ENTHALPY-BASED STEAM PRICING

The most common method for steam pricing in the process industry is based on the steam enthalpy. The method is straightforward: for a given cost of steam at generation pressure, the price of steam at lower pressure is based on the amount of heat available with this steam in comparison with the heat available for the steam at the generation pressure. In other words, HP steam is generated from boilers and the cost of HP steam can be calculated via equation (17.3). Then, MP and LP steam costs are prorated from the HP cost based on the ratio of enthalpy values. It seems the idea makes sense; however, there are fundamental flaws. The method and its limitations can be best illustrated via the example presented next.

Example 17.2   Enthalpy-Based Steam Cost Calculations

Calculate the steam prices based on enthalpy-based method for the steam system in Figure 17.1, which consists of major components of a complex steam system: boilers with deaeration and makeup, back-pressure steam turbine, condensing turbine, process steam demand, steam letdown vale, desuperheater, and so on. The economic data are provided in Table 17.1.

Solution. If BFW conditions are used as the basis to measure the heat available in steam, the heat available with a steam is equal to the enthalpy difference between the steam and BFW. Thus, the costs for MP and LP can be calculated based on the enthalpy ratio as

$$C_{MP} = \frac{h_{MP} - h_{BFW}}{h_{HP} - h_{BFW}} C_{HP}, \tag{17.5}$$

$$C_{LP} = \frac{h_{LP} - h_{BFW}}{h_{HP} - h_{BFW}} C_{HP}. \tag{17.6a}$$

Or

$$C_{LP} = \frac{h_{LP} - h_{BFW}}{h_{MP} - h_{BFW}} C_{MP}, \tag{17.6b}$$

where C_{HP} is the steam cost at the boiler, which is $13.6/klb as calculated previously based on Table 17.1.

Therefore, for given enthalpy values for HP and MP steam,

$$C_{MP} = \frac{1202.5 - 189.3}{1350.0 - 189.3} \times 13.6 = \$11.9/klb.$$

Similarly, the LP steam price can be calculated based on the enthalpy and price of HP steam

$$C_{LP} = \frac{1185.0 - 189.3}{1350.0 - 189.3} \times 13.6 = \$11.7/klb.$$

C_{LP} can also be calculated based on the MP steam as

$$C_{LP} = \frac{h_{LP} - h_{BFW}}{h_{MP} - h_{BFW}} C_{MP} = \frac{1185.0 - 189.3}{1202.5 - 189.3} \times 11.9 = \$11.7/klb.$$

17.4 WORK-BASED STEAM PRICING

Realizing the shortcomings in the enthalpy-based method, Kenney (1984) proposed to value steam based on the potential work that the steam possesses in comparison with enthalpy. Kenney pointed out that the enthalpy-based pricing method is based on the first law of thermodynamics (energy conservation), while the work-based pricing is based on the second law of thermodynamics (entropy). In thermodynamics, potential work is also termed as availability (a) implying the amount of potential work available.

In a direct comparison with the enthalpy ratio used in the enthalpy-based method, in the work-based method, the ratio of the availability of the steam at lower pressure to the availability of the steam at the generation pressure is used:

$$C_{MP} = \frac{a_{MP}}{a_{HP}} C_{HP}. \tag{17.7}$$

$$C_{LP} = \frac{a_{LP}}{a_{HP}} C_{HP}. \tag{17.8a}$$

Or

$$C_{LP} = \frac{a_{LP}}{a_{MP}} C_{MP}. \tag{17.8b}$$

Example 17.3   Work-Based Steam Cost Calculations

Calculate the steam prices based on the work-based method for the same steam system (Figure 17.1).

Solution. The availability (a) values for HP/MP/LP steam in Figure 17.1 are found to be 524/365/297 Btu/lb, respectively. Thus, the MP and LP steam prices could be calculated as

$$C_{MP} = \frac{a_{MP}}{a_{HP}} C_{HP} = \frac{365}{524} \times 13.6 = \$9.5/klb$$

and

$$C_{LP} = \frac{a_{LP}}{a_{HP}} C_{HP} = \frac{297}{524} \times 13.6 = \$7.7/klb.$$

Clearly, the steam prices show greater disparity compared with that of the enthalpy-based method (Table 17.3). The enthalpy-based steam pricing method

TABLE 17.3. Comparison Between Enthalpy and Work-Based Steam Pricing Methods

$/klb	HP	MP	LP
h-Based	13.6	11.9	11.7
W-Based	13.6	9.5	7.7

values lower-pressure steam more than that of work-based pricing. The reason is that when potential work is taken into account, the lower-pressure steam is devalued according to the potential work between steam headers. For readers who are interested in availability, you can find more detailed discussions on the subject of availability in Kenney (1984).

17.5 FUEL EQUIVALENT-BASED STEAM PRICING

Both enthalpy and work-based steam pricing methods rely on thermodynamic laws as the basis. Cooper (1989) argued that the steam pricing should reflect economic reality. Since the operating cost for a steam system mainly consists of fuel burned for steam generation, Cooper (1989) proposed to use the concept of fuel equivalent (FE) as the basis for steam pricing. In this method, the ratio of FE for steam at different pressures is used to derive the steam prices in placement of the ratios of enthalpy and availability.

The total FE for each steam header is the summation of all FEs entering the steam header via different flow paths. The specific FE for each steam header is the total FE divided by the amount of steam generated from this header, that is,

$$FE_{Header\ i} = \left.\frac{\text{Total FE consumed}}{\text{Total steam generated}}\right|_{Header\ i} \text{MMBtu/klb.} \quad (17.9)$$

A top–down approach is adopted for FE calculations. FEs based on LHV efficiencies for steam raising devices such as on-purpose boilers and waste heat boilers are calculated first. Then cascading down in the order of pressure levels, FEs for other steam headers are determined.

It is common that steam turbines are connected to steam headers. Therefore, it is logical to take power generation into account when steam price is determined. In reality, when the turbine is offline while the motor is online to run the rotating equipment, the power is imported but fuel for steam is saved. On the other hand, when the turbine is turned online while the motor is offline, import power is saved but at the expense of fuel for steam. To reflect the relative price comparison of fuel and power, the price equivalent efficiency (PEE) for power generation is defined as the price ratio:

$$\eta_{PEE} = \frac{\text{Fuel price of 1 MWh (marginal fuel)}}{\text{Power price of 1 MWh (import or export power)}}. \quad (17.10)$$

Thus, the multiplier to fuel equivalence for power defined in equation (3.5) in Chapter 3 can be adjusted based on η_{PEE}

$$MFE_{power} = \frac{3.414}{\eta_{PEE}} \, Btu/kWh. \tag{17.11}$$

The rationale behind the use of PEE is that the value of turbine exhaust steam depends on the power price relative to fuel price. In other words, the higher the power price, the lower the value of the exhaust steam, and vice versa. This fact can be reflected by defining the FE for the turbine exhaust (e.g., LP) as

$$FE_{LP} = FE_{HP} - FE_{Power}, \, Btu/kWh \tag{17.12}$$

where FE_{power} is

$$FE_{power} = \frac{\Delta W \times MFE_{power}}{M_{HP\text{-}LP}} = \frac{MFE_{power}}{m_{HP\text{-}LP}} \, Btu/lb. \tag{17.13}$$

M is the turbine steam flow (lb/h), ΔW (kWh) is the amount of power generated from the steam rate of M, and m is the specific steam rate ($m = M/\Delta W$; lb/kWh). The above equations can be readily extended to steam turbines with HP–MP and MP–LP expansions.

Example 17.4   FE-Based Steam Cost Calculations

Calculate the steam prices based on the FE-based method for the same steam system (Figure 17.1).

Solution. To determine the fuel equivalent for steam headers, the actual ways of producing steam must be identified.

17.5.1   FE for HP Steam

There are two paths for making HP steam, namely, boilers 1 and 2, respectively. The average FE for HP steam can be calculated as

$$FE_{HP} = \frac{179 + 156}{216} = 1.56 \, MMBtu/klb.$$

17.5.2   FE for MP Steam

To calculate FE for the MP header, three flows are identified:

- *Path 1:* 40 klb/h of the MP extraction from TG-1001 with 35.6 lb/kWh as specific steam rate

- *Path 2:* 21 klb/h of the letdown valve
- *Path 3:* 3.5 klb/h of BFW addition for desuperheating

To calculate the fuel equivalent for power generation in path 1, η_{PEE} and MFE_{power} are calculated first as

$$\eta_{PEE} = \frac{\$6/MMBtu \times 3.414\,MMBtu/MWh}{\$90/MWh} \times 100\% = 22.76\%,$$

$$MFE_{power} = \frac{3.414}{\eta_{PEE}} = \frac{3.414}{0.2276} = 15\,MMBtu/MWh.$$

Thus, the FE for the MP exhaust in path 1 is

$$FE_{MP}^{P1} = FE_{HP} - \frac{MFE_{power}}{m_{HP\text{-}MP}|_{TG1001}} = 1.56 - \frac{15}{35.6} = 1.14\,MMBtu/klb.$$

For path 2, $FE_{MP}^{P2} = FE_{HP} = 1.56\,MMBtu/klb$ for 21 klb/h letdown steam. This is because a letdown is an adiabatic process and the enthalpy stays the same before and after the letdown.

For path 3, FE_{BFW}^{P3} is about 0.2 MMBtu/klb BFW (see Section 3.5. 2.5 in Chapter 3 for details). Thus, the weighted average FE for the mixed MP steam becomes

$$FE_{MP}^{av} = \frac{40 \times 1.14 + 21 \times 1.56 + 3.5 \times 0.2}{(40 + 21 + 3.5)} = 1.22\,MMBtu/klb.$$

17.5.3 FE for LP Steam

LP steam price can be calculated in a similar way. There are three paths for making LP steam: (i) 70 klb/h from the TG-1002 turbine with a specific steam rate of 26.8 lb/kWh; (ii) 10 klb/h from the TG-1001 LP extraction with a specific steam rate of 22.9 lb/kWh; and (iii) 11 klb/h from the letdown valve. BFW desuperheating is not needed for the LP steam because the superheated fraction in LP steam is very small.

For path 1 involving the TG-1002 turbine, the 70 klb/h LP exhaust can be priced at

$$FE_{LP}^{P1} = FE_{HP} - \frac{MFE_{power}}{m_{HP\text{-}LP}|_{TG1002}} = 1.56 - \frac{15}{26.8} = 1.0\,MMBtu/klb.$$

For path 2 involving the TG-1001 LP extraction, the 10 klb/h LP extraction can be priced at

$$FE_{LP}^{P2} = FE_{HP} - \frac{MFE_{power}}{m_{HP\text{-}LP}|_{TG1001}} = 1.56 - \frac{15}{22.9} = 0.9\,MMBtu/klb.$$

For path 3 involving the letdown valve, the letdown steam has the same FE as the MP at 1.22 MMBtu/klb since the letdown is an adiabatic process. Thus, the average FE for the mixed LP steam is

$$FE_{LP}^{av} = \frac{70FE_{LP}^{P1} + 10FE_{LP}^{P2} + 11FE_{LP}^{P3}}{(70 + 10 + 11)} = \frac{(70 \times 1.0 + 10 \times 0.9 + 11 \times 1.22)}{91}$$

$$= 1.01 \text{ MMBtu/klb.}$$

17.5.4 Pricing of MP and LP Steam

With FEs calculated for HP/MP/LP steam headers and for the given HP price at $13.6/klb, the steam prices for MP and LP steam can be determined based on the ratio of FE as

$$C_{MP} = \frac{FE_{MP}}{FE_{HP}} C_{HP} = \frac{1.22}{1.56} \times 13.6 = \$10.6/klb, \tag{17.14}$$

$$C_{LP} = \frac{FE_{LP}}{FE_{HP}} C_{HP} = \frac{1.01}{1.56} \times 13.6 = \$8.8/klb. \tag{17.15}$$

17.6 COST-BASED STEAM PRICING

The critical question for enthalpy-based, work-based, and FE-based stem pricing methods is: Do the steam prices from these methods reflect the true costs? The question can be answered the best with the following example.

Example 17.5 Cost-Based Steam Cost Calculations

Calculate the steam prices based on the cost-based method for the same steam system (Figure 17.1).

Solution. To determine the true cost, the actual ways of producing steam must be assessed. There are three flows contributing to the MP steam header, which have been identified previously.

17.6.1 MP Steam Pricing

For path 1, which involves power generation from TG-1001, the power generation should be accounted to reflect that imported power is reduced with the same amount of power. Instead of using PEE, the actual power price is used in the cost-based method. Thus, for a path involving power generation, the steam price for the exhaust steam is

$$C_{exhaust} = C_{HP} - w \times C_P = C_{HP} - \frac{C_P}{m_w}, \tag{17.16}$$

where w is the specific power generation in MWh/klb, while m_w is the specific steam rate in klb/MWh for the turbine expansion. C_P is the power price in \$/MWh. w and m_w can be calculated using steam turbine modeling equations (15.6) and (15.8). Therefore, the price (C_{MP}^{P1}) for 40 klb/h MP steam can be calculated as

$$C_{MP}^{P1} = \$13.6/\text{klb} - \frac{\$90/\text{MWh}}{35.6\,\text{klb/MWh}} = \$11.1/\text{klb}.$$

For path 2, the price (C_{MP}^{P2}) for 21 klb/h letdown steam is the same as the HP steam, which is \$13.6/klb. This is because MP in path 2 is directly generated through HP letdown.

For path 3, the price (C_{MP}^{P3}) is the BFW cost of \$2.5/klb. Thus, the weighted average of the mixed MP steam price becomes

$$C_{MP}^{av} = \frac{40C_{MP}^{P1} + 21C_{MP}^{P2} + 3.5C_{MP}^{P3}}{(40 + 21 + 3.5)} = \frac{40 \times 11.1 + 21 \times 13.6 + 3.5 \times 2.5}{64.5}$$

$$= \$11.4/\text{klb}.$$

17.6.2   LP Steam Pricing

LP steam price can be calculated in a similar way. There are three paths for making LP steam as identified previously. BFW desuperheating is not needed for the LP steam because the LP steam is near saturated.

For path 1 involving the TG-1002 turbine, the 70 klb/h LP exhaust can be priced at

$$C_{LP}^{P1} = \$13.6/\text{klb} - \frac{\$90/\text{MWh}}{26.8\,\text{klb/MWh}} = \$10.2/\text{klb}.$$

For path 2 involving the TG-1001 LP extraction, the 10 klb/h LP extraction can be priced at

$$C_{LP}^{P2} = \$13.6/\text{klb} - \frac{\$90/\text{MWh}}{22.9\,\text{klb/MWh}} = \$9.7/\text{klb}.$$

For path 3 involving the letdown valve, the price of 11 klb/h of the letdown steam is the same as the MP price at \$11.4/klb. Thus, the weighted average price of the mixed LP steam is

$$C_{LP}^{av} = \frac{70C_{LP}^{P1} + 10C_{LP}^{P2} + 11C_{LP}^{P3}}{(70 + 10 + 11)} = \frac{(70 \times 10.2 + 10 \times 9.7 + 11 \times 11.4)}{91} = \$10.3/\text{klb}.$$

17.7   COMPARISON OF DIFFERENT STEAM PRICING METHODS

Let us press the pause button for a few minutes since we have been driving in the fast lane. The burning question requiring an answer at this point is: How do the above

TABLE 17.4. Comparison of Steam Prices

$/klb	HP	MP	LP
h-based	13.6	11.9	11.7
W-based	13.6	9.5	7.7
FE-based	13.6	10.6	8.8
Cost-based	13.6	11.4	10.3

steam pricing methods differ? Table 17.4 gives the comparison of steam prices determined by these methods. Overall, it can be observed from the table that the enthalpy-based method overvalues lower-pressure steam, while the exergy-based method underestimates lower-pressure steam. This is the key characteristic of these methods, which is caused by the nature of the steam property: enthalpy values are similar for steam at different pressure conditions, while exergy values differ significantly from pressure conditions.

The fundamental shortcoming of the enthalpy-based method is that power generation is not taken into account at all. At the other extreme, the exergy-based method gives value for any potential power generation even if there are no turbines in place, which results in undervalued lower-pressure steam. Therefore, both enthalpy- and exergy-based methods do not consider steam system configurations. This is wrong because steam system configuration affects the costs of making steam and power. A CHP (combined heat and power) configuration using highly efficient boilers and turbines can make steam and power much cheaper than a non-CHP system.

The FE-based method attempts to address these issues and is less likely to have inherent dangers in providing misleading steam prices than the enthalpy- and exergy-based methods; however, it will be more prudent to use the cost-based method if potential pricing errors are to be avoided. This is because the cost-based method can provide accounting values for steam and power, which reflect the true costs, and the best approach to economic reality.

The fundamental feature of the cost-based method is that it takes power generation into account for steam pricing. The valuation of shaft power tends to lower the average cost of lower-pressure steam. This could motivate process units to use lower-pressure steam in preference of higher-pressure steam. At the same time, it will encourage process units to consider investment on a steam generation facility for higher-pressure steam via use of waste heat and power generation facility, which will not only reduce the operating cost of the process units but also the overall energy costs. Thus, the steam pricing reflecting the true costs will avoid regretful investment. If accounting costs do not communicate the true economic conditions, business sectors such as process units may decide to install new facility, which could "benefit" the process units but will not benefit the overall entity.

17.8 MARGINAL STEAM PRICING

The above cost-based steam pricing method can be applied when determining average steam prices based on the most common production feed rates and process conditions. Thus, the steam prices can be useful for corporate reporting, energy cost allocation and budgeting for process areas, and communications with employees for awareness. However, can the average prices be used as the basis for evaluating operational improvements and energy capital investment projects? This is particularly relevant when involving important business decisions for operation improvements and implementing energy projects. In these cases, the process steam and power demands could change, which affects loadings of boilers and turbines and thus specific steam costs.

In dealing with changes in process steam and power demands, we usually consider increasing or decreasing steam usage with incremental amount and thus only interested in the incremental cost effect. This marginal effect on the operating expenditure (OpEx) is better described by the so-called marginal cost (MC) (Makwana and Zhu, 1998), which is defined as

$$C_{MC} = \frac{\text{Incremental energy cost}}{\text{Increment steam consumption}} = \frac{\Delta \text{OpEx}}{\Delta M_{steam}}. \tag{17.17}$$

ΔOpEx is incremental change in variable energy cost, which consists of fuel and BFW costs but not the fixed cost. The fixed cost does not have an effect on marginal price because the fixed cost is cancelled out in the calculation of ΔOpEx. In reality, the fixed cost has already taken place no matter how much incremental change occur in steam production.

By definition, marginal steam price is different from the average steam price as discussed previously:

$$C_{Av} = \frac{\text{Total cost}}{\text{Total steam consumption}} = \frac{\text{Variable cost} + \text{fixed cost}}{\sum M_{steam}}. \tag{17.18}$$

Let us use the next example to illustrate how marginal steam prices are determined.

Example 17.6 Marginal Steam Cost Calculations

Calculate the marginal prices for HP, MP, and LP steam if their demands are increased by 10 klb/h, respectively. Consider the steam system in Figure 17.1.

Solution.

17.8.1 Marginal Prices for HP Steam

There are two boilers in Figure 17.1. Boiler 1 is old with efficiency of 75%, and boiler 2 is relatively new with efficiency of 85%. Thus, there are two paths for HP steam generation: one from the old boiler and the other from the more efficient boiler.

Path 1: Increase HP Steam by 10 klb/h Through Boiler 1 with 75% of Efficiency   The incremental fuel consumption for making 10 klb/h of HP steam can be calculated based on equations (15.1) and (15.2)

$$\Delta Q_{HP} = \Delta m_{HP} \times (h_{HP} - h_{BFW}) \qquad (15.1)$$

and

$$\Delta Q_{fuel} = \frac{\Delta Q_{HP}}{\eta_{boiler}}, \qquad (15.2)$$

where Δm_{HP} is the incremental steam generation with additional 5–10% used internally for driving the water circulation pump and boiler fans. Internal usage of 7% is assumed here. Thus,

$$\Delta Q_{fuel}^{P1} = \frac{(1 + 0.07) \times 10 \times (1350 - 189.3)/1000}{0.75} = 16.6 \text{ MMBtu/h.}$$

With the additional BFW of 10.2 klb/h in which 0.2 klb/h is the extra blowdown, the marginal price for extra HP steam through path 1 is

$$C_{HP}^{P1} = \left.\frac{\Delta OpEx}{\Delta HP}\right|_{P1} = \frac{16.6 \times 6 + 10.2 \times 2.5}{10} = \$12.5/\text{klb.}$$

As can be observed, only the variable cost is included in the marginal price as the fixed cost is cancelled out in the $\Delta OpEx$ calculation.

Path 2: Increase HP Steam by 10 klb/h Through Boiler 2 with 85% of Efficiency   Similar to path 1, the marginal price for HP steam through path 2 can be calculated as

$$\Delta Q_{fuel}^{P2} = \frac{(1 + 0.07) \times 10 \times (1350 - 189.3)/1000}{0.85} = 14.6 \text{ MMBtu/h,}$$

$$C_{HP}^{P2} = \left.\frac{\Delta OpEx}{\Delta HP}\right|_{P2} = \frac{14.6 \times 6 + 10.2 \times 2.5}{10} = \$11.3/\text{klb.}$$

Clearly, boiler 2 can provide steam by $1.2 cheaper per 1000 lb/h steam than boiler 1.

Application Challenges: For this steam system, a revamp project requires 40 klb/h of additional HP steam and the limits of steam generation from boilers 1 and 2 are 160 and 140 klb/h, respectively. Since the current loading of both boilers 1 and 2 are at 108 klb/h respectively. Naturally, the first 32 klb/h HP steam demand will be provided from boiler 2 since it is more efficient than boiler 1. Thus, the marginal price for 32 t/h from boiler 2 is $11.3/klb, while the last 8 t/h to be made from boiler 2 will be priced at 12.5/klb.

What happens if the energy-saving project could reduce HP steam use by 40 klb/h? From which boiler should steam production be reduced and by how much? How do you value the steam to be reduced? These are the questions for you to answer.

17.8.2   Marginal Prices for MP Steam

As identified previously, there are two paths to increase the MP usage: the letdown valve and the MP extraction from TG-1001. Furthermore, there are two paths to increase HP steam: boilers 1 and 2. In combination, there are four paths connecting fuel consumption with the change in process MP demand.

Let us define the nomenclature first. The route of the HP-MP letdown is named as path 3 and the path $3 \rightarrow 1$ is to define the connecting path that links path 3 (HP-MP letdown) and path 1 (boiler 1). Similarly, path $3 \rightarrow 2$ is the connection path linking path 3 (HP-MP letdown) and path 2 (boiler 2). Let us look at each path in turn.

Alternatively, extra MP steam can be provided via TG-1001 MP extraction. If this flow path is named as path 4, the path $4 \rightarrow 1$ is the connection path link path 4 (TG-1001 MP extraction) and path 1 (boiler 1). Similarly, path $4 \rightarrow 2$ connects path 4 (TG-1001 MP extraction) and path 2 (boiler 2).

Path $3 \rightarrow 1$: Increase MP Steam by 10 klb/h Through the Letdown Valve and Boiler 1   The incremental effects include

- Boiler 1 fuel consumption increases by 16.6 MMBtu/h,
- BFW use increases by 10.2 klb/h in which 0.2 is extra blowdown rate.

Thus, the marginal MP price for path $3 \rightarrow 1$ is

$$C_{MP}^{P3 \rightarrow 1} = \frac{\Delta OpEx}{\Delta MP}\bigg|_{P3 \rightarrow 1} = \frac{16.6 \times 6 + 10.2 \times 2.5}{10} = \$12.5/klb.$$

$C_{MP}^{P3 \rightarrow 1}$ has the same value as that of C_{HP}^{P1}. Is this a coincidence? No, this is because with path $3 \rightarrow 1$, the MP steam is generated 100% by the letdown valve and through boiler 1.

Path $3 \rightarrow 2$: Increase MP Steam by 10 klb/h Through the Letdown Valve and Boiler 2   Since the fuel consumption in boiler 2 is increased by 14.6 MMBtu/h and the BFW amount increased by 10.2 klb/h for the extra 10 klb/h MP steam consumption, the marginal price for the MP steam following this path is

$$C_{MP}^{P3 \rightarrow 2} = \frac{\Delta OpEx}{\Delta MP}\bigg|_{P3 \rightarrow 2} = \frac{14.6 \times 6 + 10.2 \times 2.5}{10} = \$11.3/klb.$$

Path 4→1: Increase MP Steam by 10 klb/h Through the Steam Turbine TG-1001 and Boiler 1   TG-1001 is the steam turbine generator. Extra steam rate by 10 klb/h at MP extraction is allowed as the limit for the MP extraction is assumed to be 60 klb/h. The specific steam rate in the TG-1001 MP extraction is 35.6 lb/kWh. The incremental effects will be as follows:

- Boiler 2 fuel consumption increases by 16.6 MMBtu/h.
- BFW rate increases by 10.2 klb/h in which 0.2 is extra blowdown rate.
- TG-1001 increases power generation by 0.281 MW (=10/35.6) and thus power import reduces by the same amount with power price at \$90/MWh.

The overall incremental operating cost can be calculated as

$$\Delta OpEx = 16.6 \times 6 + 10.2 \times 2.5 - 0.281 \times 90 = 99.8\$/h.$$

Thus, the marginal MP price is

$$C_{MP}^{P4\to1} = \left.\frac{\Delta OpEx}{\Delta MP}\right|_{P4\to1} = \frac{99.8}{10} = \$9.98/klb.$$

Path 4→2: Increase MP Steam by 10 klb/h Through the Steam Turbine TG-1001 and Boiler 2   The overall incremental operating cost can be calculated as

$$\Delta OpEx = 14.6 \times 6 + 10.2 \times 2.5 - 0.281 \times 90 = 87.8\$/h.$$

Thus,

$$C_{MP}^{P4\to2} = \left.\frac{\Delta OpEx}{\Delta MP}\right|_{P4\to2} = \frac{87.8}{10} = \$8.78/klb.$$

Clearly, the MP steam from path 4→2 is the cheapest among all paths followed by path 4→1, path 3→2, and then path 3→1.

Application Challenges: What happen if extra MP demand is 40 klb/h, what sequence of paths should be followed to satisfy the extra demand with how much and at what cost? The limits for boilers and TG-1001 MP extraction have been given earlier.

What happens if MP usage is reduced by 40 klb/h? How does one implement the reduction in terms of flow paths? Please write down your answers and think it over to see whether the answer makes sense.

17.8.3   Marginal Prices for LP Steam

There are three paths to increase the LP usage: path 5 (the MP-LP letdown valve), path 6 (the TG-1001 LP extraction with specific steam rate of 22.9 lb/kWh), and path 7 (the TG-1002 LP extraction with specific steam rate of 26.8 lb/kWh), respectively. Plus two paths at the HP header, thus there will be six connection paths for linking LP steam and boilers, namely, path 5→1, path 5→2, path 6→1, path 6→2, path 7→1, and path 7→2.

The marginal prices for paths 5→1 and 5→2 are straightforward as

$$C_{LP}^{P5\to1} = \frac{\Delta OpEx}{\Delta LP}\bigg|_{P5\to1} = C_{HP}^{P1} = \$12.5/klb,$$

$$C_{LP}^{P5\to2} = \frac{\Delta OpEx}{\Delta LP}\bigg|_{P5\to2} = C_{HP}^{P2} = \$11.3/klb.$$

This is because the LP steam in these paths is generated by boilers 1 and 2 through HP letdown to the LP header.

The marginal prices for paths 6→1 and 6→2 can be determined as

$$C_{LP}^{P6\to1} = \frac{\Delta OpEx}{\Delta LP} = \frac{16.6 \times 6 + 10.2 \times 2.5 - 10/22.9 \times 90}{10} = \$8.58/klb,$$

$$C_{LP}^{P6\to2} = \frac{\Delta OpEx}{\Delta LP} = \frac{14.6 \times 6 + 10.2 \times 2.5 - 10/22.9 \times 90}{10} = \$7.38/klb.$$

Similarly, the marginal prices for paths 7→1 and 7→2 can be determined as

$$C_{LP}^{P7\to1} = \frac{\Delta OpEx}{\Delta LP} = \frac{16.6 \times 6 + 10.2 \times 2.5 - 10/26.8 \times 90}{10} = \$9.15/klb,$$

$$C_{LP}^{P7\to2} = \frac{\Delta OpEx}{\Delta LP} = \frac{14.6 \times 6 + 10.2 \times 2.5 - 10/26.8 \times 90}{10} = \$7.95/klb.$$

Clearly, the calculations of marginal LP prices determine the most economic flow paths that satisfy any extra amount of LP demand, which are paths 6→2, 7→2, 6→1 and 7→1, in order of lowest cost, while the worst paths are paths 5→1 and 5→2.

Application Challenges: Let us assume 40 klb/h of extra LP demand imposed from a revamp project. How can we satisfy this extra demand and why? Assume the flow limits of TG-1001 and TG-1002 LP extraction are 40 and 100 klb/h, respectively. The limits for boilers and TG-1001 MP extraction have been given previously.

What happens if MP usage is reduced by 40 klb/h from the base case in Figure 17.1? How can we implement the reduction in terms of flow paths and why? Please write down your answers and double check. After ensuring that your answer is correct, can you try to develop a generalized strategy for increasing and decreasing steam flows based on the concepts of flow paths and marginal prices?

17.8.4 Remarks for Marginal Steam Prices

The previous discussions demonstrate characteristics with marginal steam prices. The first characteristic of marginal steam prices is that they are path dependent.

When increasing or decreasing steam usage, there may exist multiple flow paths in the steam system to accommodate it. By calculating the marginal prices for these paths, the most economic path(s) could be selected to satisfy process variations in operation. Furthermore, the best path could be chosen for implementing an energy improvement opportunity, and the most representative energy price could be determined for cost and benefit analysis. Therefore, the marginal steam prices should be used for operation improvement and capital project evaluations.

The second characteristic of the marginal steam price is that all incremental steam change must be traced back to the fuel, BFW, and power balances so that the net effects on operating cost can be determined. This is the essence of marginal steam price.

17.9   EFFECTS OF CONDENSATE RECOVERY ON STEAM COST

In some applications such as stripping steam for separation columns, steam cannot be recovered as condensate returns to the deaerator. In other cases, condensate is lost due to either steam trap failure or lack of a recovery facility. Extra amount of freshwater must be provided as makeup to compensate for the loss. There are two major debits in costs for the makeup water: one is water treatment cost and the other is preheating cost to heat the freshwater from local temperature to condensate temperature. In the previous steam cost calculations, the costs for producing treated makeup water is evenly distributed on every pound of steam generated from boilers and cascading down to MP and LP steam.

As we know that some steam applications can have condensate return but other not, the steam for these applications must be differentiated in cost. For the example in Figure 17.1, the BFW cost is $2.5/klb for treated and preheated water in which $0.6/klb is the cost for pumping and preheating cold water to dearation temperature. In other words, if condensate is lost, the provision of makeup water is $2.5/klb. Steam costs should be revised accordingly to reflect the true costs of condensate.

17.10   CONCLUDING REMARKS

Clearly, the basis for calculating both the average and marginal steam prices is an overall steam balance including BFW and condensate balances, which is discussed in Chapter 16. The method for calculation of average steam costs follows a top–down approach, while the steam balance corresponds to the most common production scenario and steam system configuration.

In calculating the average steam costs, the focus is the average cost for meeting the total steam demand at a certain steam header. To do so, individual costs are added together to give a total cost for a certain steam header. The average steam cost for this header is equal to the ratio of this total cost over the total amount of steam demand at the same header. The average steam costs are useful for corporate reporting, energy budgeting for process units, and communications with employees.

When calculating marginal steam prices, the focus is to find the best way to meet incremental steam demand at a steam header. To do so, individual paths, which can make this incremental steam demand either fully or partially, are identified and the marginal costs for these paths are calculated. As a result, individual paths are differentiated and thus the best path(s) for meeting the incremental steam demand can be identified.

NOMENCLATURE

a availability or power potential
C_{fix} annual fixed cost
C_{var} variable cost
CapEx   capital investment
CM   maintenance and labor cost
CO   other costs including environmental control
FE   fuel equivalent
MFE   multiplier to fuel equivalence
OpEx   energy operating expenditure
PEE   power equivalent efficiency
Q heat content
q specific heat
R capital depreciation factor
x fraction of costs

REFERENCES

Cooper D (1989) Do you value steam correctly? *Hydrocarbon Processing Journal*, July, 44–47.

Kenney WF (1984) *Energy Conservation in the Process Industry*, Academic Press, Inc.

Makwana Y (1998) Energy retrofit and debottlenecking of total site, PhD thesis (X. X. Zhu is the principal supervisor), UMIST, Manchester, UK.

18

BENCHMARKING STEAM SYSTEM PERFORMANCE

For an existing steam system, if the best performance of energy efficiency can be predetermined, this could set a benchmarking basis for the purpose of monitoring and comparison. The performance gap can be readily identified by applying the methods presented in this chapter.

18.1 INTRODUCTION

The energy performance of an entire site is determined by both process energy performance (energy demand side) and steam/power system performance (energy supply side). In the previous chapters, discussions focused on the efficient use of energy from processes, which is about the energy demand side of the overall energy efficiency equation. This chapter focuses on the efficient generation and distribution of energy, which is about the energy supply side of the overall energy efficiency equation. The energy supply consists of four major components: steam generation, distribution, end-use, and recovery (Figure 18.1). The purpose of benchmarking a steam system is to provide performance measures of these four components so that the gap against the reference performance can be readily determined. The development of a steam system benchmark should provide insights and understanding for the following aspects:

- *Steam Losses:* boiler blowdown, dearation, steam distribution, steam trap and condensate discharge losses.

Energy and Process Optimization for the Process Industries, First Edition. Frank (Xin X.) Zhu.
© 2014 by the American Institute of Chemical Engineers, Inc. Published 2014 by John Wiley & Sons, Inc.

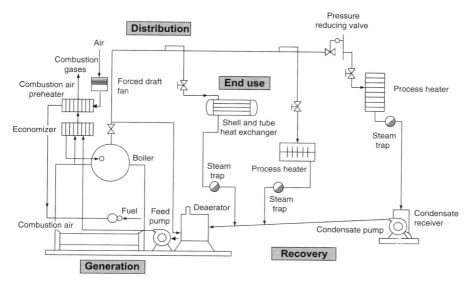

FIGURE 18.1. Four major steam components: steam generation, distribution, end-use, and recovery.

- *Equipment Performance:* boiler efficiency, turbine efficiency.
- *System Performance:* condensing power generation, steam letdown flows, venting.
- *System Configuration:* steam header pressures, steam turbine configuration, maximal cogeneration efficiency.
- Equipment operation: steam using equipment optimazation.
- Concept of *R*-curve for cogeneration (promote process heat integration).
- Energy cost distributions for process areas.
- Identify investment opportunities for on site power generation.
- Discourage energy projects with low energy efficiency.

18.2 BENCHMARK STEAM COST: MINIMIZE GENERATION COST

Benchmarking the fuel cost for steam generation is an important step in managing the steam system. For a given amount of steam production, the total fuel cost depends on boiler efficiency, steam pressure, fuel type and cost, and feedwater temperature. The benchmark can serve as a tracking tool for monitoring the boiler performance. The benchmark calculations are provided (DOE, 2012), which are briefly illustrated next.

Table 18.1 shows the heat input required to produce 1 lb of saturated steam at different operating pressures and varying feedwater temperatures, while Table 18.2 lists the typical energy content and boiler combustion efficiency for several common

TABLE 18.1. Energy Required to Produce lb of Saturated Steam, Btu[a] (DOE, 2012)

Operating Pressure (psig)	Feed Temperature (°F)				
	50	100	150	200	250
150	1178	1128	1078	1028	977
450	1187	1137	1087	1037	986
600	1184	1134	1084	1034	984

[a]Calculated based on the difference between the enthalpies of saturated steam and feedwater.

TABLE 18.2. Energy Content and Combustion Efficiency of Fuels (DOE, 2012)

Fuel Type, Sales Unit	Energy Content (Btu/Sales Unit)	Combustion Efficiency[a] (%)
Natural gas MMBtu	1,000,000	85.7
Natural gas, $\times$ 1000 ft^3	1,030,000	85.7
Distillate/No. 2 oil, gallon	138,700	88.7
Residual/No. 6 oil, gallon	149,700	89.6
Coal, ton	27,000,000	90.3

[a]Combustion efficiency is based on boiler equipped with feedwater economizer or air preheat and 3% oxygen in flue gas.

fuels. Data from these two tables can be used to determine the cost of usable heat from a boiler or other combustion unit. The calculations can also include operating costs of accessories such as feedwater pumps, fans, fuel heaters, steam for fuel atomizers and soot blowing, treatment chemicals, and environmental and maintenance costs.

18.2.1   Costs of Steam Production

Marginal steam price should be used as effective cost of stem production because the method considers the entire cycle from generation, distribution, and point of use to condensate recovery. The marginal price depends on the path it follows from the boiler to the point of use and if condensate is recovered or not. In revamp projects, when new boilers and turbines are required, total investment costs should be used as the basis for evaluation. The methods for boiler steam costs and marginal steam prices are provided in Chapter 17.

18.2.2   Fuel Selection

When boilers can handle multiple fuels, fuel selection could be very beneficial in reducing steam generation cost by taking advantage of the market variability in fuel prices. Calculations of steam cost generated by different fuels should reveal incentives for fuel switch. However, maintenance cost and emission regulations must be considered.

18.2.3 Heating Values

Fuel is sold based on its gross or higher heating value (HHV). If, at the end of the combustion process, water remains in the form of vapor, the HHV must be reduced by the latent heat of vaporization of water. This reduced value is known as the lower heating value (LHV). Fired fuel calculation should be based on LHV.

18.2.4 Benchmark Steam Cost

Make sure that the steam flow meter is installed in your boiler plant and calculate the steam cost to verity the benchmark value.

Example 18.1   Benchmark Boiler Steam Cost

A boiler fired with natural gas at $6/MMBtu produces 600 psig saturated steam. The BFW supply is 200 °F at $2.5/klb. Using Tables 18.1 and 18.2, calculate the benchmark steam cost in $/100 klb. What could be the effect on steam cost if BFW is 250 °F?

Solution. Steam cost $= 6 \times 1.034/0.857 + 2.5 = \$9.74/klb$
   If BFW is 250 °F, steam cost $= 6 \times 0.984/0.857 + 2.5 = \$9.39/klb$

18.3   BENCHMARK STEAM AND CONDENSATE LOSSES

In reality, steam generated in the boiler house is distributed through extensive steam pipelines to end users. The losses in steam distribution can be 10–30% of fuel fired in boilers and hence the net boiler efficiency is 10–30% lower than design efficiency from the user point of view. The losses do not necessarily attribute to a single cause, but are the result of a combination of various causes. It is common to observe the major steam loss caused by steam trap failure and condensate discharge problems. Steam loss could also occur due to poor insulation of steam pipes, leaks through flanges and valve seals, opened bypass and/or bleeder valves, and so on. For example, a 1/4 in. leak in a 600 psig line loses over 1000 lb/h steam. At the marginal cost of $11.9/klb for high-pressure steam, this leak could cost $95,000 per year on top of the noise and safety issues it may cause. Imagine how much money would be lost if there are more than 20 leaks, both visible and invisible, occuring in the entire plant. Simple measures such as maintenance of steam traps and monitoring of steam distribution to determine if steam generated is in accordance with steam consumed can lead to significant cost-saving benefits.

18.3.1   Steam Trap Management

A steam trap is essentially a control valve with a simple mechanism for separating steam and condensate. It is designed in such a way that it opens to allow condensate to go through but closes when presented with steam (hence "trapped"). If operating properly, steam traps have negligible loss of steam. However, when steam traps malfunction, the valve sticks either in the open or closed position. Failure in the open position can be analogous to having invisible holes in the steam system the same size as the trap's internal discharge passages where steam leaks out of the passages. Failure in the closed position results in condensate back up and causes steam heaters to cease to operate properly. There are many hundreds of steam traps all over the place in a site and many of them could malfunction if there is no best practice of maintenance in place. That is the reason why one of the most common energy wasters is steam leakage from "bad" steam traps, which is a major component of 10–30% of energy loss from steam. A world class performance is to have an annual failure rate of steam traps less than 5%. There are contractors available who can perform routine steam trap management for a site.

Failure in steam traps not only cause steam losses but also lead to damages to equipment and safety via water hammer. Two types of water hammers could be caused by steam trap malfunctioning: the cavitation occurs when a trap valve sticks in the open position as the hot steam bubbles collapse in the cold condensate, while the slug type happens when a trap valve sticks in the closed position as slugs of condensate remain in steam lines. Some traps tend to fail in an open position, while others tend to fail in a closed position. This tendency may also be influenced by the characteristics of the steam system.

The steam loss and safety issues could easily justify having an effective steam trap management program. What does this mean? Since the subject of steam traps and the management program could be mysterious to many people, it is worthwhile to discuss in some detail.

The objective of a steam trap is to remove noncondensable and condensate from steam with minimal loss of useful steam. While accomplishing this objective, a good steam trap should posses the following: does not freeze in cold weather, adaptable to the full range of loads for the given application, requires minimal maintenance, and lasts a long time.

Thus, *selection and sizing of steam traps* are critical for the management program. The steam trap should be sized for either start-up or operating conditions, depending on how often the unit will start up. If it is sized for operating conditions, then additional manual drainage should be considered for start-up. Six factors are listed in detail for selection of steam traps (Kenney, 1984):

- The anticipated condensate loads
- The pressure differential across the trap
- Start-up load
- The temperature of the condensate reaching the trap
- The need to vent air or noncondensable gases
- The proper safety factor for the application

It is important to know that steam traps should not be sized based on pipe size. A common problem is poorly sized traps that cause premature failure through excessive cycles and wear on internal parts as well as excessive steam leaks. Therefore, steam traps are selected primarily on duty; second, on duty variation for transient cases; and third, on equipment requirement (Kenny, 1989). Proper installation and maintenance together with regular vendor service can maintain good steam trap performance and long life.

Finally, a complete trap management program should consider effective condensate discharge without excessive steam leakage. This is a key part of the trap management program, which is the focus of the next discussions.

18.3.2 Condensate Discharge

The major problem with steam distribution is condensate removal. If steam traps are working properly while significant steam loss still occurs, the cause may be inadequate drainage in condensate discharge locations (CDLs). The consequence is condensate backing up in the system due to blocked traps and plugged drains. The water falls to the bottom of the pipe, which could cause water hammer and lower heat transfer efficiency.

Optimizing the CDLs requires finding ways to remove condensate without wasting steam from pipes. In a standard steam system, this means effective management of the CDLs (including the steam traps and associated piping) on the steam distribution headers, tracer lines, and equipment. The following guidelines (Broughton, 1994) can be applied for effective condensate removal:

- *Direction and slope of steam main pipes:* The recommended slope for steam mains is 1 in 200 in the direction of flow or 1 in 50 if the slope is opposite to the flow.
- *Location of drainage points:* The distance between drain points depends on the line size, location, and frequency of start-up. Much more condensate is generated during start-up because the system is initially at ambient temperature. Intervals of 45–60 m are usual if the slope of the pipe is in the direction of flow, reduced to 30 m if the slope is against the flow. Drain points are most effective where piping changes direction. Any obstruction to the free flow of condensate along the bottom of the pipe to the drainage point should be avoided. If an obstruction cannot be avoided, then a drain should be provided where the condensate is likely to collect.
- *Drain pocket:* When drainage has to be provided in a straight length of pipe, a large bore pocket should be provided to ensure the condensate will not be dragged over the opening of the pocket by the steam flow. The normal size of the drain pocket should be large enough to prevent the condensate from being whipped across the top of it.

As a best practice, CDLs should be checked once per year. For critical services such as those using superheated steam, flares, turbines, and process equipment should be checked more frequently. Failed drainage should be repaired immediately.

18.3.3 Condensate Return

For maximum efficiency, the condensate should return to boilers as much as possible. Otherwise, cold fresh water will be the makeup for any amount of condensate lost. In addition, condensate return will reduce treatment cost. The condensate is valued at the same price as BFW. To achieve maximal condensate return, a condensate system could be necessary in order to collect condensate from various steam traps. Care must be taken, however, to ensure the condensate is not contaminated. To do so, the contamination level should be monitored so it can be readily detected. Even so, condensate from many traps may not be collected if the distance is too far and the amount of condensate is too small to justify the investment. It is common to observe 20–30% of condensate lost; but 50% is excessive and should be reduced.

18.3.4 Steam Recovery

For multiple pressure steam systems, flash steam can be generated from higher-pressure condensate and used as low-pressure steam. Thus, for low pressure steam where the pipe is sized to handle the extra condensate generated in start-up, it should have the capacity to accommodate the flash steam generated.

18.3.5 Recommendations

- Steam flow meters for major users and totalizer for each process should be installed as this will provide indication of distribution losses. Install a condensate return totalizer for each process, which will identify the condensate loss.
- Establish a dedicated steam trap management program including CDL management with the objective of reducing the steam loss to a practical minimum with economic justification. A survey should be conducted on a regular basis to determine the gaps and actions to maintain the minimum level of losses.
- A monitoring tool could be helpful in tracking actual losses over time. The tool could be developed based on the steam balance with various losses incorporated. An example of such a monitoring tool is shown in Figure 18.2 in which the distribution losses, leaks, and losses from CDLs are estimated on a cumulative basis.
- Economic values for reducing these losses can be established as the costs for steam losses are the same as the marginal steam prices, while condensate loss has the same cost as BFW. Typically, 5–10% of energy savings can be achieved. If a total site spends $100 MM/year on energy operating cost, 5–10% savings is worth $5–10 MM/year.

It has been found from experience that the large energy savings usually come from minimizing steam losses.

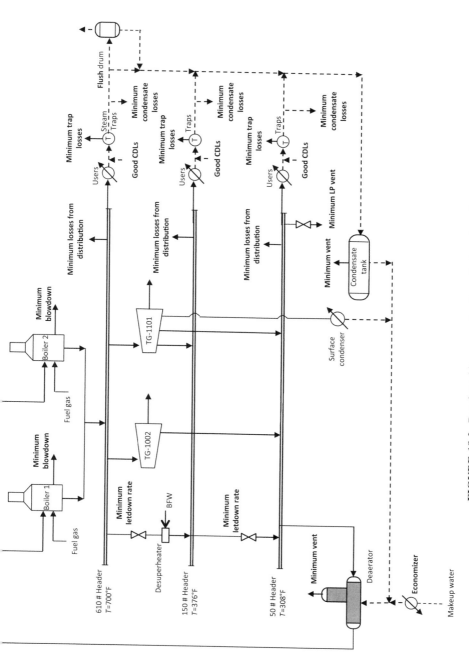

FIGURE 18.2. Benchmarking example of monitoring steam losses.

18.4   BENCHMARK PROCESS STEAM USAGE AND ENERGY COST ALLOCATION

Steam and power are used in various processes in a site. It is advisable for each process or cost center to share the overall energy operating cost, and to manage the energy operating costs as part of overall economic margin management because energy cost is the second largest variable cost, next to the raw materials, for most plants. The following illustrates how the energy allocation can be accomplished.

In conducting the energy cost allocation, the boiler steam cost is calculated based on variable cost only while capital cost is not included. The costs for lower pressure steam are determined from the boiler steam cost based on the cost-based method discussed in Chapter 17. Consider the steam system in Figure 18.3. The overall energy cost can be summarized as in Table 18.3 with the fuel price at $6/MMBtu and power at $90/MWh. Operating costs of accessories such as environmental and maintenance costs can be included in the energy costs. Total site shaft power by the process units is 20 MW of which 10.6 MW is generated on site, while 9.4 MW is imported electricity for motors. All the shaft power is billed to the process units at the variable power price of $90/MWh.

While Figure 18.3 provides an overview of the balanced steam flows, Table 18.4 gives a breakdown steam for four process areas as cost centers A, B, C, and D. Some of the steam usages have no condensate return and these users include stripping

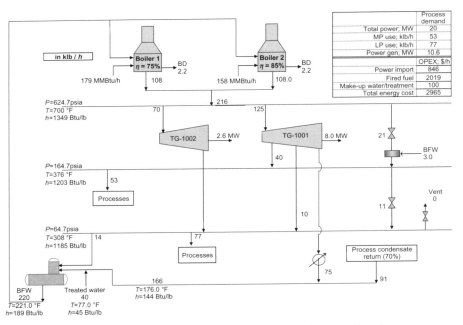

FIGURE 18.3. Steam system used as an example for cost allocation.

TABLE 18.3. Overall Steam and Power Cost Summary

Total makeup water	= 40 klb/h	= $100/h
Total boiler fuel	= 337 MMBtu/h	= $2019/h
Total import power	= 9.4 MW	= $846/h
Total cost of steam and power		= $2965/h
Cost attributed to power	= 20 MW × $90/MWh	= $1800/h
Cost attributed to steam	= $2965/h − $1800/h	= $1165/h

TABLE 18.4. Steam Summary (klb/h)

Area A MP (condensate recovery)	20
Area A LP (condensate recovery)	10
Area B LP (condensate recovery)	25
Area B LP (condensate lost)	15
Area C MP (condensate recovery)	17
Area C MP (condensate lost)	16
Area D LP (condensate recovery)	19
Area D LP (condensate lost)	8
Total steam consumption by all process areas	130
TG turbines	195
HP steam production in boilers	216

steam and locations without a condensate return facility. A power summary can be also obtained as shown in Table 18.5.

It is necessary to differentiate between steam costs for users with and without condensate recovery. It is expensive for condensate lost due to the need for makeup water, water treatment, and preheating, which costs the same as the BFW, that is, $2.5/klb in this example. For returned condensate, pumping, preheating, and deaeration LP steam are needed, which only costs $1.6/klb. Table 18.6 gives steam costs, which reflect the cost of the condensate.

With steam costs adjusted based on the statues of condensate recovery, energy cost allocation for four cost centers is established as shown in Table 18.7. Total steam and power costs are the same as the overall billing in Table 18.3 with negligible discrepancy. Thus, the overall cost is allocated for all cost centers to share the energy costs of the boiler house.

TABLE 18.5. Power Summary (MW)

Cost Centers	Electric Power	Mechanical Power	Total Power
A	1.4	2.4	3.8
B	3.0	1.8	4.8
C	2.4	3.0	5.4
D	2.6	3.4	6.0
Total	9.4	10.6	20.0

TABLE 18.6. Steam Cost Adjustments for Condensate

	HP Cost ($/klb)	MP Cost ($/klb)	LP Cost ($/klb)
Fuel cost alone	9.35	7.41	6.10
Steam cost (cond lost)	11.85	9.91	8.60
Steam cost (cond rec)	10.95	9.01	7.70
BFW cost (cond. lost) ($/klb) $= 2.5$		BFW cost (cond. rec) ($/klb) $= 1.6$	

TABLE 18.7. Cost Allocations for Cost Centers

Cost Centers	Power (MW)	Power Cost ($/h)	MP Steam (klb/h)	LP Steam (klb/h)	Steam Cost ($/h)	Steam + Power ($/h)
A	3.8	342.0	20.0	10.0	257	599
B	4.8	432.0	0.0	40.0	322	754
C	5.4	486.0	33.0	0.0	312	798
D	6.0	540.0	0.0	27.0	215	754
Total	20.0	1800.0	53.0	77.0	1106	2906

18.4.1 Recommendations and Remarks

- For cases with steam losses due to leakage, poor insulation, and so on, the costs associated with losses shall be taken into account against each cost center. In the case of condensate recovery, it is important to ensure that the quality of condensate being recovered is suitable for use as makeup water.
- The costs in Table 18.7 serve as the average costs for the cost centers. For any increased and reduced steam use from the base demand, marginal steam prices, as discussed in Chapter 17, will be used to cost the incremental amount of steam for the cost center.
- By knowing the correct energy costs for the cost centers, supervisors, engineers, and operators will include them as part of the overall margin optimization program. Benchmarking process energy cost is a very powerful tool to motivate cost centers to optimize their energy use.
- At the same time, benchmarking energy costs in the correct way will encourage each cost center to make an investment, which will not only reduce their own operating costs but also reduce the operating cost for the site as a whole. Too often an energy accounting system may look good to a cost center in its own battery limit and lead to investment in wrong equipment and result in a net penalty for the overall operating cost.

18.5 BENCHMARKING STEAM SYSTEM OPERATION

An important task of optimizing a steam system is to determine the most economical way to satisfy increasing and decreasing steam and power demands. This can be accomplished by identifying different flow paths connecting steam headers and benchmarking them based on the marginal steam prices. In other words, when increasing or decreasing steam usage, multiple flow paths may be available in the steam system to accommodate it. By calculating the marginal prices for these paths, the most economic path(s) could be selected to satisfy process variations in operation.

The second task is to benchmark the flow rates in steam letdown valves. A larger letdown flow than the benchmark value could indicate the need for a driver switch between the motor and steam turbine to minimize letdown flow. However, the driver switch must obey the trade-off between fuel and power cost. The detailed discussions and examples for the above two tasks will be provided in Chapter 19.

The third task is to benchmark steam using equipment including steam boilers, turbines, steam heaters, column stripping steam rate, steam generators, and so on. Experience shows that many steam equipment operations are not optimized. Examples are stratified heat transfer, which reduces the effective heat transfer area. Too much stripping steam causes flooding in separation columns, eroded turbine blades reduce turbine efficiency, and many other examples. With benchmark values established for the steam using equipment, we can identify equipment malfunctioning at an early stage. In some cases, it can also improve product yields while reducing energy costs.

Opened bypass valves, opened steam-bleed valves, steam leaking out of steam traps, visible waste condensate, and system water hammers are the signs that steam consuming equipment are not optimal. An opened bypass valve usually indicates equipment limitation, so operators open bypass valves to compensate. Bleeders are often visible at turbines and in steam jacket tracing applications. This could be a sign that operators may not have confidence in the installed CDL, so they bleed valuable steam to achieve the required operation.

The fourth task is to benchmark heat recovery across the process areas that are not directly related to steam usage; however, process heat recovery could lead to reduced steam usage and increased steam generation using process heat effectively. In most cases, increased process heat recovery results in the largest energy savings. The benchmark for process heat recovery can be determined by pinch analysis, which is discussed in Chapters 9, 10 and 11.

18.6   BENCHMARKING STEAM SYSTEM EFFICIENCY

No matter how much the steam system could be optimized, one relevant and important question is: What is the maximal efficiency that the steam system can achieve? The answer will be provided by the concept of cogeneration efficiency that is discussed next.

A steam system configuration implies the number of steam headers and their pressure and temperatures, the number of steam turbine drivers and generators, as well as their operating ranges and efficiency, the connections of turbines with steam headers, letdown flow rates, BFW preheat, and condensate recovery system. An efficient steam system configuration is distinguished with the following features:

  (i) Minimal condensate losses
 (ii) Minimal letdown rate through valves
(iii) Maximal BFW preheat by process waste heat
 (iv) Minimal condensing power generation
  (v) Optimal driver selection.

Processes require energy in the form of fuel, steam, and power while the primary purpose of the steam system is to produce steam to satisfy the process steam demand at different pressure levels. If the steam system has no cogeneration, power will be purchased from the electricity grid in full to satisfy process power demand. At the same time, lower pressure steam is produced from letdown of higher pressure through valves. The principle of cogeneration is to generate power based on pressure expansion from high- to lower-pressure steam, which can be accomplished via steam turbines. The measure of the cogeneration level is the ratio of power to process heat that can be supplied by steam (Kenney, 1984; Kimura and Zhu, 2000). This ratio is named as R ratio expressed as

$$R = \frac{\sum W_{\text{gen. onsite}}(\text{Btu})}{\sum Q_{\text{process}}(\text{Btu})}, \tag{18.1}$$

where W is the power generation from the steam system and Q is process heat provided by steam.

The cogeneration efficiency for the steam system is defined as

$$\eta_{\text{cogen}} = \frac{\sum W_{\text{gen. onsite}}(\text{Btu}) + \sum Q_{\text{process}}(\text{Btu})}{\sum Q_{\text{fuel}}(\text{Btu})} \tag{18.2}$$

or

$$\eta_{\text{cogen}} = (R + 1)\frac{\sum Q_{\text{process}}(\text{Btu})}{\sum Q_{\text{fuel}}(\text{Btu})}. \tag{18.3}$$

Clearly, cogeneration efficiency is a function of R ratio and the steam system configuration. Although the steam system is operated with the objective of achieving minimum cost, the maximum efficiency provides a measure of the best efficiency which the system can reach for a given configuration.

Consider improving the steam system in Figure 18.3 to operate the system in a maximal energy efficiency manner. To achieve this, three things are changed: the load for boiler 2 is increased close to the maximum limit of 160 klb/h; the condensing flow in TG-1001 is reduced to the minimum operating limit; and letdown flows are reduced to zero. In this maximal efficiency operating mode as shown in Figure 18.4, the R ratio can be calculated as

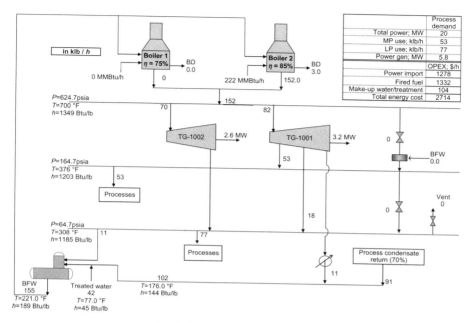

FIGURE 18.4. Maximal energy efficiency operating mode.

$$\sum W_{\text{gen. onsite}} = 2.6 + 3.2 = 5.8\,\text{MW}$$

$$\sum Q_{\text{process}} = 53 \times \Delta h_{\text{MP}} + 77 \Delta h_{\text{LP}} = 45.4 + 70.2 = 115.6\,\text{MMBtu/h}$$

where Δh_{MP} and Δh_{LP} are latent heat for MP and LP steam respectively.

$$R = \frac{5.8 \times 3.414}{115.6} = 0.17,$$

$$\eta_{\text{cogen}} = (0.17 + 1)\frac{115.6}{222} \times 100\% = 61\%.$$

The R ratio and cogeneration efficiency based on the current operation in Figure 18.3 can be calculated as

$$R = \frac{10.6 \times 3.414}{115.6} = 0.31,$$

$$\eta_{\text{cogen}} = (0.31 + 1)\frac{115.6}{336} \times 100\% = 45\%.$$

Thus, the gap in cogeneration efficiency is 16% (61–45). How much is it worth in terms of MMBtu/h? We can use net energy input to compare: fired fuel is 336 and

222 MMBtu/h, while power import is 9.4 and 14.2 MW for the current and maximal efficient operating modes, respectively. The heat rate for imported power is 9.09 MMBtu/MWh. Thus, the fuel equivalent for both cases can be calculated as

$$FE_{current} = 336 + 9.4 \times 9.09 = 421.4 \, MMBtu/h,$$

$$FE_{max \, eff} = 222 + 14.2 \times 9.09 = 351.1 \, MMBtu/h.$$

The current operating mode spends 70.3 (421.4–351.1) MMBtu/h of energy more than the maximal efficient mode, which is very significant. What is the reason for the inefficiency in the current operation?

The current operation mode spends additional fuel of 114 MMBtu/h ($336 - 222$) to make incremental power of 4.8 MW ($10.6 - 5.8$) compared with the maximal efficient operation mode. Since the power is generated through the condensing path in TG-1001, there is no credit from the turbine exhaust for process heat. Thus, the power generation alone gives efficiency of 14.6% ($4.8 \times 3.414/114$). This is very inefficient compared with two other flow paths with back pressure power generation where turbine extractions are used for process heat. Thus, the condensing power generation pulls the cogeneration efficiency down. As a general rule, the more condensing the power generation, the lower the cogeneration efficiency of the steam system.

You may ask: What is the steam system configuration that can achieve the best cogeneration efficiency for a given process steam and power demands? In the previous discussions, we know that the maximum cogeneration features maximal back-pressure power generation and minimal condensing. To achieve this, let us define the ideal R ratio, which is expressed as

$$R_{Ideal} = \frac{\sum W_{min \, Cond \, power}^{max \, BP \, power}}{\sum Q_{process}}. \tag{18.4}$$

The cogeneration efficiency reaches maximum at R_{ideal}, that is

$$\eta_{cogen}^{max} = (R_{ideal} + 1) \frac{\sum Q_{process}}{\sum Q_{fuel}}. \tag{18.5}$$

Under R_{ideal}, the existing turbines will be replaced with the more efficient ones and they can generate more power than the less efficient turbines for the same steam rates and conditions. Such an example is shown in Figure 18.5, which is developed from Figure 18.4.

R_{ideal} and η_{max} for the steam system in Figure 18.5 are calculated as

$$R = \frac{(3.6 + 3.7) \times 3.414}{115.6} = 0.21,$$

$$\eta_{cogen} = (0.21 + 1) \frac{115.6}{222} \times 100\% = 63\%.$$

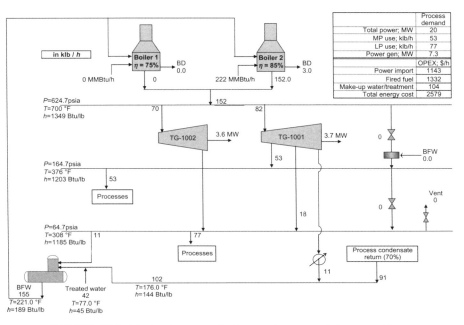

FIGURE 18.5. Maximal energy efficiency using state-of-the-art trubines.

There is only 2% (63%–61%) improvement from the steam system operation in Figure 18.4. The net fuel equivalent for the improved steam system with power import of 12.7 (20–7.3) mw is

$$FE_{Ideal} = 222 + 12.7 \times 9.09 = 337 \, MMBtu/h.$$

Compared with the net fuel equivalent of 351 MMBtu/h for the steam system in Figure 18.4, use of the most efficient turbines can only achieve energy savings of 14 MMBtu/h (351 − 337). This indicates that the optimized existing steam system can achieve cogeneration efficiency close to that by an ideal steam system configuration and thus investment on new turbines for this steam system cannot be justified. However, turbine investment could be justified for cases with very large letdown rates and expensive power. Similar evaluations can be applied to potential investment on boilers.

18.6.1 Remarks

- *R* ratio analysis with simple calculations can provide a guideline for achieving the maximal and cogeneration efficiency for a given steam system configuration. As a general rule, back-pressure power generation enhances cogeneration efficiency while condensing turbine does the opposite.
- Ideal *R*-ratio analysis can provide a high level benefit assessment for reconfiguring the steam system using more efficient equipment.

- For steam systems with boiler steam at 610 psig or lower, the operating mode with maximal cogeneration efficiency is equivalent to minimal energy lost mode for most scenarios. There are several features of maximal cogeneration efficiency mode. The most distinctive feature is the maximal power generation from back pressure turbines but minimal from condensing turbines. The second one is minimal steam rates from letdown valves as well as minimal steam losses. However, there could be exceptions when the economic operation is in conflict with energy efficient operation. This occurs when importing power is very expensive relative to onsite power generation. This could be particularly true for the systems with high power generation efficiency such as the steam systems with boiler steam at 1500 psig or combined cycle systems with gas turbines on top of steam turbines. In such scenarios with expensive power price, it could be more economic to maximize onsite power generation including condensing turbines. If possible, the plant wants to capture the spot opportunity via exporting power to the grid. In contrast, when power is cheaper relative to onsite power generation, the plant wants to import as much power as possible and select motors as process drivers but operating back pressure steam turbines to balance the steam system. The question is how can we identify this kind of opportunities? The analysis method is provided in Chapter 19.

NOMENCLATURE

FE   fuel equivalent
Q heat duty
R power to heat ratio
W power
η efficiency

REFERENCES

Broughton J (1994) *Process Utility Systems: Introduction to Design, Operation and Maintenance*, Institution of Chemical Engineers.(IChemE).

DOE (2012) Energy tips: Benchmark the fuel cost of steam generation, January, DOE.

Kenney WF (1984) *Energy Conservation in the Process Industry*, Academic Press, Inc.

Kimura H, Zhu XX (2000) R-Curve concept and its application for industrial energy management, *Ind. Eng. Chem. Res.*, **39**, 2315–2335.

19

STEAM AND POWER OPTIMIZATION

The ultimate goal in managing a steam and power system is to satisfy the process energy demand at the least cost. Flow paths available in the system provide major degrees of freedom to satisfy incremental steam and power demands. Power and fuel purchases are additional options that should be compared with existing flow paths available in the steam system to minimize energy operating cost. A steam balance based optimization approach can play an important role for optimizing the overall steam and power system.

19.1   INTRODUCTION

Within a site comprising process units, the interconnections of steam pipes can be frighteningly complex. Regardless of the complexity of a steam system, the overall objective is to meet the steam and power demand with minimal operating cost shown in the annual utility bill. Steam system optimization can significantly reduce overall operating cost and improve profit, while reducing total fuel consumption and greenhouse gas emissions.

What are the characteristics of an optimized steam system? The main characteristics can be summarized as follows:

(i)   Minimum steam and condensate losses
(ii) Minimum letdown rate through valves
(iii) Maximal boiler feedwater (BFW) preheat by process waste heat

Energy and Process Optimization for the Process Industries, First Edition. Frank (Xin X.) Zhu.
© 2014 by the American Institute of Chemical Engineers, Inc. Published 2014 by John Wiley & Sons, Inc.

(iv) Optimized steam header pressure and good control of header temperature

(v) Optimized steam equipment loadings

(vi) Optimized trade-off between on-site power generation and power import/export.

The first three items are discussed in detail in Chapter 18 while the last three items are the focus of this chapter.

19.2 OPTIMIZING STEAM HEADER PRESSURE

Optimizing steam header pressure could significantly improve turbine power generation efficiency as the inlet and back pressure of the turbine are the major parameters affecting power generation. The result is reduced steam usage required for making the same amount of power. The general direction of optimizing header pressure is increasing the high (HP) header pressure and reducing condensing pressure. The medium (MP) and low (LP) header pressures depend on the trade-off of power generation between HP–MP turbines and MP–LP turbines.

In one steam system as shown in Figure 19.1, the current HP header pressure is 610 psig and 700 °F, which had been recognized as a comfort zone by the mechanical engineers in a plant. When the new supervisor of the boiler house took the position, he knew that the HP header pressure was a key operating parameter in reducing operating cost and thus asked the unit engineer to conduct an assessment. The

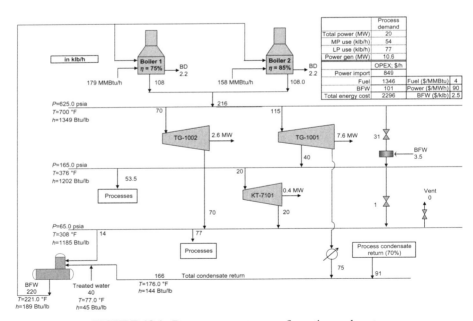

FIGURE 19.1. Base steam system configuration and costs.

engineer came back with the assessment result: If the HP header can be operated at the design conditions, that is, 650 psig and 750 °F, an extra 1 MW of power can be generated without additional HP steam required. The net benefit could be worth $480,000 per year of utility cost savings. By knowing the high value at stake, the supervisor issued an inspection project to check the pressure rating and metallurgy integrity for the HP steam header and related equipment. The inspection identified a few items in the steam pipeline to be upgraded, which was fixed during the turnaround maintenance. Afterward, the steam header pressure was able to be increased to 650 psig and the benefit was realized.

19.3 OPTIMIZING STEAM EQUIPMENT LOADINGS

Energy can be distributed to processes via different routes. Thus, energy supply optimization is about distributing energy via the least cost route. For example, steam can be provided via letdown valves, letdown steam turbines, or purchased from the grid a nearby power plant. On the other hand, a process pump or compressor could be run by a steam turbine or motor. These different routes of providing energy give degrees of freedom on increasing energy efficiency and reducing costs.

The loadings of steam equipment such as on-purpose boilers and turbine generators must be optimized. When multiple boilers are available, optimizing boiler loading will reduce fuel consumption, while optimizing the loading of turbine generators will reduce steam consumption for a given power generation. When varying steam loadings, the allowable limits must be observed as they could affect equipment operability and reliability. Loading optimization for boilers and turbo-generators should be conducted based on efficiency curves of the boiler and turbines. This strategy allows for the fastest dynamic response while always trending to the most economical steady state operating position.

Consider the steam system operation with the loading profile as shown in Figure 19.1 but with the HP header pressure increased as discussed previously. The operating policy for the plant was to run two boilers at equal loadings. However, the new supervisor of the boiler house wanted to know if the current loading profiles were the best possible. He asked the unit engineer to calculate the efficiency for boilers and turbines, which are given in Table 19.1. If you were the engineer, what suggestion would you like to give to the supervisor for the loadings?

Based on the specific fuels for boilers and steam rates for steam turbines calculated as shown in Table 19.1, the engineer suggested that directionally, the boiler 2 capacity should be increased while boiler 1 should be decreased. Steam rates for various turbine paths should be adjusted based on specific steam rates. One step further, the engineer made a simulation of loading changes for the steam system to see the effects of these changes. He shared the results with the supervisor as in Figure 19.2 indicating the cost savings of $360,000 per year by changing the boiler loading only while satisfying the minimum loading limit. This value proposition convinced the supervisor and operators that the boiler loadings should be determined based on the boiler efficiency, which is very different from the current boiler operating policy of equal loading.

TABLE 19.1. Steam Equipment Efficiency and Limits

	Specific Fuel Rate (MMBtu/klb)	Maximum Limit (klb/h)
Boiler 1 steam generation	1.7	160
Boiler 2 steam generation	1.5	140
	Specific Steam Rates (klb/MWe)	Maximum Limit (klb/h)
TG-1001		
HP–MP expansion	32.6	70
HP–Cond expansion	10.5	100
TG-1002	25.0	80
KT-7101	56.9	20

On top of changing boiler loadings, the engineer also suggested the loading change to turbines as shown in Figure 19.3. The incremental cost savings from turbine loading change alone is $787,000 per year, which is very significant.

As a recap, we have identified three changes for the same steam system:

- Increasing HP header pressure saves steam cost of $480,000 per year.
- Optimizing boiler loading leads to cost savings of $360,000 per year.
- Optimizing turbine loading results in cost savings of $787,000 per year.

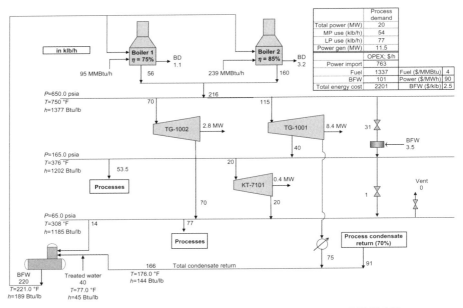

FIGURE 19.2. Optimized boiler loadings resulted in cost savings of $360,000 per year.

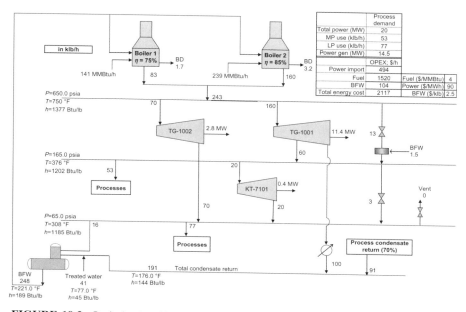

FIGURE 19.3. Optimized turbine loadings resulted in cost savings of $787,000 per year.

Thus, these three changes give a total benefit of $1.63 MM/year, which is 8% of total cost savings for the steam system, not mentioning other opportunities available.

19.4 OPTIMIZING ON-SITE POWER GENERATION VERSUS POWER IMPORT

For plants that are allowed to import/export electricity to the grid, it adds an additional degree of freedom to consider as to when to import or export and by how much. Electricity price could vary due to the market balance between supply and demand. More on-site power generation is the right thing to do if electricity price is relatively higher than the cost of on-site generation and vice versa. For some process plants, opportunities arise to export electricity when the electrical grid is in sort of supplying. Thus, changes in on-site power generation could be an important operating parameter to influence operating cost.

Let us revisit the steam system in Figure 19.2 with the purpose of assessing the effects of changing on-site power generation on cost savings and revealing the fundamentals behind on-site power generation versus power import/export. To assess the effects, the engineer made two changes:

- The first change is to increase the power generation from TG-1001: If the steam rate for the condensing path is pushed to the limit (Figure 19.4), an extra 2.4 MW of power can be generated from TG-1001 at the expense of extra fuel

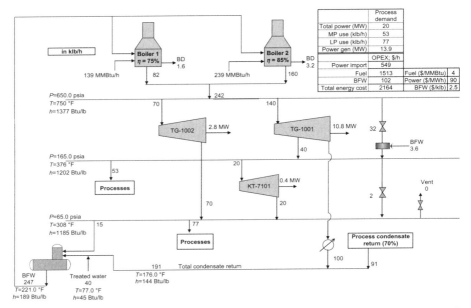

FIGURE 19.4. Increasing the power generation from TG-1001 resulted in cost savings of $330,000 per year.

fired in the boiler compared with that of Figure 19.2. The net cost savings is $330,000 per year.

- The second change is to increase the power generation from TG-1002 via raising the steam rate to the limit at 80 klb/h (Figure 19.5). Although an extra 0.4 MW of power is generated from TG-1002, additional fuel in the boiler outweighs the benefit from reduced power import. Thus, this change results in a net cost penalty of $400,000 per year.

Therefore, the million dollar question is: Why would increasing power generation from one turbine reduce cost savings while from another turbine increase cost? You may be smart enough to say: this is because power generation efficiency is different between the two turbines. The answer is partially correct because it does not take into consideration of price of fuel and electricity. In other words, if the fuel price is much cheaper than $4/MMBtu and power price more expensive than $90/MWh, increasing power generation from both turbines could give cost savings. So what is the fundamental and quantative explanation?

19.4.1 Market Power Equivalent Efficiency and Heat Rate

Qualitatively, on-site power generation could be favorable if it is more cost-effective compared to power import and vice versa. The truth is that various steam turbines

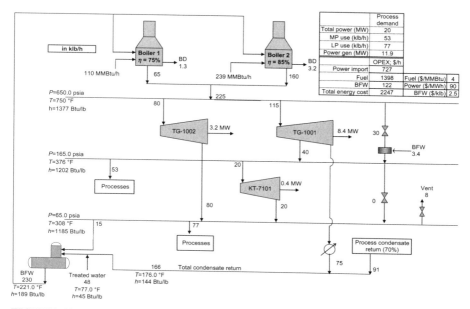

FIGURE 19.5. Increasing the power generation from TG-1002 resulted in cost penalty of $400,000 per year.

have different power generation efficiency. It is possible that one turbine could make power cheaper than power import, while the other turbines may not be able to. As steam is made from fuel fired in boilers, the relative price of power to fuel is more relevant to the comparison of on-site power generation cost. On the other hand, on-site power generation cost is a function of turbine efficiency, boiler efficiency and fuel cost. With this qualitative understanding, we are ready to tackle the issue of comparing on-site power generation with power import.

To reflect the relative price comparison of fuel and power, the concept of price equivalent efficiency (PEE) for power generation defined in Section 17.5 is used here to express the price ratio of fuel to power:

$$\eta_{PEE} = \frac{\text{Fuel price of 1 MWh (marginal fuel)}}{\text{Power price of 1 MWh (import or export power)}}. \tag{17.10}$$

Or

$$\eta_{PEE}^{import} = \frac{C_f}{C_p}, \tag{19.1}$$

where C_f is marginal fuel price ($/MWh-fuel) and C_p is imported or exported power price ($/MWh-power). The discussions below focus on power import, which can be applied to the case of power export.

To understand the physical meaning of the price ratio defined in equation (19.1), we can revise the fuel and power prices as how much Q_{fuel} in MWh can be purchased by \$1 versus how much W_{power} in MWh can be purchased per \$1. Thus, the price ratio in equation (19.1) can be redefined as

$$\eta_{PEE}^{import} = \frac{\$1/Q_{fuel}}{\$1/W_{power}} = \frac{W_{power}}{Q_{fuel}}. \tag{19.2}$$

Equation (19.2) converts the price ratio to the ratio of power to fuel, each amount of which can be purchased with \$1. The expression in equation (19.2) is the same as that of power generation efficiency and thus η_{PEE}^{import} can be called import power generation efficiency, which will be compared against on-site power generation efficiency.

The fuel equivalence for power import can be calculated based on equation (17.11) with a revised version as

$$FE_{PEE}^{import} = \frac{3414}{\eta_{PEE}^{import}} \text{ Btu/kWh}. \tag{19.3}$$

In the power industry, the fuel equivalent for power generation is also called heat rate for power generation. Since the fuel and power prices are market driven, the expression in equation (19.3) can be named as market heat rate for power import. With the market heat rate defined as such, it can be compared to heat rates for on-site power generation.

There could be several turbines in the steam system for power generation. For some turbines, there could be multiple extraction points. Each turbine extraction is a power generation path with a specific steam rate ($m_{steam,i}$) and connects with boiler efficiency ($\eta_{boiler,j}$). Based on equation (15.11), on-site power generation efficiency is

$$\eta_{PEE,ij}^{on\text{-}site} = \frac{3414}{m_{steam,i} \times FE_{inlet\ steam,ij}}, \tag{19.4}$$

where

$$FE_{inlet\ steam,ij} = \frac{h_{inlet\ steam,i}}{\eta_{boiler,j}}.$$

$m_{steam,\ i}$ is the specific steam rate of power generation for the turbine path i, $h_{inlet\ steam,i}$ is the enthalpy for the steam inlet to the turbine path i and $\eta_{boiler,j}$ is the efficiency for boiler j.

The heat rate for on-site power generation can then be defined as

$$FE_{PEE,ij}^{on\text{-}site} = \frac{3414}{\eta_{PEE,ij}^{on\text{-}site}} \text{ Btu/kWh}, \tag{19.5}$$

where $\eta_{PEE,ij}^{on\text{-}site}$ and $FE_{PEE,ij}^{on\text{-}site}$ are the power generation efficiency and heat rate for on-site power generation from turbine extraction i and boiler j.

Introduction of power generation efficiency and heat rate for both power import and on-site power generation makes us ready to answer the fundamental question raised previously from Figures 19.4 and 19.5: Under what price ratio of fuel to power is on-site power generation more economical than power import?

Consider the three options we discussed previously: power import, power generation from TG-1001 condensing path, and TG-1002. Determining the preference of these options becomes a matter of calculating PEE and heat rates.

For power import, we have

$$\eta_{PEE}^{import} = \frac{C_f}{C_p} = \frac{\$4/MMBtu \times 3.414\,MMBtu/MWh}{\$90/MWh} = 0.152.$$

This implies that importing power has equivalent power generation efficiency of 15.2%. Thus, the heat rate for power import is

$$FE_{PEE}^{import} = \frac{3414}{\eta_{PEE}^{import}} = \frac{3414}{0.152} = 22,460\,Btu/kWh.$$

This implies that importing power requires fuel equivalent of 22,460 Btu to produce 1 kWh of power.

For TG-1001 condensing path, the power generation efficiency and heat rate are

$$\eta_{PEE}^{on\text{-}site} = \frac{3414}{m_{steam} \times h_{steam}/\eta_{boiler,1}} = \frac{3414\,Btu/kWh}{10.5\,lb/kWh \times 1377\,Btu/lb/0.75} = 0.177,$$

$$FE_{PEE,ij}^{on\text{-}site} = \frac{3414}{\eta_{PEE,ij}^{on\text{-}site}} = \frac{3414}{0.177} = 19,288\,Btu/kWh.$$

For TG-1002,

$$\eta_{PEE}^{on\text{-}site} = \frac{3414}{m_{steam} \times h_{steam}/\eta_{boiler,1}} = \frac{3414\,Btu/kWh}{(25.0\,lb/kWh \times 1377\,Btu/lb)/0.75} = 0.074,$$

$$FE_{PEE,ij}^{on\text{-}site} = \frac{3414}{\eta_{PEE,ij}^{on\text{-}site}} = \frac{3414}{0.074} = 46,135\,Btu/kWh.$$

The reason for using boiler 1 as the basis is that boiler 2 has reached the limit after boiler loading optimization in Figure 19.2.

Clearly, from heat rate point of view, TG-1001 condensing path beats power import, while power import option comes second, with TG-1002 as the worst among the three options. That explains why increasing power generation from TG-1001 condensing path will reduce operating cost (Figure 19.4), while increasing power generation from TG-1002 will make operating cost go up (Figure 19.5).

These discussions for power import can be extended to the case of power export. If the plant is allowed to export power to the grid, the question then becomes: Under what price ratio of fuel to power is it beneficial to export power? Similar to power import, answering this question is a matter of calculating heat rate for power export using equation (19.3) and on-site power generation heat rate using equation (19.4).

So far, we discussed power import and export with their PEE and heat rates. As a matter of fact, the PEE and heat rate of power import are the same as that of power export. To avoid confusion, we use the terms *market power equivalent efficiency* and *market heat rate* for both power import and export.

After all, as the simplest way to conclude, on-site power generation beats power import when on-site heat rate from certain turbine flow path beats market heat rate. In this case, maximize power generation through this turbine path while observing steam balance. If possible, surplus power generated is exported to grid.

19.4.2   Cut Point of Power Price

What is the cut point for the power price in competing with TG-1001 condensing path if the fuel price is maintained the same as $4/MMBtu/h? This could be an interesting question. Since at the cut point of power price, the power generation from the condensing path equals the power import. Thus, the condition for the cut point can be expressed as

$$\eta_{\text{PEE}}^{\text{import}} = \eta_{\text{PEE,Cond}}^{\text{on-site}}. \tag{19.6}$$

Since $\eta_{\text{PEE,Cond}}^{\text{on-site}} = 0.177$,

$$\eta_{\text{PEE}}^{\text{import}} = \frac{C_f}{\hat{C}_p} = \frac{\$4/\text{MMBtu} \times 3.414\,\text{MMBtu/MWh}}{\hat{C}_p\$/\text{MWh}} = 0.177.$$

Solving for $\hat{C}_p$ yields

$$\hat{C}_p = \frac{4 \times 3.414}{0.177} = \$77/\text{MWh}.$$

This leads to a powerful observation that the power generation from TG-1001 condensing path will beat power import as long as power price is above $77/MWh for the fixed fuel price at $4/MMBtu and the opposite is true. When the fuel price changes, the cut point of power price varies accordingly.

19.5   MINIMIZING STEAM LETDOWNS AND VENTING

In many steam systems, steam letdowns are used to satisfy the steam demands in lower-pressure headers as the demands could increase due to production variation from time to time. However, providing steam through a pressure letdown valve is at the expense of lost power potential. Two major handles can be used in operation to minimize the letdown flows, which are loading optimization and driver switch. In case excessive letdown flows are still remaining after conducting these two steps, installing a new turbine generator could be considered.

In a process plant, the steam system is operated with a large amount of steam letdowns from 610 to 150 psig and from 150 to 60 psig in summer and winter operation, respectively, as well as 60 psig venting in summer.

To capture the energy loss from steam going through these letdown valves, the process plant considered to install a new steam turbine with multiple extractions to capture HP–MP and MP–LP letdowns. The steam turbine will operate in varying loads depending on the amount of letdown flows available. On average, a minimum of 4 MW power using the letdown steam flows can be generated, which reduces power import and results in cost savings of $3 MM/year and fuel equivalent savings of 36 MMBtu/h. The installed cost for a steam turbine is $7 MM and thus the simple payback is slightly higher than 2 years, which could be a good investment project.

In another process plant, the steam system operates with a large amount of LP vent. To prevent loss of valuable condensate due to LP vent, a condenser was installed to cool down the LP steam and return LP condensate back to the deaerator. Although this solution saves condensate, it did not resolve the LP long issue. It was later identified that a driver switch could help to reduce the LP vent. In the boiler house, there are three forced draft (FD) fans currently run by MP–LP extraction turbines but fans have spare motor drivers. The operation policy acceptable to the plant was to use steam turbines for reliability reasons. The engineer wanted to establish the value of the driver switch to minimize the LP dump.

When switching to motor for one FD fan, 100 t/d of MP use for the turbine is eliminated and the LP dump is reduced by the same amount. The net benefit comes from the HP steam saved since there is a large amount of HP letdown, which is worth about $890,000 per year. The significant value triggered the debate to challenge the current operating policy, which was developed many years ago when the electricity supply from the public grid, was not reliable. But the grid reliability has been improved significantly. It only triped once in the last five years.

After discussions with deliberation, the operating policy was revised so that one FD fan can be electrically driven when three boilers are online. The reason for this practice is that, in the event of a loss in electricity supply, the steam load can be handled by the two remaining boilers. This change was evaluated after a few months later. Since the motor worked fine, it was decided by the utility supervisor that two FD fans were run by motor while one fan run by turbine. As a result of change in the operating policy, the majority of the above benefit was captured.

19.6 OPTIMIZING STEAM SYSTEM CONFIGURATION

The primary purpose of the steam system is to produce steam to satisfy the process steam demand at different pressure levels. If the steam system has no cogeneration, power will be purchased completely from the electricity grid to satisfy process power demand. At the same time, lower-pressure steam is produced from letdown of higher pressure through valves. However, if there are steam turbines placed between steam headers, both power and steam heat can be cogenerated at the same time. This is a kind of cogeneration from the steam system alone. Another kind of cogeneration is that of a combined gas and steam turbine system. We use the term combined heat and power (CHP) to cover both cogeneration systems above. The question is: How do we assess the cogeneration efficiency for a CHP system?

The parameter influencing cogeneration is the ratio of power to process heat that can be supplied by a steam system (Kenney, 1984). This ratio is named as R ratio, which is expressed as

$$R = \frac{\sum W_{\text{on-site}}(\text{Btu})}{\sum Q_{\text{process}}(\text{Btu})}, \tag{19.7}$$

where $\sum W$ is the maximal power generation potential from the steam system and $\sum Q$ is the sum of direct-steam-supplied process heat.

The cogeneration efficiency is defined as

$$\eta_{\text{cogen}} = \frac{\sum W_{\text{on-site}}(\text{Btu}) + \sum Q_{\text{process}}(\text{Btu})}{\sum Q_{\text{fuel}}(\text{Btu})}. \tag{19.8}$$

Or,

$$\eta_{\text{cogen}} = (R+1)\frac{\sum Q_{\text{process}}(\text{Btu})}{\sum Q_{\text{fuel}}(\text{Btu})}. \tag{19.9}$$

Clearly, cogeneration efficiency is a function of R ratio and the CHP system configuration. Although the CHP system is operated with the objective of achieving minimum cost, the maximum efficiency provides a measure of the best efficiency the system can reach for a given configuration. Calculation of cogeneration for a CHP system is illustrated through an example problem below.

Example 19.1   Evaluate Cogeneration Efficiency

Consider the steam system in Figure 19.1. Tasks:

a. Calculate the cogeneration efficiency for the current operation.
b. Determine the most efficient operation mode.
c. Compare the two operation modes.
d. Compare the maximal efficient operating modes with minimal cost operation. Note that the fuel and power prices for this example are $6 per MMBtu and $70 per MWh.

Solution.

19.6.1   Calculate the Cogeneration Efficiency for the Current Operation

$$\sum W = 2.6 + 7.6 + 0.4 = 10.6 \text{ MW} = 36.2 \text{ MMBtu/h},$$

$$Q_{MP} = 53 \times \Delta h = 53 \times (1202 - 189)/1000 = 53.7 \text{ MBtu/h},$$

$$Q_{LP} = 77 \times \Delta h = 77 \times (1185 - 189)/1000 = 76.7 \text{ MMBtu/h},$$

$$R = \frac{36.2}{53.7 + 76.7} = 0.28,$$

$$\eta_{cogen} = (0.28 + 1)\frac{(53.7 + 76.7)}{158 + 179} \times 100\% = 49\%.$$

19.6.2   Calculate the Maximal Cogeneration Efficiency

To achieve this, the following changes have to be made:

- Condensing flow to the minimum limit.
- Minimum letdown flows and zero LP vent.
- Increased power generation from back pressure extractions.
- Steam generation from more efficient boiler.

The steam system loadings are shown in Figure 19.6 with the above tasks accomplished.

However, what is the improved cogeneration efficiency from these changes? Let us do the numbers.

$$\sum W = 2.7 + 3.1 + 0.4 = 6.2 \text{ MW} = 21.2 \text{ MMBtu/h},$$

$$Q_{MP} = 53 \times \Delta h = 53 \times (1202 - 189)/1000 = 53.7 \text{ MMBtu/h},$$

$$Q_{LP} = 77 \times \Delta h = 77 \times (1185 - 189)/1000 = 76.7 \text{ MMBtu/h},$$

$$R = \frac{21.2}{53.7 + 76.7} = 0.16,$$

$$\eta_{cogen} = (0.16 + 1)\frac{(53.7 + 76.7)}{226} \times 100\% = 67\%.$$

Thus, the gap in cogeneration efficiency between the maximal cogeneration efficiency (Figure 19.6) and the current operation (Figure 19.1) is 18% (67%–49%). How much is it worth in terms of MMBtu/h? Does it tell something else?

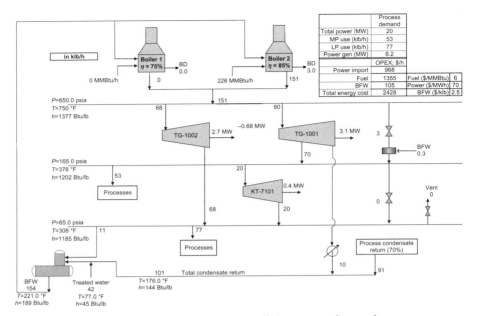

FIGURE 19.6. Maximal energy efficiency operating mode.

19.6.3 Compare the Current Operation with the Maximal Cogeneration Efficiency Operation

We can use total net energy input to compare. For the current and maximal efficient operating modes, fired fuel is 337 and 226 MMBtu/h, while power import is 9.4 (20–10.6) and 13.8 (20–6.2) MW, respectively. Assume fuel equivalent (FE) of 9.09 MMBtu/mWh for power import, fuel equivalent for both cases can be calculated as

$$\text{FE}_{\text{current}} = 337 + 9.4 \times 9.09 = 422.4 \, \text{MMBtu/h},$$

$$\text{FE}_{\text{max eff}} = 226 + 13.8 \times 9.09 = 351.4 \, \text{MMBtu/h}.$$

Therefore, the current operating mode spends 71 (351.4–422.4) MMBtu/h of energy more than the maximal efficient mode, which is very significant. The reasons for the gap are self-explanatory from the features of maximal efficiency mentioned above. In particular, the single largest contributor to the current inefficient operation is the large condensing flow. This is because the power generation efficiency for the condensing path is only 20% (see calculation below), which is much lower than the boiler steam efficiency of 85%. In other words, 65% of heat is lost on the surface condenser.

$$\eta_{\text{PEE}}^{\text{on-site}} = \frac{3414}{m_{\text{steam}} \times h_{\text{steam}} / \eta_{\text{boiler},2}} = \frac{3414 \, \text{Btu/kWh}}{10.5 \, \text{lb/kWh} \times 1377 \, \text{Btu/lb} / 0.85} \times 100\% = 20\%.$$

In the maximal efficient operation mode, the condensing path flow is reduced to the minimum limit. The reduced power generation from on site is compensated by increased power import. The heat rate for power import is credited at 9.09 MMBtu/MWh, which is based on 37.6% of power generation efficiency for a typical coal-fired power plant. In comparison, this import power generation efficiency is much higher than that of the condensing power generation. That is the reason why reduced condensing power generation boosts energy efficiency of the steam system.

19.6.4 Maximal Efficient Operation = Minimal Cost Operation?

The burning question becomes: Is the maximal efficient operation mode also the minimal cost mode? The answer is not simply yes or no but with a necessary and sufficient condition to be satisfied.

For the current operation in Figure 19.1 with the boiler fuel of 337 MMBtu/h at $6/MMBtu, import power of 9.4 MW at $70/MWh and BFW of 40 klb/h at $2.5/klb, the total energy cost is $24.3 MM/year versus that of $21.3 MM/year for the maximal operation in Figure 19.6. Thus, the maximal efficient mode saves $3 MM/year of energy cost compared with the current operation. The fundamental reason behind this can be explained with the concept of PEE introduced previously.

The PEE for power import can be calculated based on equation (19.1) as

$$\eta_{PEE}^{import} = \frac{C_f}{C_p} = \frac{\$6/MMBtu \times 3.414\ MMBtu/MWh}{\$70/MWh} \times 100\% = 29\%.$$

The PEE for the condensing path is only 20% as calculated before. Therefore, economically, power import beats the condensing path in this case.

Until now, we can safely derive the conclusion: It is not always true that maximal efficient operation is the most economical operation. It is true *only when* $\eta_{PEE}^{import} > \eta_{PEE}^{on\text{-}site}$. In plain words, the necessary and sufficient condition can be stated as: When the PEE is higher than on-site power generation efficiency, the maximal efficient operating mode is also the minimal cost operating mode. Under this condition, the problem of optimizing the steam system is converted to a simplified problem to maximize the cogeneration efficiency.

That could explain the reason why the minimal cost operation in Figure 19.3 favors maximal condensing flow, where the fuel price is $4/MMBtu and power price is $90/MWh. This is because the above condition does not hold for this operation as the PEE for power import is 15.2%, which is worse than that of the condensing path. Thus in this case, the condensing path beats the power import in cost but at the expense of cogeneration efficiency.

19.7 DEVELOPING STEAM SYSTEM OPTIMIZATION MODEL

Energy supply optimization is about generating steam and power at the lowest cost via optimizing boilers and turbines. At the same time, it is about optimizing power import versus on-site generation. The overall goal is to achieve minimal energy operating cost. There are several methods that could be employed for steam system optimization, which include: experience based, maximal cogeneration based, and mathematical optimization based. No matter what methods are used, the basis is the steam balance simulation. Mathematical modeling for steam system simulation and optimization is discussed the next.

19.7.1 Building Steam Balance Simulation Model

Although the details of establishing steam balances are discussed in Chapter 16 while steam equipment modeling equations are given in Chapter 15, the intention here is to combine all the related equations to develop a complete simulation and optimization model for the steam system. For specific applications, users can revise the model by providing specific configuration and conditions for the steam system.

19.7.1.1 For Boilers (Including Waste Heat Boilers) Subsystem for HP Steam Generation   For complex steam systems, boiler (i) could burn different fuel (r). Fuel could be fuel gas, LPG, and natural gas. Boilers could have different efficiencies. Even for the same boiler, efficiency depends on steam production and the conditions (temperature and pressure) of the steam generated:

$$M_{\text{BFW},i} = M_{\text{steam},i} + M_{\text{blowdown},i}. \tag{19.10}$$

$$Q_{i,r} = \frac{M_{\text{steam},i,r} \times (h_{\text{steam},i,r} - h_{\text{BFW},i})}{\eta_{i,r}}. \tag{19.11}$$

$$\eta_{i,r} = f(M_{\text{steam},i,r}, M_{\text{fuel},i,r}). \tag{19.12}$$

19.7.1.2 For Steam Header Subsystem   Steam enters a header (j) via boiler steam, steam from PRV (pressure reduction valve), turbine extraction, process waste heat boiler steam, and recovered steam from a flash drum. Process steam generation and use depend on feed rate and process conditions. Steam leaves a header to go to steam turbines, processes, PRV as well as steam losses due to leaks, and trap losses. For the balance closure, steam input to a header must be equal to the output:

$$M_j^{\text{in}} = \sum_i M_{\text{boiler } i,j} + \sum_s M_{\text{PRV } s,j} + \sum_k M_{\text{turbine } k,j} + \sum_l M_{\text{WHB } l,j} + \sum_m M_{\text{FS } m,j}. \tag{19.13}$$

$$M_j^{\text{out}} = \sum_n M_{\text{process } n,j} + \sum_s M_{\text{PRV } s,j} + \sum_k M_{\text{turbine } k,j} + \sum_q M_{\text{Loss } q,j}. \tag{19.14}$$

$$M_{\text{process } n,j} = f(\text{feed rate}, \text{process condition}, \text{losses}). \tag{19.15}$$

$$M_{\text{WHB } l,j} = f(\text{feed rate}, \text{ process condition}). \tag{19.16}$$

$$M_j^{\text{in}} = M_j^{\text{out}} \qquad j = \{\text{VHP}, \text{HP}, \text{MP}, \text{LP}\}. \tag{19.17}$$

19.7.1.3 For Power Generation Subsystem Power is generated from a steam turbine (k), theoretic steam rate ($m_{\text{turbine } k}$), and isentropic efficiency ($\eta_{\text{turbine } k}$). The isentropic efficiency depends on the inlet and outlet steam conditions and steam rate ($M_{\text{turbine } k}$). For a given total power demand, power import can be determined from on-site power generation:

$$W_k = \frac{M_{\text{turbine } k}}{m_{\text{turbine } k} / \eta_k}. \tag{19.18}$$

$$\eta_k = f(M_{\text{turbine } k}, T_{\text{in}}, P_{\text{in}}, P_{\text{out}}). \tag{19.19}$$

$$W_{\text{power}}^{\text{on-site}} = \sum_k W_k. \tag{19.20}$$

$$W_{\text{power}}^{\text{import}} = W_{\text{total}}^{\text{demand}} - W_{\text{power}}^{\text{on-site}}. \tag{19.21}$$

19.7.1.4 For Condensate and Deaeration Subsystem Condensate return and treated water together with LP steam to the deaerator provide BFW for boilers. LP steam is injected to the deaerator for removal of oxygen and other dissolved gases (α) from the treated water and condensate:

$$M_{\text{BFW}} = M_{\text{Cond}} + M_{\text{TW}} + M_{\text{LP}}^{\text{D}}(1 - \alpha). \tag{19.22}$$

$$M_{\text{LP}}^{\text{D}} \times (h_{\text{LP}} - h_{\text{TW}}) \times (1 - \alpha) = M_{\text{BFW}} \times (h_{\text{BFW}} - h_{\text{TW}}) - M_{\text{Cond}} \times (h_{\text{Cond}} - h_{\text{TW}}). \tag{19.23}$$

19.7.1.5 Steam Desuperheater Desuperheater is used to saturate superheated steam by using BFW because saturated steam is preferred for process use:

$$M_{\text{Sat},j} \times h_{\text{Sat},j} = M_{\text{Sup},j} \times h_{\text{Sup},j} + m_{\text{BFW}} \times h_{\text{BFW}}. \tag{19.24}$$

$$M_{\text{Sat},j} = M_{\text{Sup},j} + m_{\text{BFW}}. \tag{19.25}$$

19.7.1.6 For Steam Recovery from Blowdown A flash drum could be used to recover steam from boiler blowdown or high-pressure condensate. In the former case, the recovered LP steam ($m_{\text{FS}}^{\text{LP}}$) comes from the overhead of the drum while the bottom is sent to the sewer:

$$M_{\text{FS}}^{\text{LP}} \times h_{\text{FS}}^{\text{LP}} = M_{\text{BD}} \times h_{\text{BD}} - M_{\text{bttm}} \times h_{\text{Cond}}^{\text{LP}}. \tag{19.26}$$

$$M_{\text{FS}}^{\text{LP}} = M_{\text{BD}} - M_{\text{bttm}}. \tag{19.27}$$

19.7.1.7 For Steam Pressure Reduction Valve Letdown valves (PRV) are used to supplement steam demands at lower pressures for operational flexibility. Modeling of PRV is based on the valve characteristics, which are expressed in the general form as shown next. In most applications, the valve position is measured online. Thus, the following equations can be converted to a relationship between letdown steam flow and valve position (see Section 15.6):

$$m_j = K_1 \mathrm{Cv}\, P_i \text{ for choke flow when } \frac{P_o}{P_i} \le 58\%. \tag{19.28}$$

$$m_j = K_2 \mathrm{Cv}\sqrt{(P_i + P_o)(P_i - P_o)} \text{ for nonchoke flow.} \tag{19.29}$$

19.7.1.8 Initial Point and Convergence Criteria   The recommended commutation strategy is to use total boiler steam generation as initial trial rate and repeated until the convergence criteria is met:

$$\frac{\left| M^{\mathrm{D}}_{\mathrm{LP},i+1} - M^{\mathrm{D}}_{\mathrm{LP},i} \right|}{M^{\mathrm{D}}_{\mathrm{LP},i}} \le \varepsilon, \tag{19.30}$$

ε is a small number, for example, as 0.001.

19.7.1.9 Total Operating Cost   The total cost of the steam system can be back calculated from the converged steam balance based on fuel fired, treated water used, and power imported, which can be expressed as

$$C_{\mathrm{T}} = \sum_r \sum_i C_{\mathrm{fuel}\, r} Q_{\mathrm{boiler}\, i,j} + C_{\mathrm{BFW}} \sum_k M_{\mathrm{BFW},k} + C^{\mathrm{import}}_{\mathrm{power}}\, W^{\mathrm{import}}_{\mathrm{power}} - C^{\mathrm{export}}_{\mathrm{power}}\, W^{\mathrm{export}}_{\mathrm{power}}. \tag{19.31}$$

19.7.1.10 Application Remarks   You do not need to feel intimidated with the set of equations above. You will find the task simple when you follow the top–down approach to establish steam balances from boilers to each steam header, to steam turbines and letdown valves, and to deaerator and blowdown flash drum. The top–down approach for steam balance is discussed in detail in Chapter 16. During the process of setting up the steam balances, you could apply some of the equations above for modeling the equipment and subsystems. The steam balances can be conducted readily in a spreadsheet environment.

19.7.2 Establishing Optimization Criteria and Constraints

19.7.2.1 Objective Function   The objective function is used in optimization to drive the overall operating cost to minimal with all potential independent variables (or manipulated variables) available for optimization.

Minimize objective function f:

$$f = \sum_r \sum_i C_{\mathrm{fuel}\, r} Q_{\mathrm{boiler}\, i,r} + C_{\mathrm{BFW}} \sum M_{\mathrm{BFW}} + C^{\mathrm{import}}_{\mathrm{power}}\, W^{\mathrm{import}}_{\mathrm{power}} - C^{\mathrm{export}}_{\mathrm{power}}\, W^{\mathrm{export}}_{\mathrm{power}}. \tag{19.32}$$

19.7.2.2 *Constraints*   Optimizing the steam system is to know what knobs to turn, which can change operating costs. First, the steam system should avoid any foreseeable losses such as steam trap losses, leaks, and poor insulation. This issue has been discussed in detail in Chapter 18. Second, the pressure of steam headers could be optimized. This is usually done once within a time period and they are maintained to the optimized values after optimization.

Once the above tasks are managed in good order, minimizing operating cost of the steam system becomes optimizing loadings of steam equipment such as on-purpose boilers and turbine generators in the equipment limits. At the same time, trade-off between power import/export versus on-site power generation must be optimized. This optimization involves optimizing both equipment loading and driver selection. Loading optimization is continuous optimization within the bounds or limits; but driver selection is discrete (driver on or off) optimization as a turbine driver is switched to motor or vice versa.

Subject to the following constraints:

- Boilers are constrained in limits:

$$M_{\min r} \le M_{\text{boiler},r} \le M_{\max r}. \tag{19.33}$$

- Turbine generators are constrained in limits:

$$M_{\min k} \le M_{\text{turbine},k} \le M_{\max k}. \tag{19.34}$$

- Turbine drivers are constrained as either on or off:

$$z_k M_{\min k} \le M_{\text{turbine},k} \le z_k M_{\max k} \quad z = (0, 1). \tag{19.35}$$

- Power import is modeled with turbine generator on a continuous basis and turbine driver on a discrete basis:

$$W_{\text{power}}^{\text{driver}} = \sum_k z_k W_k^{\text{driver}}. \tag{19.36}$$

$$W_{\text{power}}^{\text{gen}} = \sum_k W_k^{\text{gen}}. \tag{19.37}$$

$$W_{\text{power}}^{\text{import}} = W_{\text{power}}^{\text{demand}} - W_{\text{power}}^{\text{driver}} - W_{\text{power}}^{\text{gen}}. \tag{19.38}$$

- And equations (18.10)–(18.30).

19.7.2.3 *Application Remarks*   The whole set of equations 19.10–19.38 forms the optimization model for steam and power. This optimization strategy allows the fastest dynamic response while always trending to the most economical steady state operating position. Furthermore, the steam marginal prices will be automatically determined from the optimization model (Varbanov et al., 2004).

NOMENCLATURE

C_f	marginal fuel price
C_p	imported or exported power price
C_v	valve orifice coefficient
f	objective function
FE	fuel equivalent
h	enthalpy
m, M	mass flow
P_i, P_o	inlet and outlet pressure of the control valve
PEE	price equivalent efficiency
Q	heat duty
R	power to heat ratio
W	shaft work
z	binary variable

Greek Letters

α	fraction of deaeration steam vented
ε	a very small number
η	efficiency

REFERENCE

Varbanov P, Perry S, Makwana Y, Zhu XX, Smith R (2004) Top-level analysis of site utility systems, *Chemical Engineering Research and Design,* Trans IChemE, Part A, **82**(A6), 784–795.

PART 5

RETROFIT PROJECT EVALUATION AND IMPLEMENTATION

20

DETERMINE THE TRUE BENEFIT FROM THE OSBL CONTEXT

Energy improvement projects are usually identified from the process level, and benefits are estimated based on marginal steam prices and local energy and water balances. When selecting energy projects for implementation, these projects must be evaluated in the context of OSBL (outside system battery limit), that is, in the context of site-wide steam/power/fuel balances, to determine the true benefits in reducing operating cost and greenhouse gas emissions. This method can prevent bad investment decisions.

20.1 INTRODUCTION

It is common that decisions for implementing beneficial improvement ideas are made with the evaluation based on the inside process battery limit (ISBL). It is not surprising to find, in some cases, that the implemented energy-saving projects deliver far less benefits and require much high capital compared to what are predicted from evaluation. The reason is that energy savings at the process level can only be realized at the site-wide balances when the utility bills come at the end of the year.

Therefore, it is critical that improvement ideas must be assessed in-line with overall balances of steam, power, and fuel. In typical cases, energy-saving projects could affect steam marginal prices, and hence impact the net benefits that the energy

Energy and Process Optimization for the Process Industries, First Edition. Frank (Xin X.) Zhu.
© 2014 by the American Institute of Chemical Engineers, Inc. Published 2014 by John Wiley & Sons, Inc.

projects could generate. In extreme cases, energy-saving projects could cause fuel gas and steam to become long.

A method is discussed here, which is based on OSBL conditions. This method can have a profound influence on evaluating project economics and thus determining the project selection. Improper project evaluation could lead to bad investment decisions: bad projects are selected and good projects are discarded, which, regrettably, is common in energy optimization evaluations. To avoid the bad investment decisions and determine the true benefits for energy-saving projects, the effects on incremental steam, power, boiler feed water (BFW), and fuel balances must be assessed. A practical example is provided the next to illustrate how this method is applied.

20.2   ENERGY IMPROVEMENT OPTIONS UNDER EVALUATION

In the site wide energy optimization project, five energy improvement ideas among many others were generated and evaluated at stand-alone basis. The benefits and capital costs for these improvement ideas are given in Table 20.1. These assessment results were obtained at the process ISBL.

20.2.1   Idea 1: Driver Switch for Boiler Fans

There are three boiler forced draft (FD) fans, which are run by medium-pressure to low-pressure (MP–LP) extraction turbines currently with motors in spare. It was identified that motor driving is more economical than turbine due to relatively low electricity price. Based on steam marginal pricing, it is estimated that if one fan can be switched to motor, the benefit would be $738,144 per year. The cost savings would be $2.2 MM/year if three fans could be switched to motors.

20.2.2   Idea 2: Reduce High-Pressure Steam in the Hydrogen Plant

The steam reformer in the hydrogen generation unit (HGU) is to make hydrogen from natural gas via carbon reacting with steam. The reaction temperature by design is 825 °C, while the running temperature is 800 °C due to the mechanical integrity issue in the reaction cooler. The lower reaction temperature results in higher steam to carbon ratio (S:C) and thus requires more steam for making the same amount of hydrogen. With the feed composition remaining relatively constant, the S:C ratio can be reduced by increasing the reaction temperature up to 825 °C by design. The cost for fixing this issue was initially estimated to be very expensive. Without knowing the benefit of fixing it, the resolution was to reduce the reactor temperature to 800 °C and maintain it at this low temperature. This operating policy has been adopted without challenge for many years with the mechanical issue remaining unresolved but at the expense of increased steam rate.

TABLE 20.1. Project Economics Obtained on Stand-Alone Basis

	Driver switch for one boiler fan	Reducing HP steam in HGU	K-301 motorization	Extra LP stripping steam in HCU	F304 convection revamp for extra MP gen	Total
Steam saving (Mt/d)	100 MP	200 HP	625 MP	−150 LP	85 MP	
Power use (kW)	160	n/a	2,000	None	None	
Net energy cost saving (US$/yr)	738,144	1,800,000	4,051,800	−1,026,000	703,800	6,267,744
Capital cost (US$)	None	550,000	6,000,000+	n/a	1,900,000	2,450,000

Basis: marginal steam prices: HP = $25/ton, MP = $23/ton, LP = $19/ton, Electricity = $65/MWh, 360 operating days/year.

427

In the energy optimization project, this mechanical issue was raised again. To justify the case of fixing it, a model was built for the relationship between the reactor temperature and steam-to-carbon ratio and thus the benefit of increasing the reactor temperature can be estimated. The calculation indicated a reduction in high-pressure (HP) steam by 200 t/day and thus $1.8 MM/year in cost savings if the reaction temperature could be increased to 825 °C. To give a more realistic capital estimate, a detailed mechanical inspection was conducted. The cost of fixing the mechanical issues was estimated at around half a million dollars, which was much cheaper than what was expected.

20.2.3   Idea 3: K-301 Driver Motorization in Reformer

Currently, the recycle gas compressor K-301 in the reforming process is driven by a partial condensing turbine KT301 with MP as the inlet steam, which is drawn off at atmospheric pressure. At very low power generation efficiency, the turbine consumes a large amount of steam, which is around 625 t/day MP steam on average costing $5.1 MM/year. Replacing the existing turbine with a variable speed motor requires 2 MW of electrify costing $1.1 MM/year. Thus, switching to motor will give a net benefit of $4.0 MM/year in cost savings due to the relatively low cost of electricity compared to steam and efficiency elevation. However, the capital cost for a large motor of 2 MW is expensive with an equipment installation cost of $6 MM.

20.2.4   Idea 4: Stripping Steam for HCU Fractionation Revamp

A yield improvement option was identified from the hydrocracking unit (HCU) as a part of the energy optimization project. It was found that a large amount of diesel was lost into the unconverted oil due to poor fractionation efficiency in the main fractionators. To recover this lost diesel with an estimated value of $12 MM/year, a new stripper column will be installed. The stripper column will require stripping LP steam of 150 t/day costing $1.0 MM per year. The cost for related piping would be estimated as part of HCU revamp project.

20.2.5   Idea 5: Increase of F304 Steam Generation

F304 is the common convection bank for two fired heaters in the reforming unit. Currently 1300–1900 t/day of flue gas from F304 leaves the stack at temperatures around 390 °C. This is considerably higher than the typical value of 220 °C seen in peer units within the industry. One of the options proposed is to add surface area in the convection section for extra MP steam generation of 85 t/day with a cost savings of $703,800 per year at the expence of $ 1.9 MM capital cost.

20.3   A METHOD FOR EVALUATING ENERGY IMPROVEMENT OPTIONS

The evaluation results in Table 20.1 show large benefits of these improvement ideas based on the ISBL level. If decisions for project selection were made at this ISBL level, these options could be selected for implementation due to large benefits compared with relatively low capital costs. To avoid the bad investment decisions and determine the true benefits for these ideas, the effects on incremental steam, power, BFW, and fuel balances are assessed in detail as follows.

20.3.1   Step 1: Predetermine the Implementation Sequence

Once the improvement ideas are selected for assessment, the first thing to do is to determine the sequence of potential implementation because different sequences of project implementation will have different effects on OSBL. As a general guideline, quick hits of improvement with no and low costs should be considered for implementation first at the early stage. Capital projects are ranked based on benefits and costs and then prioritized. Other criteria such as easy implementation and turnaround schedule are also considered in determining the implementation sequence.

For this example problem, the improvement ideas will be implemented according to the following sequences:

- The driver switch to the motor for boiler fans is considered as the first implementation item as it can generate large benefit without capital cost. With this choice made, two implementation sequences are formed as the basis for OSBL assessment.
- *Implementation Sequence One:* Increasing reaction temperature in HGU $\rightarrow$ motor driver for K301 $\rightarrow$ HCU revamp $\rightarrow$ F304 revamp.
- *Implementation Sequence Two:* Increasing reaction temperature in HGU $\rightarrow$ HCU revamp $\rightarrow$ F304 revamp. K301 motorization project is not considered in this sequence as its mutual exclusivity with F304 revamp, which will become clear later.

The following evolutions for these two implementation sequences will determine which options should be implemented in which sequence and why.

20.3.2   Step 2: Establish the Base Case of Steam Balance

The base utility (steam, power, water, and fuel) balances used for this assessment purpose should represent the most common operation production scenario. The common production scenario corresponds to the most frequent feed rate and compositions, product rates and quality, and process conditions. The utility consumption could be the most representative under this operating scenario. Furthermore, the utility balances must achieve high balance precision to avoid misleading investment decisions.

For the example problem, the steam balance is given in Figure 20.1. The main feature of this steam balance is the large HP steam letdown for meeting the MP demand while there is a relatively large amount of LP dump. By LP dump in this case, it implies that when LP steam is long, it is condensed and then the condensate is sent back to the deaerator to preserve the condensate. Thus, LP dump is better than LP venting as condensate is lost to the atmosphere by LP venting.

20.3.3   Step 3: Evaluate Operational Changes (Quick Hits)

One quick hit opportunity was identified in the boiler house, which is the driver switch for the boiler FD fans. This opportunity is used here for illustration purposes for how to assess operation opportunities in the context of steam balance. Other operational changes that could affect steam balance would be assessed in a similar manner.

Although the fan drivers require MP steam as inlet into the turbines, this MP is generated via HP letdown valve. Thus, shut down of the fan turbines will result in less HP letdown rate. At the same time, shut down of the fan turbines will reduce LP dump as the fan turbines have LP extractions. In the current operation, the LP dump is 148 t/day, which is sufficient to accommodate the driver switch to motor for one fan as the steam turbine requires 100 t/day of MP steam. The entries in bold in Figure 20.2 are the new steam rates after implementing the driver switch for one fan.

From Figure 20.2, it can be seen that switching one driver to motor could give cost savings of \$810,144 per year. It can be observed that the savings benefit from the steam balance in Figure 20.2 is higher than that from the ISBL evaluation as shown in Table 20.1. The reason is that MP steam price was used as the basis for benefit calculations in the ISBL evaluation, while true benefit comes from HP savings, which can only be observed from the overall steam balance.

When considering the driver switch to motor for the second FD-fan, the benefit is not twice of that of a single driver switch. The reason is that there is only 48 t/day of LP dump available after implementing a driver switch for the first fan. Thus, the incremental HP savings for the second fan is only 48 t/day compared to 100 t/day for the first fan, but it is still beneficial.

The total cost savings from the driver switch for two fans is \$1.14 MM/year with benefit for the second fan as \$335,000 per year. The steam balance after implementing the second driver switch is given in Figure 20.3. Obviously, driver switch to motor is the way to go. Based on the current LP dump, one fan should be switched to motor at the earliest time possible. It is possible for the driver switch for the second fan as well, but it depends on electricity supply reliability.

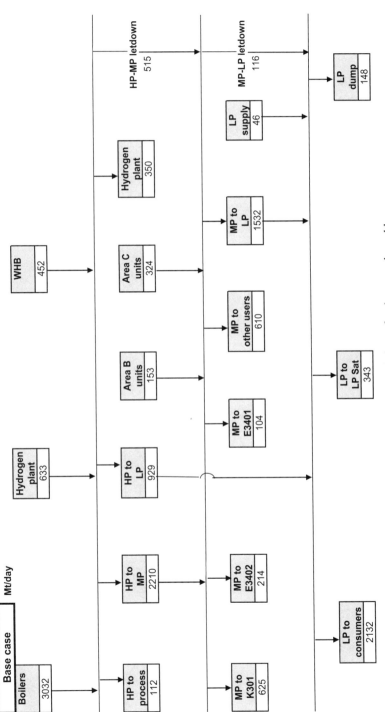

FIGURE 20.1. The base steam balance for the example problem.

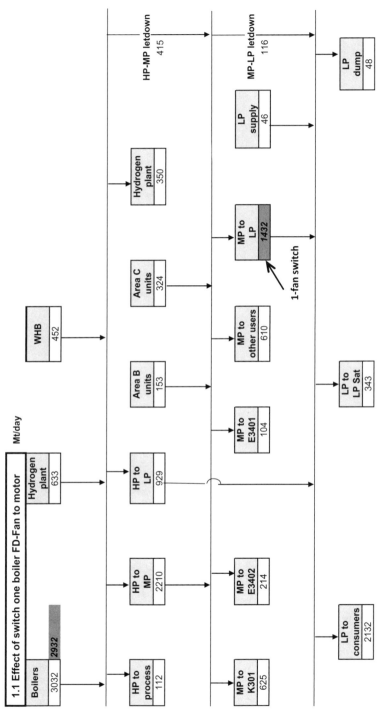

FIGURE 20.2. Effects of driver switch for one boiler fan.

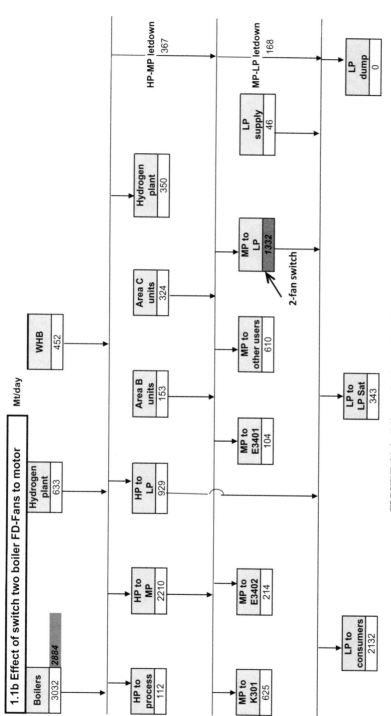

FIGURE 20.3. Effects of driver switch for two boiler fans.

433

In the past, electricity supply trips were quite often and thus the operating policy required operators to use steam turbines for critical services as preference to motors. However, the electricity supply has been very reliable for the past 5 years and only one trip occurred. By knowing the benefit of using motor over turbine, the plant technical personnel reviewed the current operating practice and recommended to the plant management to implement the option for driver switch for one fan to start with. The second fan could be switched to motor if the first implementation went well without trip in one year.

20.3.4 Step 4: Evaluating Energy Capital Projects for Options in Sequence 1

The steam balance from implementing operational changes becomes the basis for evaluating energy capital projects. Figure 20.2 is the basis for evaluating the stripping steam for HGU revamp and the resulting balance from implementing the HGU energy option will become the basis for evaluating the K301 motor option and so on.

20.3.4.1 Reducing HP Steam Use in the Hydrogen Plant (HGU) It requires mechanical work to fix some of the cracks in the reaction cooler, which can allow the reaction temperature to increase to 825 °C. From the steam balance point of view, this option will reduce 200 t/day of HP generation from the boilers. This option is assessed in the steam balance as shown in Figure 20.4, which indicates that implementation of this project is expected to save energy operating cost of $1.8 MM/year on top of driver switch for boiler fan. The capital cost is estimated around $500,000. It is a very promising project.

20.3.4.2 Compressor K301 Motorization Following the HGU Steam Reduction When the compressor K301 driver is changed to motor, there could be a knock-on effect on steam balance as shut down of the turbine could reduce MP use by 625 t/day. The steam balance basis for assessing this option is Figure 20.4.

After implementing the options for boiler driver switch and steam reduction in the hydrogen plant, the HP letdown will be at 415 t/day, which is the major part of the supplier for the KT301 turbine. Thus, implementing the motorization for K301 driver with shutting down of KT301 turbine will save 415 t/day of HP steam. At the same time, 210 t/day surplus MP steam (625–415) will add to the MP letdown valve and eventually go to the LP dump. The steam balance after K301 motorization is shown in Figure 20.5. This results in boiler fuel saving of $3.7 MM/year, but requiring 2 MW electricity for the motor, which is at the expense of annual electricity cost of $1.1 MM/year. Thus, the net energy savings is $2.6 MM/year. With the capital cost of $6 MM for the installed equipment without including other related costs such as demolish cost and downtime, this project should be treated as a borderline project because the actual capital cost could be much higher than $6 MM if other costs are considered. This warrants a more detailed evaluation for infrastructure and downtime costs.

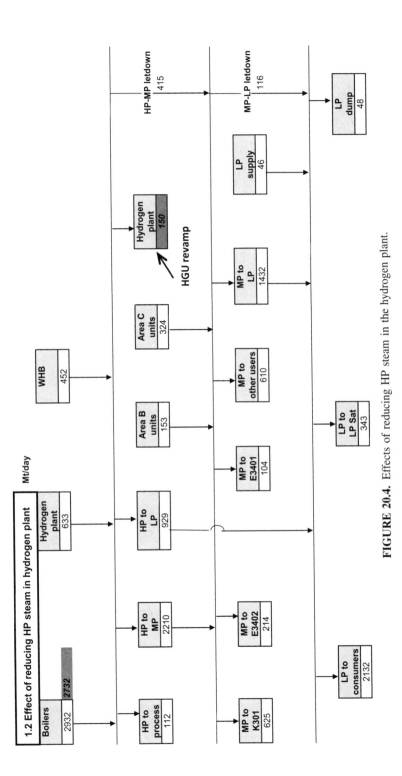

FIGURE 20.4. Effects of reducing HP steam in the hydrogen plant.

The above OSBL-based evaluation gives a very different benefit estimate for the K301 motor project compared with that of $4 MM/year net benefit obtained from stand-alone basis (Table 20.1). Clearly, the stand-alone evaluation is misleading as it provides a too high benefit estimate, which cannot be realized. If implementation decisions were made based on stand-alone evaluation, a green light could have been given for implementation.

From Figure 20.5, several key observations on steam balance can be observed:

- This option will capture part of the benefit compared with 625 t/day MP required for the turbine KT301. Capture of the total benefit is limited by the amount of HP letdown available, which is 415 t/day on average. A statistical analysis is warranted to analyze the means and variations of the true letdown flows, which can provide a basis for more realistic benefit estimates (see Chapter 21).
- If implementing the K301 motor option, the steam balance will be impacted significantly. First, the HP letdown will become the pinch point for other steam saving projects. Second, there will be a large amount of MP surplus, which has to go through MP–LP letdown and cause a large LP dump.

20.3.4.3 HCU Revamp Following HGU Revamp and K301 Motor Project

The HCU is expected to be revamped to recover extra diesel product. In this case, a stripper column will be installed, which requires stripping LP steam of 150 t/day. This extra LP steam can be completely satisfied by use of LP dump. Thus, there is no extra requirement from boilers. This result is in sharp contrast with that from the stand-alone assessment (Table 20.1), which indicates the cost of $1.0 MM/year for supplying this 150 t/day of LP steam.

The incremental effects on steam balance are shown in Figure 20.6. This steam balance is better than the base steam balance in Figure 20.1 in that boiler load is reduced by 23%. In addition, HP letdown is shifted to MP letdown. However, there is still a large MP letdown and LP dump, which are the sources of inefficiency. The potential solution for reducing the LP dump is to consider a driver switch to motor for HP–LP extraction turbines in process units. Such opportunities should be the tasks for further investigation.

20.3.4.4 F304 Revamp Following HGU/K301 Motor/HCU Revamp

Although the stand-alone assessment indicated the moderate benefit for extra MP steam generation from the F304 revamp, the steam assessment (Figure 20.7) determines that there is no benefit of F304 revamp in this implementation sequence. Extra 85 t/day MP steam generated from F304 revamp will have no proper home to go and end up in LP dump via MP letdown valve. Another important finding from this steam assessment is that F304 revamp and K301 motorization are mutually exclusive. In other words, F304 revamp for steam generation should not be considered after the K301 motor option is implemented and vice versa.

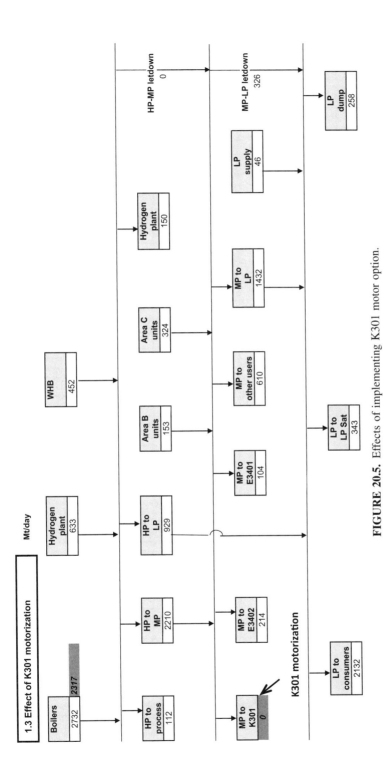

FIGURE 20.5. Effects of implementing K301 motor option.

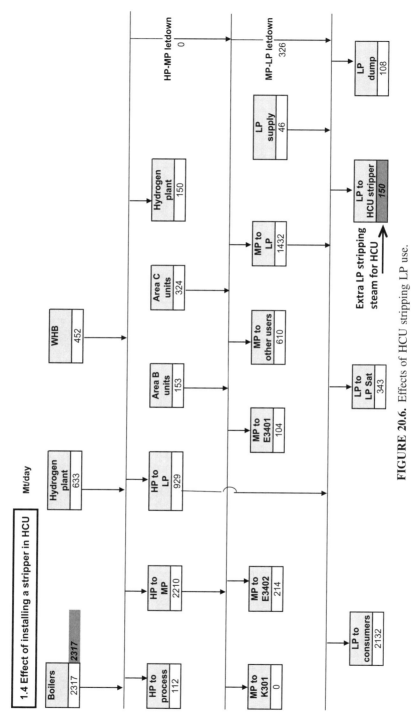

FIGURE 20.6. Effects of HCU stripping LP use.

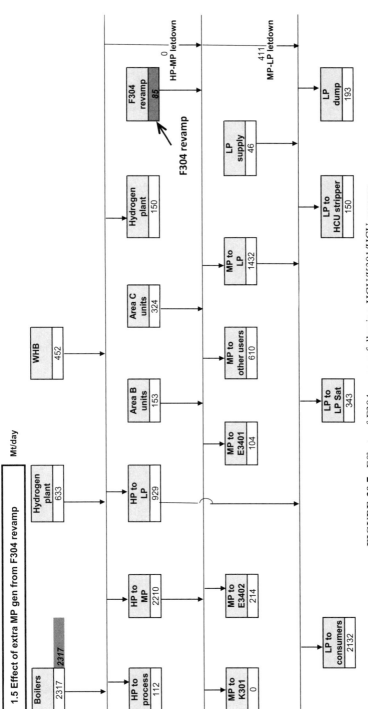

FIGURE 20.7. Effects of F304 revamp following HGU/K301/HCU revamp.

20.3.4.5 Overall Summary for Implementation Sequence 1   The evaluation
results for sequence 1 are summarized in Table 20.2. What observations can be
highlighted?

- The driver switch for one boiler fan should be implemented at the earliest time
  possible. The driver switch for two fans is also possible.
- The HGU revamp project is highly recommended as it can reduce operating
  cost with small capital cost. However, close motoring of the HGU steam
  reformer is required after fixing the cracks in the reaction cooler.
- There will be no cost for providing LP stripping steam for a new stripper
  column in the HCU revamp following this implementation sequence.
- Installing a motor for the K301 compressor is on borderline. To justify the
  implementation, a more detailed capital cost estimate including infrastructure
  cost and downtime should be conducted to identify any showstoppers and
  hidden cost for implementation.
- Last but not least, the K301 motor option is mutually exclusive from the F304
  revamp for MP generation. In other words, F304 revamp does not generate any
  value following the option of K301 motorization. Before throwing this option
  out, it will be assessed next through implementation sequence 2, which is
  discussed next.
- By the end of this implementation sequence, there is a large amount of MP
  letdown and LP dump.

What can we learn from the assessment of the options in the context of steam
balances? The most important lesson is that the benefit estimates obtained from
steam balances reflect the true benefits. The true benefits can differ significantly from
that calculated on a stand-alone basis. The second lesson is that implementing
energy-saving options could have big impacts on steam balances, which must be
taken into account in making decisions for project selection.

20.3.5   Step 5: Evaluate Energy Projects for Options in Sequence 2

Now let us look at what could happen if we take the route of sequence 2 of the
implementation. This sequence starts with operational change, which is driver switch
for boiler fan followed by steam reduction in the hydrogen plant, which is the same as
the sequence 1 discussed previously. Thus, the focus is placed on the remaining
options in the implementation sequence: providing extra LP steam in HCU revamp,
and F304 revamp for MP generation.

20.3.5.1 HCU Revamp Following Boiler Driver Switch and HGU Revamp   -
After boiler driver switch and HGU steam reduction, the steam balance (Figure
20.4) features a small amount of LP dump at 48 t/day. Thus, to accommodate extra
LP use in HCU at 150 t/day, 102 t/day ($150 - 48$) LP has to be provided by
additional boiler HP generation, which is cascaded through letdown valves, thus at

TABLE 20.2. Summary of Individual Energy Projects for the Implementation Sequence 1

Project implementation sequence 1	Driver switch for one boiler fan	Reducing HP steam in HGU	K-301 motorization	Extra LP stripping steam in HCU	F304 revamp for extra MP gen
Incremental cost saving ($/year)	810,144	1,796,630	2,611,800	0	0
Incremental capital cost ($)	None	550,000	6,000,000+	n/a	1,900,000
Incremental payback (years)	n/a	0.3	>2.3	Quick (qualitatively)	No payback
Accumulative FE energy saving (MMBtu/h)	9.1	30.0	55.5	55.5	55.5
Accumulative cost saving ($/year)	810,144	2,606,774	5,218,574	5,218,574	5,218,574
Steam balance comments	From Figure 20.2: 1. Will reduce boiler HP by 100 t/day 2. Will reduce LP dump	From Figure 20.4: Will reduce boiler steam by 200 t/day and has no effect on the rest of steam balance	From Figure 20.5: 1. Will increase MP letdown and LP dump. 2. Switch off HP–MP turbines and switch on MP-Cond turbines can reduce both MP letdown and LP	From Figure 20.6: HCU LP stripping steam will reduce LP dump	From Figure 20.7: Will increase MP letdown and LP dump
Recommendations	Way to go	Way to go	Boardline	Most likely	Cancel
Remarks for implementations	Need to revise operating policy for drivers	Conduct inspection after mechanical work for fixing the cracks in the reaction cooler	1. More detailed capital cost estimate needs to be conducted. 2. The benefit depends on the amount of HP letdown available. Thus, a statistic analysis for HP letdown is needed. 3. Too large LP dump cold limit the benefit as the condenser for LP cooling will be too small	1. The high value from diesel recovery justifies the HCU revamp. 2. Extra LP will be provided from LP dump	Not beneficial to implement F304 revamp if K301 motor project is implemented

an extra cost of $916,133 per year. This is different from the no-cost solution if HCU revamp was conducted at the end of sequence 1.

From the steam balance point of view (Figure 20.8), there is a large amount of HP and MP letdown flows, which is the source of inefficiency. Let us see if the F304 convection revamp project could help to mitigate this inefficiency.

20.3.5.2 *F304 Revamp for More MP Generation*   In the end of this sequence, F304 convection section revamp will generate cost savings of $765,000 per year. This is because 85 t/day MP generation from F304 will reduce boiler steam generation by the same amount (Figure 20.9). The more MP generation from F304, the more HP reduction from boilers.

20.3.5.3 *Overall Summary for Implementation Sequence 2*   What new observations can the assessment of sequence two reveal to us? This question can be answered by Table 20.3 and the key observations are highlighted as follows:

- Extra LP stripping steam for a new stripper column in the HCU will cost close to $1 million per year.
- In view of the high capital and uncertainty associated with implementing the K301 motorization project, F304 is relatively cheap and easy to implement than the K301 motorization project. If F304 revamp is selected, the sequence 2 would be selected for implementation and Figure 20.9 can represent the steam balance after the implementation.
- From the steam balance point of view, by the end of this implementation sequence, there is no LP dump, but large amount of HP and MP letdown flows. In this case, marginal prices for MP and LP are the same as HP steam. This may call for optimizing process steam use to reduce both MP and LP steam usage.

20.4   FEASIBILITY ASSESSMENT AND MAKE DECISIONS FOR IMPLEMENTATION

From the above assessment in the context of steam balances and also the effects of different implementation sequences, the pros and cons for each sequence are identified. Feasibility analysis should be conducted to identify hidden costs, mechanical integrity issues, as well as operation and reliability issues. The details for feasibility assessment will be discussed in Chapter 22. If the feasibility analysis shows strong incentives, it is the time for the plant management to make decisions for approving the project implementation, and the plant engineers can then proceed to work with engineering vendors to provide FEED packages.

As a final comment, energy improvement ideas must be assessed in OSBL in terms of potential effects on steam, fuel, and water balances to obtain the true benefits and capital costs. Another comment is that this assessment is based on average values for steam balances and does not take into account variations. For

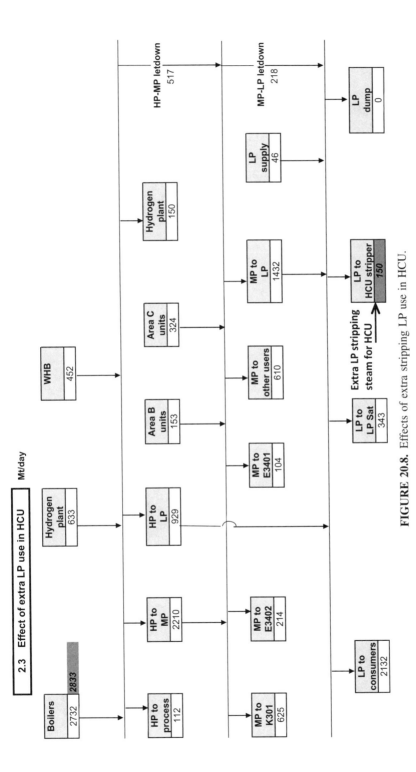

FIGURE 20.8. Effects of extra stripping LP use in HCU.

444

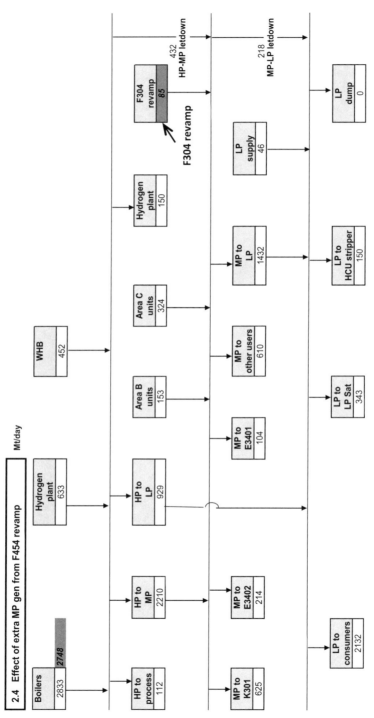

FIGURE 20.9. Effects of extra F304 MP generation following HGU/HCU revamps.

TABLE 20.3. Summary of Individual Steam Projects for Implementation Sequence 2

Project implementation sequence 2	Driver switch for one boiler fan	Reducing HP steam in HGU	Extra LP stripping steam in HCU	F304 revamp for extra MP gen
Incremental net cost saving ($/year)	810,144	1,796,630	−916,133	765,000
Incremental capital cost ($)	None	550,000	n/a	1,900,000
Incremental payback (years)	n/a	0.3	Quick (qualitatively)	2.5
Accumulative FE energy saving (MMBtu/h)	9.1	30.0	19.3	28.3
Accumulative net cost saving ($/year)	810,144	2,606,774	1,690,642	2,455,642
Steam balance comments	From Figure 20.2: 1. Will reduce boiler HP by 100 t/day 2. Will reduce LP dump	From Figure 20.4: Will reduce boiler steam by 200 t/day and has no effect on the rest of steam balance	From Figure 20.8: HCU LP stripping steam demand will increase HP and MP letdown flows	From Figure 20.9: Will reduce HP letdown by 85 t/day with the rest of the steam balance remained the same
Recommendations	Way to go	Way to go	Most likely	Likely
Remarks	Need to revise operating policy for drivers	Conduct inspection after mechanical work for fixing the cracks in the reaction cooler	1. The high value from diesel recovery justifies the HCU revamp. 2. Extra LP will cost close to 1 million dollars	Extra MP generation has the same value as HP steam

example, the availability of HP letdown could determine the true benefit of the K301 motorization project. The average of this letdown rate sets the primary test for economic benefits of this motorization project while the variation of this letdown rate sets the secondary test. Large variation could indicate that the letdown rate is not high enough over majority of time for realizing the benefit of K301 motorization. This fact alone could kill the motorization project. Detailed discussions on statistic analysis will be given in Chapter 21.

21

DETERMINE THE TRUE BENEFIT FROM PROCESS VARIATIONS

Operating conditions may vary from time to time. Economic assessment estimated based on average values must be revised to take variations into account so that more accurate benefit estimates can be obtained.

21.1 INTRODUCTION

In Chapter 20, K301 motorization was identified as a promising project for improving energy efficiency and reducing energy cost. It was mentioned that the key parameter for justifying this project is the high-pressure (HP) letdown rate on a yearly basis. This is because the turbine driver receives medium-pressure (MP) steam through HP letdown, which is then drawnoff at atmospheric condition. Thus, shutdown of the turbine will reduce MP intake, and thus HP steam. Naturally, the larger the letdown rate, the larger the fuel savings from the boilers when the driver is replaced with a motor. In the previous assessment, the average HP letdown rate was used as the basis for benefit calculations. When an average value is used, there is an underlying assumption that the letdown valve opens at the same position all the time and provides a constant flow rate at the average value. Use of average could act as the first test for techno-economic assessment. If an improvement idea cannot pass this test, the option may be cancelled right there. However, for the case of K301 motorization, this option passed the test, but it was recommended to conduct a second test based on statistical variations of the HP letdown rate.

Energy and Process Optimization for the Process Industries, First Edition. Frank (Xin X.) Zhu.
© 2014 by the American Institute of Chemical Engineers, Inc. Published 2014 by John Wiley & Sons, Inc.

The reason is that, use of average does not address the variation, which could significantly affect the economic evaluation of improvement. In reality, the letdown rate varies due to the fact that production rates change as well as weather changes and thus the MP requirement varies for MP users including the turbine to run the K301 compressor. The first question is: What is the variation of the HP letdown rate? This question can be answered by measurement for the letdown valve shown in the distributed control system.

If it is found that the variation is significant by eye inspection; the next question of interest will be: How often can the letdown valve produce flows larger than the average? It could be possible that a high average rate is the result of peak MP demands that occur infrequently. Thus, the frequency of the letdown valve opening above the average is the deciding factor to justify the K301 motorization project. If the letdown flows are higher than the average in an overwhelming majority of time over a year, the true benefit will be close to that estimated based on the average flow. Otherwise, the true value could be much lower and thus there is a very slim chance to implement the K301 motorization project. Whichever case it is, let us find out. The calculation procedure will be illustrated with examples that follow.

If the online data of steam rates resembles the normal distribution, we can readily determine the mean or average, the distribution frequency, standard deviation, and lower and upper bounds of the steam rates.

21.2   COLLECT ONLINE DATA FOR THE WHOLE OPERATION CYCLE

Online data are the direct measurement of the HP letdown flows and such data can be obtained from the historian; Figure 21.1 gives the 360 data points per year. The data indicate large variations of letdown steam rates. This could be caused by operation mode changes due to seasonal product requirements.

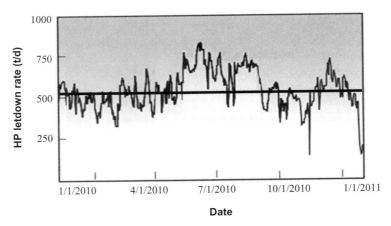

FIGURE 21.1. Online data for the HP letdown rate.

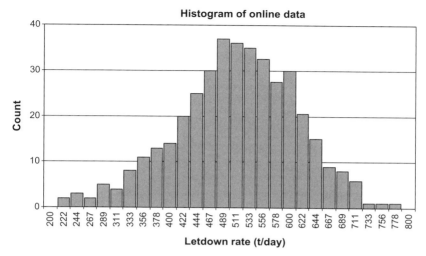

FIGURE 21.2. Frequency plot for HP letdown rate.

The first lesson is that we should collect data for the whole operation cycle to obtain the true average flow. If the data for one season are used as the basis, the average flow could only represent this season but not the other. However, the turbine driver actually sees the HP letdown flows over the whole operating cycle of the years. Using the average flow derived for one season could be either underestimating or overestimating the true average flow.

To understand the frequency of letdown flow at different rates, the online data in Figure 21.1 are represented by the flow ranges and the frequency of steam rates in terms of counts in each range as shown Figure 21.2. Note that the counts for all intervals add up to 360, which is equal to the total data points in Figure 21.1. The profile in the figure resembles a bell-shaped normal distribution. Assume the normal distribution fits the actual distribution. This assumption can allow us to readily determine HP letdown average, flow distribution and variance. The calculations of these parameter are explained in the steps that follow.

21.3 NORMAL DISTRIBUTION AND MONTE CARLO SIMULATION

The normal distribution is the most commonly known as it represents the typical variation pattern in the process industry: Most data are close to the average representing the normal operation while a small fraction of data corresponds to campaign operation or turndown operation. Therefore, the normal distribution is widely used to describe variations of real data that cluster around the single mean value as a simple model for complex phenomena. The normal distribution can be described by a probability density function as

$$f(x) = \frac{1}{\sigma\sqrt{2\pi}} e^{-\frac{1}{2}([x-\mu]/\sigma)^2}, \tag{21.1}$$

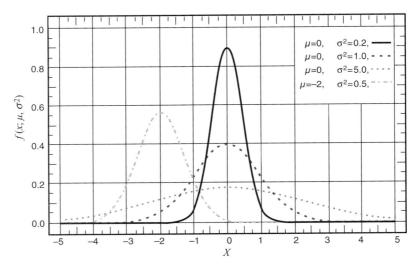

FIGURE 21.3. Probability density function of normal distribution.

where μ is the mean or average, σ is the standard deviation, and σ^2 the variance. The probability density function of a normal distribution features a bell-shaped curve as shown in Figure 21.3.

The mean determines the location of the peak of the distribution in relation to the y-axis. If the mean is zero, the central line of the profile overlaps with the y-axis. The standard deviation determines the width of the distribution profile. A large σ implies a wide variation featuring a short and fat profile, while a small σ indicates a narrow variation corresponding to a long and thin profile.

Let us use the letdown steam rates as an example to walk through how to create a normal distribution model to fit real data.

21.3.1   Step 1: Calculate the Average and Standard Deviation

The average for the HP letdown flows obtained from the historian can be calculated as

$$\mu = \frac{1}{N}\sum_{i=1}^{N} x_i = 515 \, \text{t/d}, \tag{21.2}$$

where N is the total number of data points and x_i is the letdown rate for the ith data point. Standard deviation is calculated as

$$\sigma = \sqrt{\frac{1}{N}\sum_{i=1}^{N}(x_i - \mu)^2} = 85 \, \text{t/d}. \tag{21.3}$$

One thing that needs to be emphasized again is that the sample data must cover the whole cycle of variations in operation. For example, a production cycle could be one year if it covers production seasonal change and this cycle is repeated year-by-year. In this case, the data for a typical production year should be collected and are used as the basis for calculating the average and deviation. Spurious outliers' data due to instrumentation errors must be removed from the sample data set before calculating μ and σ.

21.3.2 Step 2: Generate Normal Distribution via Monte Carlo Simulation

Based on the average and standard deviation calculated, a Monte Carlo simulation can be used to generate random numbers that mimic a normal distribution with the peak as the average value and the spread defined by the standard deviation. Monte Carlo simulation can be conducted readily in an Excel spreadsheet. Within a databook in the spreadsheet, select "tool" in the main menu and then click "data analysis" in the pull-down menu. In the dialog box, select "random number generation" and a data entry window opens. For generating one Monte Carlo simulation run, enter 1 for "Number of variables"; otherwise, type 2 for two runs, and so on. When specifying the "number of random numbers," it is recommended to enter 5000 for a Monte Carlo simulation as a too small number of data points could give a misleading profile. Then select "normal distribution." By providing the values of mean (average) and standard deviation as calculated from equations (21.2) and (21.3), a data set of normal distribution is generated, which features the same mean and standard deviation as specified.

There are cases when trustable online for an operating parameter are not available, but the minimum and maximum operating limits are known from experience. In this case, the sample data set of normal distribution can be generated via

$$x = x_{min} + \text{Rand}() \times (x_{max} - x_{min}), \tag{21.4}$$

where Rand() is the random number generator between 0 and 1. It is recommended to generate a relatively large set of data with typically 5000 data points. Then you can conduct statistical assessment for x based on the succeeding steps.

Another common scenario is to understand the trend of a variable, which is a function of other independent variables. If the independent variables are of normal type, the dependent variable features a joint normal distribution.

Let us consider a dependent variable, z, which is equal to $a \cdot x + b \cdot y$, where x and y are independent variables.

In the case where online, data are available, then μ and σ can be calculated for both x and y. Assume $x \sim f(\mu_x, \sigma_x^2)$ and $y \sim f(\mu_y, \sigma_y^2)$. Then z corresponds to a bivariate normal joint distribution as $z = a \cdot x + b \cdot y \sim f(\mu_z, \sigma_z^2)$, and μ_z and σ_z are

$$\mu_z = a\mu_x + b\mu_y, \tag{21.5}$$

$$\sigma_z^2 = a^2\sigma_x^2 + b^2\sigma_y^2 + 2abR\sigma_x\sigma_y, \tag{21.6}$$

where R is the correlation coefficient between x and y.

This can be done readily in a spreadsheet. First, a Monte Carlo simulation is employed to generate normal data for x and y respectively based on their μ and σ. After the data for z are generated based on $z = a \cdot x + b \cdot y$, μ_z and σ_z can be calculated for the set of data z via equations (21.2) and (21.3). In this way, the calculation is simpler as there is no need to calculate correlation coefficient R. Otherwise, equations (21.5) and (21.6) are used to calculate μ_z and σ_z with R estimated based on correlation of x and y.

21.3.3 Step 3: Generate Histogram from Monte Carlo Simulation

Generating a histogram is an important step as the data from the Monte Carlo simulation will be organized to provide a visual representation. To do so, the lower and upper limits of steam letdown rates from the simulation data are determined, which are 200 and 800, respectively. Then the intervals (or bins named in spreadsheet) of letdown rates are calculated as

$$x_i - x_{i-1} = x_{min} + (x_{max} - x_{min})/M, \qquad (21.7)$$

where M is the total number of intervals.

The frequency (percentage of occurance) or the number of Monte Carlo simulation data points with values falling within each interval can be calculated in the spreadsheet using the function "frequency $(x, bins)$" where x is the 5000 data points of Monte Carlo simulation and $bins$ is the set of intervals as calculated in equation (21.7). Twenty-six intervals are used in this example with minimum and maximum as 200 and 800 t/d, the function *frequency* $(x, bins)$ will calculate the count (the number of occurance) of letdown rates that belong to each interval. The counts for all intervals will add up to 5000 as Monte Carlo simulation generates 5000 data points. To generate a histogram similar to Figure 21.2 for a like-to-like comparison, scaling is required to reduce the number of counts from 5000 to 360, the same as the online data points. To do so, the scaling factor of 360/5000 is applied to the count for each interval.

The histogram is generated for the example problem as shown in Figure 21.4 based on 26 intervals as x-axis and frequency as y-axis. With the histogram, we transfer the original data from scattered data as shown in Figure 21.1 to the organized data in Figure 21.2 (based on online data) and in Figure 21.4 (based on normal distribution) from which we can visualize how much the letdown rates might be expected to vary in operation. To simply put it, the histogram allows us to see the operation variation of letdown rates over an entire year.

The histogram reveals the fact that the uncertainty of letdown rate is large varying from 200 to 800 t/d. But in many cases, we wish to know the probability of above or below a certain value, or between specified limits. For this example problem, we would like to know how often the letdown rates are below 300 t/d, which is the threshold deciding the economic payback of the K301 motorization project. This question will be answered in step 4.

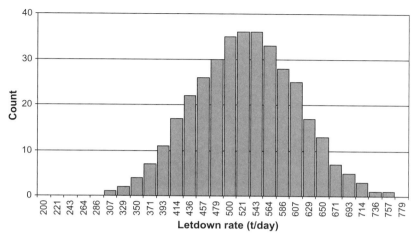

FIGURE 21.4. Histogram of Monte Carlo simulation results.

21.3.4   Step 4: Cumulative Probability and Percentage

The objective of using the Monte Carlo simulation based on the hypothesis of normal distribution is to predict how often (percentage) the letdown flow rates are below 300 t/d for this example problem. This percentage is an essential parameter as it could make or break the K301 motorization project. But first, let us take a look at the cumulative probability.

For example, we wish to know the percentage of counts (the number of occurrences) with letdown rates less than 415 t/d. This cumulative probability is determined as the percentage of the accumulative counts of steam rates to the left of 415 t/d, as shown in Figure 21.5, over the total counts in the entire profile. The counts

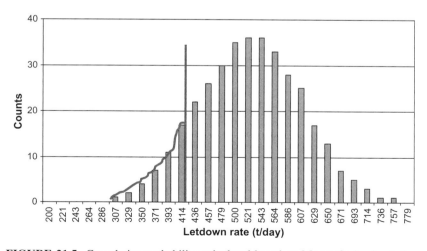

FIGURE 21.5. Cumulative probability calculated based on Monte Carlo simulation.

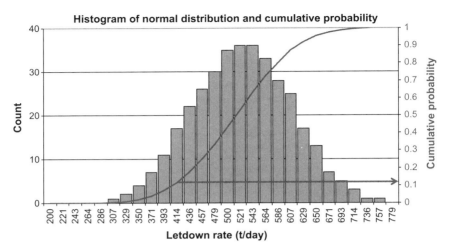

FIGURE 21.6. Cumulative probability curve based on Monte Carlo simulation.

of letdown rates to the left of 415 t/d (Figure 21.5) are cumulatively summed up and then the cumulative sum is divided by the total count (360 for this example) to give the cumulative probability or the percentage.

In this way, we can build a cumulative probability curve alongside the histogram of normal distribution as shown in Figure 21.6. The reason for showing the cumulative probability curve along with the histogram is to show the percentage of the letdown rates to the left of the letdown rate of interest.

Now we can answer some of the questions as follows for how often the letdown steam rates are less than a certain amount.

Question 1: What percentage of the time are the steam rates less than 415 t/d?
Answer: It is 12% of the time (Figure 21.6). This can be calculated directly using PercentRank (array, 415) in the spreadsheet where array is the 5000 data points obtained from Monte Carlo simulation.

Question 2: What percentage of the time are the steam rates larger than 650 t/d?
Answer: It is 5% of the time. This is calculated as {1 − PercentRank (array, 650)} in the spreadsheet. Letdown rates of 650 t/d or higher are used for peak demand.

Question 3: What percentage of the time are the steam rates between 415 and 650 t/d?
Answer: It is 83%, which is calculated as {1 − (12% + 5%)}.

Question 4: What is the true benefit of the K301 motorization project?
Answer: It is $2.16 MM/year (0.83 × 2.6), which can be used as the benefit basis for the K301 motorization. $2.6 MM/year was calculated in Chapter 20 based on 415 t/d of letdown average with the assumption of 415 t/d of HP letdown available at 100% of the time.

Question 5: What are the 90% interval limits for the letdown flow rates?
Answer: This question is about the spread or variation of letdown rates. The lower and upper limits are 5 and 95%, respectively, which makes 90% interval. The limits can be calculated using Percentile(Array, p) where p is the cumulative probability. For the present example, Percentile(Array, 0.05) = 375 t/d, while Percentile(Array, 0.95) = 650 t/d. This implies that the letdown rate varies between 375 t/d and 650 t/d in 90% of time. Clearly, it is a pretty wide spread.

The accuracy of the results will depend upon the number of data points and how far out on the tails of the distribution and on how realistic the statistic model is, how well the input distributions represent the true uncertainty or variation. Rerunning Monte Carlo simulation a few times will give you an idea of how much the results may vary between simulation runs.

21.3.5 Step 5: Verification

Figure 21.7 compares the observed distribution of online data with the normal distribution created by Monte Carlo simulation, which fits well together. However, it can be observed that there are a few points of actual distribution that distorts the normality. It may cause a concern if it is correct to use a normal distribution to model the online data. Before jumping into any conclusions, it is recommended that process engineers need to look into the data to understand why these outlier data occur: Is it due to instrumentation error or something else? It should be figured out from looking into production data such as feed rate and weather change to see if there is any cause for the abnormal situation. In most cases, these outlier data are caused by instrumentation errors.

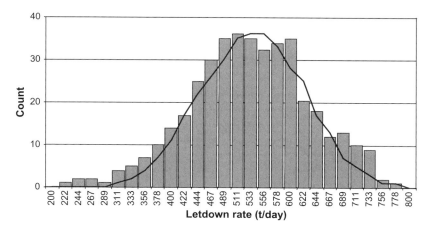

FIGURE 21.7. Comparison of online data with normal distribution.

If the abnormality is not caused by the bad data, the fundamental question then becomes: Is it justified to use a normal distribution to model the actual distribution? In general, it is rare that normal distribution fits the online data perfectly. As long as the variation pattern mimics the bell curve of normal distribution, it is reasonable to use the normal distribution to model the variation of real data. Use of normal distribution enables us to determine the true benefit based on variance.

Statistical methods (such as F-test, Student's t-test, and χ^2-test) can be used to prove the justification of using normal distribution to model online data. Usually, F-test is applied first to determine if a data set follows the normal distribution. Student's t-test is applied to determine the means or average assuming a common variance between a small set of sample data points and a large data pool. On the other hand, χ^2-test can be applied to determine the variance for a small set of sample data points by assuming a common means between a small set of sample data points and a large data pool. For readers who are interested in this topic, please read Schmuller (2009).

21.4 BASIC STATISTICS SUMMARY FOR NORMAL DISTRIBUTION

For more general applications, you may want to know a bit more about statistical metrics. As a brief summary, the basic statistics include mean, standard deviation, quartiles, standard error, and confidence intervals. Let us explain the physical meaning for each parameter in turn.

21.4.1 Central Tendency: Mean and Median

Mean is the most important metric, which describes the peak location of the distribution. It is calculated the same as the arithmetic average expressed in equation (21.2).

Median is the middle value of the Monte Carlo simulation data when they are sorted from low to high. In other words, 50% of the data is less than the median. If there is an even number of data points, then the median is the average of the middle two points. In spreadsheet, the sample mean = Average (array) and the sample median = Median (array).

Extreme values can have a large impact on the mean, but the median only depends upon the middle point(s). This property makes the median useful for describing the center of skewed distributions. In concept, the mean and median are different. If the distribution is symmetric (like the Normal distribution), then the values of mean and median are identical.

21.4.2 Spread: Standard Deviation

This is the second most important metric as it is the measure of dispersion: deviation from mean in root squared form (see equation (21.3)). In other words, the standard deviation describes the spread of the data.

Another measure of spread is called the quartile in spreadsheet. If the data are sorted from lowest to highest, quartiles can be established in a spreadsheet:

The minimum value $=$ Quartile(array, 0) or $=$Min(array).
The first quartile or 25th percentile $=$ Quartile (array, 1).
The median or 50th percentile $=$ Quartile (array, 2) or $=$Median(array).
The third quartile or 75th percentile $=$ Quartile (array, 3).
The maximum value $=$ Quartile (array, 4) or $=$Max(array).

21.4.3 Nonequal Distribution: Skewness

Skewness describes the asymmetry of the distribution relative to mean. In other words, when a distribution is skewed, it does not have equal probabilities above and below the mean. A positive skewness indicates that the distribution has a longer right-hand tail (skewed toward values larger than mean). A negative skewness indicates that the distribution is skewed to the left. In spreadsheet, skewness $=$ Skew (array).

21.4.4 Standard Error

It is a measure of uncertainty of the sample mean from the population true mean. It is the difference between a sample mean and the population mean: standard deviation of the means of different samples from the same population. For example, if Monte Carlo simulation for the same population is repeated and the sample mean is recorded each time, the distribution of the sample means would resemble a normal distribution. The standard error is a good estimate of the standard deviation of the distribution of multiple mean values. Assuming that the sample is sufficiently large ($n \geqslant 50$), the standard error can be calculated as

$$StdErr = \frac{\sigma}{\sqrt{N}}.$$
(21.8)

In spreadsheet, StdErr $=$ Stdev(array)/Sqrt(Count(array)), where Count gives the number of sample data.

21.4.5 Confidence Intervals for the True Mean

Alternative to standard error, confidence interval can be used to describe the uncertainty of sample mean. The sample mean is only an estimate of the population mean. How accurate is the estimate? Nobody knows. However, the concept of the confidence interval can make people feel good with the interpretation: We can be 95% confident that the true mean of the population falls somewhere between the lower and upper limits. In reality, the population of all samples is seldom known. In that sense, standard error may be a better measure of uncertainty of mean than the confidence interval.

In principle, the larger the sample size, the smaller the confidence interval (or more confident with the estimate). For example, it is advisable to generate a sufficiently large number of data points (at least 5000) in Monte Carlo simulation.

The 95% confidence interval is the interval produced in such a way that there is a probability of 95% that the true population mean lies inside it. For the 95% confidence interval (two-sided), the upper confidence limit and lower confidence limit are calculated as

$$U_{95\%} = \text{Mean} + 1.96 \, \text{StdErr} = \mu + 1.96 \frac{\sigma}{\sqrt{N}}, \tag{21.9}$$

$$L_{95\%} = \text{Mean} - 1.96 \, \text{StdErr} = \mu - 1.96 \frac{\sigma}{\sqrt{N}}. \tag{21.10}$$

To get a 90% or 99% confidence interval, 1.96 in the above equations should be replaced by 1.645 or 2.575, respectively.

Similarly, the upper and lower limits for 50% confidence interval can be derived as

$$U_{50\%} = \mu + 0.674 \frac{\sigma}{\sqrt{N}}, \tag{21.11}$$

$$L_{50\%} = \mu - 0.674 \frac{\sigma}{\sqrt{N}}. \tag{21.12}$$

NOMENCLATURE

L lower limit of a confidence interval
M the number of intervals
N the number of sample data points
StdErr   standard error
U upper limit of a confidence interval

Greek Letters

μ mean value in normal distribution
σ variance in normal distribution
ρ correlation coefficient

REFERENCE

Schmuller J (2009) *Statistical Analysis with Excel® for Dummies®*, 2nd edition, John Wiley & Sons, Inc.

22

REVAMP FEASIBILITY ASSESSMENT

The purpose of feasibility assessment is to identify technical and operating constraints for modifications. More importantly, the assessment should aim to find ways to reduce capital costs for promising options via exploiting spare capacity available in design margins and interactions of process design, equipment limits, and process conditions. Effective methodology and guidelines introduced here could help engineers find ways to reduce the level of modifications to key equipment, which is a main attribute to make a successful retrofit project.

22.1 INTRODUCTION

The previous chapters explained the methods that can help identify and evaluate improvement ideas in terms of cost and benefit as well as the effect on the utility system. If an option survived from the evaluation, there is one more step to go for identifying critical constraints and determining ways of overcoming them before being implemented in design. This step is called feasibility assessment, which is the focus of this chapter.

During feasibility assessment, improvement options will be evaluated in greater detail with more accurate estimates of costs and benefits based on rigorous process and equipment simulations. The traditional approach for feasibility assessment is to investigate modification options without extensively exploiting interactions between modifications, process conditions, process redesign, heat integration, and equipment

Energy and Process Optimization for the Process Industries, First Edition. Frank (Xin X.) Zhu.
© 2014 by the American Institute of Chemical Engineers, Inc. Published 2014 by John Wiley & Sons, Inc.

constraints. Thus, the traditional approach takes a narrow view of feasibility assessment with the focus on the negative part of assessment, that is, identifying showstoppers.

In contrast, this chapter will present a feasibility assessment methodology that takes a more proactive view of the feasibility with focus on searching for innovative ways to overcome critical constraints. It is not the intension of this chapter to provide a manual of feasibility assessment, but instead to introduce the methodology and share key lessons. These lessons are gained from experience of conducting numerous feasibility assessment studies.

22.2   SCOPE AND STAGES OF FEASIBILITY ASSESSMENT

A retrofit design is generally more challenging than grassroots design. For grassroots design in engineering firms, standard design templates are available. Designers make adjustments to the templates based on new feed rates, product specifications, and process conditions following a standard design procedure. In retrofit feasibility studies, one needs to be innovative to minimize modifications to existing equipment while maximizing potential improvements. At the same time, practical experience and engineering judgment also play a critical role during feasibility assessment.

The essence of a feasibility project is to investigate the effects of modifications toward existing processes and key equipment and identify critical constraints in existing equipment. Prior to feasibility assessment, the improvement options are usually identified in the idea discovery stage by using the methods discussed in the previous chapters. The million dollar question requiring an answer from the feasibility assessment is: Is it feasible to implement the options from the economic, technical, and operational points of view?

It usually goes through two stages in feasibility assessment to find an answer. The first stage is called feasibility scoping study, while the second one is feasibility study, after which the retrofit design phase follows.

22.2.1   Feasibility Scoping Study

A feasibility scoping study enabled by process simulation is the preliminary evaluation of improvement ideas in the context of overall process with the main objectives to perform cost and benefit analysis and determine if major equipment can accommodate the improvement ideas. The major equipment includes furnaces, reactors, main fractionators and separators, compressors, special pumps, and major heat exchangers. Usually, the scoping study is not concerned with relative ancillary equipment such as receivers, drums, heat exchangers, pumps, piping, instruments, relief valves, and so on.

In the scoping study, alternatives for improvement options will be identified and their effects will be evaluated with the aim of narrowing down alternatives and finding the best option. Why do we need to consider alternatives in the scoping study?

In reality, there are a number of ways for achieving an improvement. These alternatives could be identified in the energy study phase, but the effects to major equipment are yet to be determined. For example, the reboiling duty for a separator could be reduced via increasing the feed temperature, reducing reflux ratio, installing a side reboiler, and so on. Each option has pros and cons, which must be identified and compared for selection of the best option to go to the retrofit design phase. Otherwise, it could be very time consuming if evaluation of alternatives is left for the design stage because design is an iterative process and requires tremendous effort to make a rigorous simulation correctly. It could be a waste of time if alternative designs turn out to be infeasible and not cost-effective from the retrofit design phase.

In the scoping study, the accuracy of cost estimation is on the order of ±30. The main deliverables of the study include building a process simulation, evaluating the effects of changes, as well as conducting high-level cost and benefit estimates.

22.2.2 Feasibility Study

The purpose of a feasibility study is to determine what modifications should be selected for implementation, what constraints the modifications could face, and what solutions can be used to overcome the constraints. In this level of feasibility assessment, more details than the scoping study are considered in the process simulation as all the major and minor equipment are taken into account. Thus, potential effects on process and equipment as well as infrastructure from modifications will be identified and evaluated.

To identify equipment constraints, equipment details must be modeled in simulation. For example, column geometry and internal data need to be provided so that column loadings, for example, jet flooding under maximum capacity operation and weeping for turndown operation, can be obtained. Compressors and turbines are modeled with performance curves, which predict head, speed, and efficiency. The design performance curves must be "tuned" to mimic test run data. This allows a more realistic prediction of performance at different loads and/or process conditions.

In some cases, an improvement idea looks very promising in the scoping stage, but it may fail to pass the feasibility study because major equipment could pose constraints for the improvement idea. These constraints or restrictions are called the showstoppers in the industry.

The accuracy of cost estimates is on the order of $20 \pm 30\%$. The deliverables include basic engineering design with process flow diagram (PFD) markup of modifications and specifications for existing equipment, which require modification and for new equipment.

22.2.3 Retrofit Design Phase

The design phase can be divided into two stages: conceptual and final retrofit design. In the conceptual retrofit design, process redesign to incorporate modifications will be obtained. A complete PFD will be provided together with process conditions, heat and mass balances, as well as utility requirements, which are additional deliverables on top of the feasibility study. The accuracy of cost estimates is on the order of 10–20%.

The final retrofit design delivers P&ID (piping and instrumentation diagram) and generates a complete basic design package including complete process design, engineering design information, datasheets for all new and existing equipment, utility summaries, complete hydraulics, together with piping and instrumentation. The process design and equipment evaluation are conducted based on equipment-rated simulation.

22.3   FEASIBILITY ASSESSMENT METHODOLOGY

22.3.1   Traditional Approach for Feasibility Assessment

Traditional feasibility assessment investigates modification options more from identifying showstoppers instead of searching for opportunities to overcome them. The main reason is that traditional methodology does not simultaneously optimize process redesign, equipment performance, and operating changes that can reduce investment and operating costs. As a consequence, a revamp package often has a high capital cost.

22.3.2   Systematic Approach for Feasibility Assessment

Major changes to infrastructure and installing key equipment such as a new reactor, main fractionation tower, and/or gas compressor could form a major capital cost component in a retrofit project. In some cases, it is possible that the level of modifications to major equipment could be reduced or even avoided by exploiting design margins for existing processes and optimizing degrees of freedom available in the existing design and equipment. It is the goal of the retrofit methodology (Zhu and Martindale, 2007) to achieve minimum investment cost, which is shown in Figure 22.1 and explained later. Applying this feasibility assessment approach could give results in three categories. First, alternative options for each improvement idea will be provided. Second, any potential limitations, either in process conditions or equipment, will be flagged out. Third, solutions for overcoming or relaxing these limits will be obtained via exploiting interactions between process conditions, equipment performance, process redesign, and heat integration. This approach provides a pathway to help engineers find an answer to the challenging question: How to make improvement options work? In other words, what does it take to make the improvement options techno-economically feasible?

This methodology consists of four core components, namely, process simulation development, equipment rating analysis, heat integration analysis, and opportunity interaction optimization. This methodology has been applied to numerous process energy retrofit projects (Zhu et al., 2011) with the following common features:

- Clearly defined objectives, scope and basis.
- Reduced capital investment and operating cost to achieve the objectives.
- Increased possibility of retrofit approval.

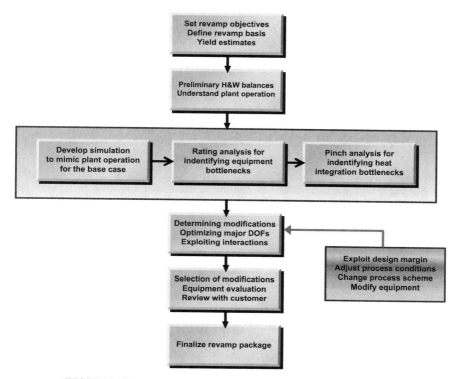

FIGURE 22.1. Retrofit methodology (Zhu and Martindale, 2007).

The purpose of process simulation development is to gain a good understanding of the plant design and operation in terms of key operating parameters and their interactions. The process simulation can mimic the plant operation or test run for the base case defined and agreed upon by the plant engineers. Thus, the simulation can provide the specifications for equipment rating.

The key role of equipment rating analysis is to assess equipment performance and identify equipment spare capacity and limitations. Utilization of spare capacity can allow capacity expansion up to 10–20% in general and accommodate improvement projects with low capital cost. When equipment reaches hard limitations, for example, a fractionation tower reaches the jet flood limit, or a compressor at the flow rate limit, or a furnace at the heat flux limit, it could be expensive to replace them or install new ones. The important part of a feasibility study is to find ways to overcome these constraints, which is accomplished in the succeeding steps.

The third step is to apply the heat integration methods (Chapters 9 and 10), the process integration methods (Chapter 11), the fractionation capability diagram (Chapter 12) and utility optimization (Chapter 19) to exploit interactions and hence determine changes to process conditions, equipment, process redesign, and utility system with the purpose of shifting plant bottlenecks from more expensive to less expensive equipment. By capitalizing on interactions, it is possible to utilize equipment spare capacity and push equipment to their true limits to avoid the need for replacing the existing equipment or installing new ones. This is a major feature of this retrofit methodology.

A simple example is fractionation tower feed preheat. A tower reboiler could reach a duty limit. With a tower feed preheater, it reduces the required reboiling duty. The column assessment may show the effects on separation with increased feed preheat and reduced reboiling at the bottom. If the effects are acceptable by the column, this modification could avoid replacing the existing reboiler with a larger surface or installing a new reboiler. If the tower requires refrigerant for cooling, feed cooling shifts colder utility use from the condenser to a warmer level in the feed chiller. As a result, adding a new feed cooler finds more capacity in the refrigeration compressor, beneficially reduces the vapor and/or liquid traffic within the column, and may unload refrigeration load in the condenser.

The fourth step is integrated optimization and the driver is optimizing process conditions to utilize the spare capacity available in the existing assets. Process redesign provides another major degree of freedom as it can increase heat recovery and relax equipment limitations. Integrated optimization can identify synergetic opportunities in achieving retrofit objectives with greatly reduced capital cost.

Process redesign includes process flow scheme redesign and equipment internal redesign. Redesign of process flow scheme could change material flows through equipment. This could help utilize spare capacity in some equipment to relax the limitations in other major equipment. An example of equipment redesign is changing column internals, which could allow an existing column to process extra capacity with relatively lower cost than installing a new column.

22.3.3 Process Models and Yield Estimates

It must be pointed out that there is a critical role that the process simulation model and the yield estimating model can play in exploiting interactions between process conditions and yields. The effects of changing process conditions on equipment and utility can be evaluated in the process simulation model while the effects of changing process conditions on yields can be assessed in the yield estimating model. Having both yield estimating models and process simulation models allows engineers to conduct what-if analysis effectively to determine optimal revamp solutions.

However, it is rare that the yield estimating models are available to operating plants. That is one of the reasons why it is recommended by the author that a retrofit project for critical technology be conducted by the technology licensors as they have yield estimating capability. The criteria of selecting the vendor to conduct a retrofit project is desired process and operating knowledge, yield estimating capability and design experience, and track record in technology development and in conducting retrofit projects.

22.4   GET THE PROJECT BASIS AND DATA RIGHT IN THE VERY BEGINNING

In most cases, feasibility assessment is conducted under a tight timeline. When the plant management approves a feasibility project, they want to move things as soon as possible in order to see the benefit to be captured quickly. However, it happens from time to time that feasibility assessment cannot generate correct results on time or is unable to address the key concerns sufficiently and hence turns out to be unsuccessful. In majority of such cases, it can be traced back to one source: failure to define the basis correctly and get data right. Thus, the feasibility assessment embarks on the wrong track until it is too late to meet the project deadline.

It is true that getting the basis and data right could take a lot of time and effort in the early stage. People may have a concern on the project expenditure and schedule as the feasibility project usually faces a tight timeline and budget. However, the time is worth spending to identify the objectives and requirements first before finding ways to achieve it. Getting the basis right is like "look before you leap" if simply putting it. The benefit is avoidance of rework down the road of the project.

Getting project design basis right implies that we need to determine and agree on:

- *Processing Objectives*: There could be different processing objectives such as throughput increase, technology upgrade, changes in feeds and/or products, energy reduction, environmental regulations.
- Specifications for feeds and products.
- *Operating Cases as the Basis*: A normal operating case for benefit estimate and a control case for equipment sizing.
- Utility provision.
- Economic criteria.
- The scope, deliverables, and schedule.

Project basic data consist of the following:

- Plant data for selected case scenarios agreed in the basis definition.
- Equipment datasheets.
- PFD and P&ID.
- Preliminary functional description.
- Utility systems balances for the base case.
- Revamp history.

Due to the importance of getting the design basis and data right for a complex retrofit project, the following guidelines could be used:

- Good communications internally and externally is the key to get the basis and data right. To achive this, assign a dedicated coordinator who can ask the right questions and extract the required information from plant personnel management, marketing engineer, commercial planners, process engineering, maintenance engineer, utility engineer, and energy engineer for issues pertinent to their responsibilities.

- Engineers with different responsibilities will provide their input of basis data: The process engineer will need to generate test run data if needed as well as explain process steps and postulate functional descriptions. The commercial engineer will provide input from business planning reality and help develop the project objectives and product specifications. The utility engineer will provide data for utility and infrastructure and bring the local plant situation into focus as it relates to the newly postulated facilities.

- Assign a project engineer with good experience for the process unit of interest who can provide input and feedback in developing a simulation for the process, and prepare tables and graphs that can summarize the design basis effectively for the engineering vendor who provides process design. The engineer should be dedicated to the task of gathering data and be allowed to be free from other responsibilities for the period of data extraction and reconciliation. This engineer should proceed to be involved in the feasibility assessment and revamp design.

- Bring out all possible surprises (good or bad) as early as possible. Do not bring surprises to management without proper answers (reasons and solutions).

22.5   GET PROJECT ECONOMICS RIGHT

It is critical that correct project economic data must be provided for determining justification of improvement options and selecting between alternatives, which include the following:

- Feed and product prices, including by-products.
- Utility costs, including fuel, electricity, steam, and cooling water.
- Basis for equipment capital cost estimates.
- Criteria of return on investment, namely, simple payback, net present value (NPV), and internal rate of return (IRR).

22.5.1   Uncertainty of Fuel Prices

There are significant challenges to get the project economics right. This does not simply mean "agree economic data" but on "how to get it right." The first challenge is: What are the energy prices to be used for evaluating the benefits of improvement options? In the late 1990s, the price for natural gas was about $3/MMBtu; in 2007 the price was about twice as high, and then in 2012 the price dropped to lower than $3/MMBtu in the United States, but it was still very high in Asia. An energy retrofit design for increased heat recovery in 2007 may now be economically unjustified. However, who can predict the future?

To deal with the first uncertainties, the method of cost range analysis (Zhu and Martindale, 2007) could be used to determine a flat cost range in which the total costs (energy plus capital) for different heat recovery levels are similar and close to the minimum total cost. In this flat cost range, energy prices are insensitive to the total cost because the effect is cancelled out by capital cost. A maximum heat recovery level within this flat cost range is then selected as the basis for the revamp project. This approach provides effective mitigation of the risk of a design being rendered suboptimum due to changes in energy costs.

22.5.2 Selection of Steam Prices

The second challenge is the selection of correct steam price. In most companies, the reported cost of boiler steam is the average cost. The total operating costs, including fuel, power, water, chemical additives, labor, maintenance, depreciation, interest, and administrative overheads, are divided by the total amount of steam produced. Then the costs of medium- and low-pressure steam are usually determined based on their enthalpy values in relation to the boiler steam. However, these prices do not reflect the true costs of steam in operation. Use of these steam prices as the basis can lead to wrong estimates of benefits from energy projects. As a consequence, many good energy projects may be rejected, and bad projects may be approved for implementation, which is the main reason for investment regret. Therefore, how do we determine the correct steam prices?

For tackling this challenge, the method based on marginal analysis can indicate true steam costs at point of use. The reason why the marginal price method (Chapter 17) can provide the true steam cost is that it is based on the last incremental amount of steam saved or generated. Determination of marginal steam prices relies on an overall steam and power balance model, which takes into account the effects from steam balances.

22.5.3 Capital Cost Estimates

Another key challenge is to achieve accurate capital cost estimates. Capital costs for a typical energy-saving project include costs for heat exchangers, columns, compressors, heaters, and coolers. Improper estimates of capital cost can lead to suboptimal trade-offs between capital and utility costs. One important aspect is estimates of heat transfer coefficients for individual process streams since heat transfer coefficients determine the size and surface area of heat exchangers. It is advisable to store a database of heat transfer coefficients for a variety of different processes from experience, which are used for feasibility assessment. Another aspect for accurate capital costs is the use of installation factors for revamp. In general, the values of installation factors in revamp could be several times higher than that used in grassroots design. This is because in revamp scenarios, demolishing, removal, piping, and other infrastructure costs could be a major part of the cost component. In addition, the penalty due to shutdown or reduced production should be also considered.

The common mistake observed is the capital cost estimate for a retrofit project being too low compared with reality. The main reason is that it does not include installation and the outside system battery limit (OSBL) cost, which could be as expensive as the inside system battery limit (ISBL) cost. This topic is discussed in Section 22.6.

22.5.4   Economic Analysis

Economic analyses at early stages in a retrofit project are primarily useful as a screening tool. More detailed economic analyses can be completed when the project is more clearly defined. Different economic approaches are used for evaluating investment benefits, which include simple payback, NPV, and IRR (internal rate of return).

An investment project involves spending an investment (I, in $), and then gaining a return over time. If an energy project is implemented with money spent now, the plant hopes to see a return on investment from a subsequent net energy-saving benefit (B, in $/year). The simple payback approach is to calculate the payback time (P_B in years) as

$$P_{\mathrm{B}} = \frac{I}{B}.\tag{22.1}$$

Investment costs are typically installed equipment cost. Net benefits are the annual increase in profits from the project, based on increased benefits minus increased costs. Payback is usually expressed in years, with a value of 2–3 usually a reasonable investment threshold for an energy retrofit project. Some companies use 4–5 years as the payback criteria for energy project selection. However, this payback indicator does not take into account the time value of money for which the NPV accounts for.

The NPV approach involves assigning a rate of return (r, in %) that is specific to the project and then computing the present value of the expected stream of revenues (R_i, in $). It provides an indication of what the present value of the investment (I) is considering the investment cost while realizing the future revenue from the investment at the present time. The NPV is simply the present value of future cash flows minus the investment price, taking inflation and returns into account. Since the investment is initially expended, it is counted as negative revenue. An NPV formula is expressed as

$$\mathrm{NPV} = -I + \frac{R_1}{(1+r)} + \frac{R_2}{(1+r)^2} + \frac{R_3}{(1+r)^3} + \cdots + \frac{R_N}{(1+r)^N},\tag{22.2}$$

where R_1 is the first year revenue, R_2 the second year revenue, and so on. N is the last year to account for revenue (R_N) from the investment (I). The analysis uses a minimum acceptable return on investment, r. This NPV indicator is defined as the difference between the present value of the expected returns (r) and the initial investment required to generate the returns. If the NPV is positive, the NPV will add value to the business. If it is negative, the investment will subtract value from the business.

Two tasks are required to conduct an NPV analysis. First, investment and revenues must be estimated. For a retrofit project, a total capital cost required becomes the investment. The total cost should include all possible major capital cost related items such as equipment, installation, infrastructure, downtime, and so on. The revenues are the net benefits expected in the future. Second, an appropriate rate of return must be identified. Most investments undertaken by companies are financed with retained earnings with profits from previous activities instead of borrowing. Thus, once a company approves and undertakes one investment, it cannot execute other investments at the expense of the approved investment, and the interest rate has to account for the internal corporate value of funds. As a result of these factors, interest rates of 10–20% are common for evaluating the NPV of projects.

If we assume a constant benefit over time, equation (22.3) will reduce to

$$\text{NPV} = -I + R \times \sum_{i=1}^{N} \frac{1}{(1+r)^i} = n \times R - I, \tag{22.3}$$

where n is the factor converting benefits across N number of years into present value,

$$n = \sum_{i=1}^{N} \frac{1}{(1+r)^i} = \frac{1}{r}\left(1 - \frac{1}{(1+r)^N}\right). \tag{22.4}$$

Typically, n takes a value within the range of 5 to 10. For example, if $N = 10$ years and $r = 10\%$, $n = 6.14$. If $N = 20$ years and $r = 10\%$, $n = 8.51$. If $N = 10$ years and $r = 15\%$, $n = 5.02$.

A retrofit project reduces net operating expenditure (ΔOpex) by \$10 MM/year with capital expenditure (CapEx) of \$30 MM. Assuming capital is spent initially, the present value (PV) for the Opex savings across 10 years at a discounted rate of 10% is

$$\text{PV}(\Delta\text{Opex}) = 6.14 \times \Delta\text{Opex} = \$61.4\,\text{MM},$$

$$\text{NPV} = \text{PV}(\Delta\text{Opex}) - \text{CapEx} = 61.4 - 30 = \$31.4\,\text{MM}.$$

This project should go ahead based on the NPV approach as it has a high NPV. However, the simple payback is 3 years, which would put this project borderline.

For the above retrofit project, if the CapEx is spent over the next 3 years in four installments, \$10 MM initially (year zero), 8 in the first year, 7 in the second year, and the last 5 in the third year,

$$\text{PV}(\text{CapEx}) = 10 + \frac{8}{(1+10\%)} + \frac{7}{(1+10\%)^2} + \frac{5}{(1+10\%)^3} = \$26.8\,\text{MM},$$

$$\text{NPV} = \text{PV}(\Delta\text{Opex}) - \text{PV}(\text{CapEx}) = 61.4 - 26.8 = \$34.6\,\text{MM}.$$

In some cases when a grassroots design is improved via design changes, these changes could lead to reduction in both Opex and CapEx. In these cases, CapEx takes a negative value. Thus, CapEx savings contribute to NPV as well.

For example, an improvement idea reduces both Opex by \$3 MM/year and CapEx by \$5 MM. Assume this Opex savings spreads across 10 years with discount rate of 10% and CapEx savings is obtained initially,

$$PV(\Delta OpEx) = 6.14 \times \Delta Opex = \$18.4 \, MM$$

$$NPV = PV(\Delta Opex) - \Delta CapEx = 18.4 - (-5) = \$23.4 \, MM.$$

The IRR approach is similar to NPV. The difference is that rather than assume an acceptable return on investment, the IRR approach solves the equation (22.3) with NPV $= 0$ to determine an effective interest rate (r) or IRR. In other words, IRR is the interest rate at which the net present value of all the cash flows (both positive and negative) from an investment equal zero,

$$0 = -I + \frac{R_1}{(1 + IRR)} + \frac{R_2}{(1 + IRR)^2} + \frac{R_3}{(1 + IRR)^3} + \cdots + \frac{R_N}{(1 + IRR)^N}. \quad (22.5)$$

If the IRR determined from equation (22.5) exceeds a company's required rate of return, that project is desirable. Otherwise, if IRR falls below the required rate of return, the project should be rejected.

When the question is whether a company should invest a project or not, NPV is a stronger metric than simple payback and IRR for making investment decisions. This is because NPV determines the present value of making the investment and not just the amount of time needed to realize the investment. For this reason, NPV has become the most common approach to investment decisions. The IRR approach gets the profit-maximizing answer only if it agrees with NPV.

The payback approach indicates how many years an investment project will be run before the investment is realized and profitability is reached. The problem with the payback period is how to decide between projects. The payback approach would select the project with the faster payback—it may make a small amount of money very quickly, but it is not obvious that this is a good choice. However, the choice is clearer with NPV—select the project with the higher NPV because it has the highest value and produces the strongest benefit for the company.

When performing a retrofit study, there are a lot of things unknown to the engineers. This is especially true if the effects on equipment, infrastructure, downstream processes, utility system, and tank farms are unknown. Some of the uncertain issues could be dealt with by doing various cases (scenarios) and then assigning probabilities to them. The result gives probability weighted economic scenario analysis.

22.6   DO NOT FORGET OSBL COSTS

One opportunity for driver replacement was identified, which is replacing the steam turbine driver for a recycled gas compressor with a motor. The reason was that electricity from the grid is much cheaper compared with high-pressure (HP) steam generated on site. Based on the cost estimate for installing the new motor within the ISBL, it showed a good payback of less than 2 years. But is it really good investment?

I was asked by the plant management to evaluate this option. I made field visits to the process unit and went to the location where the new motor could be installed. One of my questions to the process engineer who accompanied me was: from where will this 5 MW electricity be supplied? The answer was there is a substation nearby. Then I asked the question that hit the bottom line: Does the substation have enough spare capacity to accommodate electricity supply for an extra 5 MW? The process engineer could not answer this question and he had to call the electrician on site. The answer was a big NO! This fact instantly killed the motorization project as it would cost several million dollars to build a new substation plus cable runs, supporting structure, and all the tie-ins. It was a surprise to everyone involved as it was ranked as a great project. I felt partly smug as I found the truth of the matter and partly guilty as I felt I was responsible for killing this project.

The lesson learned was that the OSBL cost could be a major cost component for a retrofit project. It should be included in the cost estimate at the early stage because this could avoid rework and surprise in the later stage.

22.7 SQUEEZE CAPACITY OUT OF DESIGN MARGIN

Design margin is a magic word to many people in the process industry because it implies a great potential: It is known in the industry that additional capacity can be squeezed up to 10–20% out of design margins with aggressive operation and little capital cost! This fact raises very high expectations out of retrofit projects from plant management: Achieve the most out of the existing assets with the least investment.

To understand the plant design margins and constraints, it is recommended to conduct a rating survey for key equipment, and the results can be shown in a plot with rankings and indications of which equipment are limiting and which ones have spare capacity with how much.

For example, a product fractionator is usually designed with 80–85% jet flood as the limit, which gives around 10–15% spare capacity for use in revamp. For processes with reaction effluent used for feed preheating, if this heat recovery is high, the feed heaters are usually designed with large spare capacity for start-up, control, and end of run purposes. This spare capacity could be exploited as normal capacity in revamp. For many process pumps and compressors, they are designed based on rated flow rates (110% of normal); therefore, a 10% spare capacity is available for use in revamp. For the off-site utilities and the boiler house, design margins are applied to BFW pumps, storage tank pumps, and so on. There are many such examples of design margins built into the grassroots, which are available for grab in revamp.

Some design margins are hidden, but can be discovered. For example, many process furnaces are designed based on radiant flux or tube wall temperature (TWT) limit. If the furnace of interest is operating below the design limit, the gap provides the room for extra capacity through the furnace. For reactions, if reduced performance is acceptable, it may be able to avoid addition of new reactors in revamp. These examples include reduced conversion, reduced catalyst run length, and a reduction in severity at the desired cycle length.

However, everything has two sides. One must remember that when design margins are exploited in revamp, the related equipment will be operated tight or close to the limits. Something may need to be compromised. For example, if a fractionator is operated too close to the flooding condition, product quality may suffer due to reduced fractionation efficiency. This risk could be acceptable if the tower provides a rough separation and the net stream is recycled within the process unit. A furnace operating close to the flux or TWT limit leaves less room for safe operation. However, this risk could be managed by adopting better monitoring and control and advanced process control techniques. The bottom line is that you must know what objective you want to accomplish and what it takes to achieve it and how to mitigate risk.

22.8   IDENTIFY AND RELAX PLANT CONSTRAINTS

Finding ways to relax plant limitations is one of the most important tasks in feasibility assessment. The question is: What can be done in reality to overcome plant limitations? Let us take a tour on this important topic.

Three general limitations for equipment are pressure, temperature, and metallurgy as each equipment is designed with the limits for these parameters. If it is identified that equipment will operate at higher pressure than the design limit, it is necessary to modify the existing equipment so that it can be rated for higher design conditions or be replaced; however, both equipment modification and replacement come with a high cost. This is similar to the case when operating temperature will exceed the design limit. These constraints could be resolved if process conditions could be changed. However, if metallurgy is found less than required for a new unit design, this could be a major showstopper and it needs to be flagged out for the metallurgist's attention as early as possible. In some cases, the equipment could still be usable if the metallurgist agrees, and the plant takes actions for routing inspection.

For fired heaters, the limitations could be heat flux or TWT. The former is applied to heaters with pressure drop larger than 20 psi, which is the most common. The latter is for low-pressure drop heaters. When the heater duty must be increased to handle duty much larger than design, the existing heater may be insufficient in meeting the limits. Installing a new heater could be very expensive. The most effective way to avoid this is to increase feed preheating via process heat recovery.

For separation/fractionation columns, the major limitation is the tray loadings. Typically, the column is designed with 80–85% jet flooding. In revamp cases, the column can run harder and flooding limit can be relaxed to 90% or even 95%. It is possible to push extra throughput in the range of 10–20% at the column flooding limit if split specifications can be relaxed. If the revamp capacity expansion objective can still not be accomplished, more efficient trays could be used to replace the old ones partially or completely. Although it is costly, it is cheaper than erecting a new column.

For reactors, insufficient catalyst volume is the major limitation in revamp. Increasing space velocity could be effective but at the compromise of a shorter cycle length. The alternative solutions are to replace the existing reactor or install a new reactor, but they are expensive. Changing reactor internals could be effective to resolve the issue. Flow maldistribution is another common constraint that could be resolved by alternating loadings. The third constraint is increased pressure drop due to new catalyst or increased flow rate. In this case, reducing pressure drop upstream could help relax this constraint.

For compressors, when it reaches the capacity limit in flow or head, the direct solution is reducing the gas flow. The alternative solutions include increasing speed by gear change or adding wheels to the rotor or adding a compressor in series. The worst is to replace the major compressor, which is very expensive. Similarly, for pumps, the major constraint is insufficient capacity in flow or head. The cheap solution is replacing the impeller with a larger-sized one. The alternative solution is to add a third pump with two operating in parallel.

For heat exchangers, the most common constraint is insufficient surface area. In resolving this constraint, adding surface area to the existing exchanger could help, but area addition could only be up to typically 20%. Beyond this, adding a shell in parallel may be required. Although adding a shell in parallel gives lower ΔP, it can cause problems in flow distribution. For boiling service, replacing bundles with UOP High Flux tubing can increase heat transfer capacity. Replacing bundles with enhanced heat transfer tubes such as UOP High Cond tubing applies to condensing service. Installing a new exchanger unit with a new foundation is the last resort.

The above constraints occur in the ISBL. However, it is important to identify OSBL limitations as well such as off-site utility, layout, piping, substations, and so on. For example, a retrofit project requires additional HP steam for process use but existing steam system may reach the boiler capacity limit of HP steam generation. Installing a new boiler could kill the retrofit project. Another retrofit project requires installing a new exchanger, which is well justified from ISBL conditions; however, the piping to connect two process streams in the exchanger is too long in distance, which becomes cost prohibitive. A recent retrofit project determined the great benefit of installing a new motor with 5 MW to run a recycled gas compressor in a reforming unit, but it was deemed infeasible because the electricity requirement of the new motor is beyond the capacity of the nearby substation. Installing a new substation is a multimillion dollar project. Numerous ISBL and OSBL limitations could occur, which are not listed here in detail.

22.9   INTERACTIONS BETWEEN PROCESS CONDITIONS, YIELDS, AND EQUIPMENT

In operation, the operator can keep the plant running optimally by taking complex interactions into consideration under constantly changing conditions such as changes in feed quality, the product slate, process heat recovery heat exchanger performance, heaters performance and reliability, catalyst deactivation, and turndown. Some examples of key parameters and their interactions can be highlighted:

- For a fractionation column, the key operating conditions are the temperature and pressure of operation, vapor-to-liquid ratios, feed temperature, reflux rate, and pump-around duties. These parameters have significant effects on energy use and fractionation efficiency.
- For a reactor, the key operating parameters are the temperature and pressure, space velocity, and potentially H_2/HC ratio. These parameters greatly influence product yields.
- For heat exchangers, the key operating parameter is the U value, which affects heat transfer duty.
- For process heaters, heat flux and tube wall temperature are the key operating parameters, which affect not only fuel efficiency but also throughput and reliability.
- There are complex interactions between heater, reaction, fractionation, and heat recovery systems.

In revamps, we could do the same by exploiting these parameters and interactions to reduce the capital investment required for new equipment and retrofitting existing equipment. The following are some of the examples:

- Change to the existing reactor internals to enhance mixing and vapor distribution may avoid the need to add a new reactor.
- Change to the existing fractionator internals with enhanced mass transfer capability may avoid the addition of a new tower.
- Providing more feed preheat to a tower could save an existing reboiler.
- Increasing the pressure for a separator may save an existing recycled gas compressor.
- Adding a few trays to the bottom section of a main fractionator could save a new separator.
- Increasing process heat recovery could save a feed heater.
- In many processes, compressors are very expensive in capital. In revamp situations, if it is found that the new operating condition is above the choke point, changing operating pressure or reducing recycle gas rate could avoid the need to replace the compressor.

Discovering these opportunities requires a good understanding of these interactions and the technical skills to exploit them for meaningful use. The skills come from experience and creativity.

22.10   DO NOT GET MISLED BY FALSE BALANCES

In an oil refining plant, opportunities were identified for installing three steam turbines for power generation. Based on the estimates from a consulting firm, these turbines could produce large amounts of power due to high letdown steam rates in the steam system and generate very significant benefit due to the high electricity price with a quick payback of less than 2 years as a whole. The plant technical engineers were very excited with this finding and reported it to the management for approval. As a common practice in the capital approval process in the company, the management requests an independent review before making a decision for the large capital project.

I was called upon to conduct the review. Since the letdown steam rates are the key parameters that determine the benefits, my first question was how the steam letdown rates were estimated in the opportunity evaluation. This question implies: On what production basis was the steam balance developed?

To findout, I agreed with the plant technical manager that the base steam balances for estimating the benefit of new steam turbines should be developed according to the most common production mode(s). This is because the steam balances generated in this manner can be used as the basis for conducting the assessment to avoid under estimation or overestimation of the benefit. From the analysis of online data over an entire year, it was found that production rate varied widely and it featured a pattern: 20% of the time for campaign production, 70% of the time as normal production, and the rest of the time had low feed rates. Thus, the normal production was chosen as the basis to develop steam balances. At this stage, I realized that the base steam balance used by an outside consultant was developed according to the maximal production rate! Clearly, the benefit for the new steam turbines was overestimated.

This finding made the management feel relieved. If a green light was given for the options to be implemented, it would lead to regretful investment. The lesson learned is that in evaluating benefits for yield and/or energy improvements, it is common that people use maximal capacity mode as the basis, which could lead to overestimating the benefit. As a best practice, the common operating mode should be used to give a more realistic benefit estimate. Maximal capacity could be considered as the basis for sizing the new equipment in revamp assessment. Thus, it is advisable to have multiple reviews to check for critical parameters. The importance is not just to find the difference but lies in searching for reasons behind the difference.

22.11   PREPARE FOR FUEL GAS LONG

A process plant is in fuel gas long when fuel gas production is more than the consumption. When surplus fuel gas cannot be exported, it may have to be flared, which will undo energy improvements and also produce CO_2 emissions to the environment. Fuel gas long could happen when energy efficiency improves.

There is a real story of fuel gas long, which was caused due to the implementation of several large energy-saving projects. The fuel gas long scenario was not predicted in advance as the plant had a poor fuel gas balance. The plant fuel gas balance underestimated the production of refinery fuel gas but overestimated fuel consumption! During the period of fuel gas long, the plant had to cut down feed rates for major process units, which hurt its economic margins. This was not enough to curb fuel gas long and thus the plant had to send large amounts of fuel gas to flare. Eventually fuel gas balance was obtained after two gas turbines were installed for power generation.

The lesson learned from this painful event is that the plant fuel gas balance needs to accurately model the amount of fuel gas produced in individual processes as well as the amount consumed in individual fired heaters and boilers. The balance model should account for variations in feed rates and process conditions as well as weather changes. The variation could be very significant and thus modeling of variation in fuel gas production is a major challenge in building the plant fuel balances.

The main feature of a good fuel gas balance is to be able to balance fuel gas to 97% closure across variations in feed and product slates, processing conditions, and seasons. For example, processing heavy feeds could have a significant impact on fuel gas production and hence the fuel gas balance. A similar effect occurs for dramatic seasonal weather changes.

In modeling fuel gas balance, it is important to understand both the rate and composition of the fuel gas. As changes in feed and product slates are implemented, the relative proportions of fuel gas produced from the different process units can change, thus causing the overall fuel gas composition to vary. Different fuel gas composition has different heating value. For example, fuel gas with higher $C_3/C_4/C_5$ concentration has much higher heating value than the fuel gas containing C_1 and C_2 only.

Therefore, it is important for a plant to have a good fuel gas balance available for review of the impact of feed changes and energy-saving projects.

If a fuel gas balance model predicts that the plant will be in fuel gas long, a solution roadmap detailing what the plant can do in the short and long term should be developed. Efforts to reduce fuel gas production must attempt to understand the root causes and determine the best choice for reducing it via both operational changes and capital projects.

The first thing to consider is operational changes starting from increasing column pressure, reducing column overhead temperature, and reducing reaction severity, then reducing feed rate as the last resort. Another option is to use more fuel gas or steam, which can enhance production of desirable products and improve product quality.

The plan for capital projects should be developed as well if operational changes cannot bring the fuel gas long back into balance. Examples include removing column overhead cooling limits, which could minimize LPG lost to fuel gas. This could be achieved by better cooling using advanced heat transfer equipment.

A fuel gas long situation also provides the opportunity to recover valuable products from the fuel gas. For example, a refinery plant might consider recovery of hydrogen from a hydroprocessing gas source. Options for H_2 recovery include pressure swing adsorption (PSA) and membrane systems based on the H_2 purity requirement. Alternatively, ethane and ethylene could be recovered from certain fuel production sources such as fluid catalytic cracking (FCC) off-gas. On the other hand, a fuel gas long scenario could provide an opportunity to install a cogeneration facility as a long-term investment project to generate steam and power that might either reduce the plant's dependence on outside electricity or generate excess power that could be sold to the grid in a premium price.

Overall, prevention of fuel gas long is the best strategy. In the scenario of fuel gas long, a mitigation road map must be developed well in advance so that the plant has options in short, middle, and long terms, and can be successful in turning an unfortunate fuel gas long situation into a profitable opportunity.

22.12 TWO RETROFIT CASES FOR SHIFTING BOTTLENECKS

22.12.1 Case 1: Retrofit Project for a Refiner

Zhu et al. (2011) reported a successful retrofit feasibility assessment for a North American refiner during 2010. The global economy was still in recession but there was increased diesel demand in the market. The refinery wanted to capture the market opportunity, but it was so difficult to obtain loans from investors. The refinery management decided to investigate the possibility of capacity expansion by 15% in the hydrocracking unit with a stretch goal of 20% at minimum capital cost. If it can be done, they wanted to implement the modifications fast. Therefore, a feasibility project was conducted to look into the possibility.

In the kick-off meeting, the plant management emphasized the retrofit objective that increasing the throughput by 15% must be met with minimal capital investment because it was difficult to obtain large investments during the recession.

In this project, energy optimization was used as an enabler to move the bottleneck from "expensive-to-fix" locations to cheaper ones, allowing optimization of process conditions and designs. Actually achieving and implementing this synergy is difficult as one needs to pay attention to every detail of practical feasibility.

The assessment first looked into the opportunities associated with design margin. By optimizing process conditions and pushing the equipment to the absolute limits, the plant could make majority of the spare capacity and push plant capacity by 10% with some minor modifications. However, beyond a 10% expansion, several major limitations were identified from detailed simulations, which included the feed heater flux limit before the main fractionator column, jet flooding limit in the main fractionation column, and reactor space velocity limit in the hydrocracking unit.

The retrofit methodology (Zhu and Martindale, 2007) was adopted and UOP proprietary tools were applied to identify ways to shift bottlenecks from expensive to cheaper equipment. First, by adding two exchangers in the feed preheat circuit before the main fractionation feed heater, the feed preheat was increased by 20% and thus the need for retrofitting the existing feed heater was avoided, which could be very expensive.

At the same time, the opportunity to recover hydraulic power from the high-pressure separator liquid by installing a power recovery turbine was identified. Other energy-saving opportunities included installing a combined convection section for high-pressure steam generation for two reactor charge heaters, which had radiant sections only. A large amount of flue gas heat is wasted to the atmosphere as the stack temperature is higher than 1000 °F.

These modifications removed the throughput bottlenecks in the heaters and thus made it possible for a 15% throughput increase with minimal capital requirements.

There were however other aspects to consider beyond removing the bottlenecks in process heaters to actually achieve the 15% increase in throughput. First, the original reactor internals were already unable to provide uniform gas/liquid flow distribution let alone accommodate the increase in throughput. Hence, new state-of-the-art reactor internals were selected. Second, the cold flow properties of the product, which resulted from vaporizing the fractionator feed more than before, were unacceptable. This was solved by specifying a new catalyst that can handle cold flow properties. Last, after assessing the vapor/liquid equilibrium and hydraulics for the debutanizer and main fractionator columns, it was clear that these may not be able to handle 15% increase in throughput. New fractionator internals to allow for high liquid rates were implemented.

With the mentioned modifications, the energy-saving opportunity was around 100 MMBtu/h with capital payback of less than 2 years. The overall energy study captured energy savings and enabled the desired throughput increase worth $14 million per year. More importantly, the plant could achieve 20% capacity expansion with minimal capital investment. Proper application of pinch analysis to explore the synergy between heat integration and process changes was critical for shifting bottlenecks from expensive locations to cheaper ones.

22.12.2 Case 2: Retrofit Project for an Ethylene Plant

Lee et al. (2007) reported a successful retrofit project for an Asian ethylene plant. This retrofit case went through every single phase of feasibility assessment and retrofit design. Initially, the plant management had a two-phase retrofit strategy for capacity expansion, namely, 10% increase in the first phase (to be implemented in the forthcoming turnaround) and 20% increase in the second phase. The limit for 20% expansion was set by the cracking furnace capacity.

A previous feasibility study, without exploiting the interactions between process conditions and design and equipment changes, had concluded that, among other modifications, a new cracking heater and associated equipment, replacing of the low-pressure casing of the cracked gas compressor (CGC), rerotoring of the CGC turbine, a new depropanizer and associated equipment, and rerotoring of the ethylene and propylene compressors/turbines would have been required to achieve new production capacity. This retrofit proposal was rejected by the plant as it was too expensive. The new strategy was to revamp the ethylene and propylene refrigeration compressors with minimal cost and to push plant capacity by 20%, which was set by cracking furnace limitations.

In the new revamp study (Alanis and Sinclair, 2002), a pinch analysis study considering both process changes and heat integration was conducted to identify process design changes that can reduce refrigerant compressor shaftwork. Pinch and column analyses revealed interesting interactions between the process and the utility systems. By capitalizing on these interactions, it is possible to more effectively use existing equipment. A good example of this is distillation column feed conditioning. Column analysis may find that thermal conditioning of a column feed is beneficial. Feed preheat recovers refrigeration at a colder level than the reboiler, or, conversely, feed cooling shifts colder utility use from the condenser to a warmer level in the feed chiller. In both cases, adding a new feed exchanger finds more capacity in the refrigeration compressor, beneficially reduces the vapor and/or liquid traffic within the column, and could also unload the condenser and/or reboiler.

After modifications were identified, equipment rating assessment showed that the charge-gas compressor turbine already worked at the capacity limit under the current rates and needed to be modified to achieve higher production capacity. Increasing the CGC suction pressure was an option to reduce turbine's power. This, however, would only achieve a modest increase in plant capacity. Second, the propylene compressor and its turbine were also at their limit and had no additional spare capacity. A predictive process simulation model was used to estimate the compressor loading at the increased production rate of +20%. At this stage, the compressor manufacturer was contacted to assess the required compressor modifications. The compressor manufacturer's initial study concluded that the compressor required a new, larger casing.

At the project team's request, the compressor manufacturer advised that the maximum increase in capacity attainable with internal modifications while retaining the existing casing was between +10 and +15%. The compressor manufacturer also advised the feasible operating region of the compressor's second- and third-stage discharge pressures of the rerotored compressor. A number of process design and operating changes were identified to reduce the loading of the compressor. A few iterations with the compressor manufacturer were needed to define the optimum set of design and operating changes required to achieve the target production rate of +20% without replacing the propylene compressor.

Other major modifications in the final revamp package include the retraying of some towers, rerotoring the ethylene compressor, rerotoring the ethylene and propylene compressor turbines, and installing two new heat exchangers and three new pieces of equipment as part of process modifications that reduced the propylene compressor loading.

The turbine vendor indicated that a 13 month lead time was needed to install the required turbine modifications, which could be too late to be accommodated in the initial schedule for the turnaround. On the other hand, the market showed a strong regional demand for ethylene. As a result, the plant owner decided to delay the plant turnaround by 1 year and adopt a single-stage revamp strategy to push the plant capacity by 20% so that the revamp package could be implemented in the coming turnaround.

Most of the modifications were implemented during the 2005 turnaround. Tie-ins were also put in place to install the remaining modifications early in 2006 without a plant shutdown. The plant is now fully operating at the target production rate of 470 Mt/year. The economic success has also been very satisfying. The total annual profits were estimated by the plant engineers from production increase and energy reduction at $16.1 million with a 15.7-month payback.

22.13   CONCLUDING REMARKS

What are the lessons learned from successful retrofit projects? A few important lessons can be highlighted as follows:

- Redefine processing objectives based on process and equipment constraints.
- Avoid replacing key equipment and installation of new expensive equipment by exploiting interactions of process conditions, process, and equipment designs.
- Use equipment rating to identify equipment limitations and spare capacity and exploit spare capacity for extra capacity.
- Apply pinch analysis together with process simulation to identify process changes that can shift bottlenecks from expensive to cheap equipment.
- Work with equipment vendors to find the cheapest ways to retain or retrofit key equipment.
- Modify the original design and establish new process operating conditions that can accommodate the changes in the most economical manner.

NOMENCLATURE

CapEx	capital expenditure
I	investment
IRR	internal rate of return
NPV	net present value
Opex	operating expenditure
PV	present value
R	revenue
r	interest rate

REFERENCES

Alanis FJ, Sinclair IJC (2002) Understanding process and design interactions: the key to efficiency improvements and low cost revamps in ethylene plants, Fourth European Petrochemicals Technology Conference, June 26–27, Budapest, Hungary.

Lee JJ, Ye BH, Jeong HY, Alanis FJ, Sinclair I, Park NS (2007) Reduce revamp costs by optimizing design and operations, *Hydrocarbon Processing*, April issue, 77–81.

Zhu X, Maher G, Werba G (2011) Spend money to make money via revamp, *Hydrocarbon Engineering*, September issue, 33–38.

Zhu XX, Martindale D (2007) Energy optimization to enhance both new projects and return on existing assets, NPRA, San Antonia.

23

CREATE AN OPTIMIZATION CULTURE WITH MEASURABLE RESULTS

Process engineers are accountable for keeping products on specification and mechanical engineers accountable for maintaining equipment reliability. A good energy management system can provide tools and enablers to process engineers, mechanical engineers and operators to put the operating costs on check via effective strategy, actions, and measures to cut down the operating costs.

23.1 INTRODUCTION

In many organizations, cutting energy operating costs has been an on-going effort for a long time. However, "cherry-picking" has been the dominant approach. In other words, projects to reduce costs are usually selected from stand-alone evaluations without investigating synergetic opportunities from yields, capacity, and reliability. These kinds of projects typically have limited benefits and a long-time payout.

To achieve greater benefits and be able to sustain them, we need a systematic approach consisting of a solution methodology, supporting structure, and activities. The objective of applying this approach is to identify energy-saving projects, capturing hidden opportunities in process operation, closing of various loose ends in steam system and offsite utility, and optimizing utility system operation.

Energy and Process Optimization for the Process Industries, First Edition. Frank (Xin X.) Zhu.
© 2014 by the American Institute of Chemical Engineers, Inc. Published 2014 by John Wiley & Sons, Inc.

23.2   SITE-WIDE ENERGY OPTIMIZATION STRATEGY

There is no single solution for energy improvements. As a matter of fact, improving energy efficiency within the process can be achieved in a variety of ways ranging from little or no capital cost operational improvements to optimizing the process, recovering more heat, or adopting new process technology that fundamentally improves the efficiency of the operation. To manage and sustain the improvement benefits, a continuous improvement culture must be established supported by enablers such as performance matrices, key indicators, roles, and responsibilities. The strategy consisting of five steps to achieve the best energy efficiency are discussed next (Sheehan and Zhu, 2009).

23.2.1   The First Step: Minimize Waste and Losses via Diligence

In reality, steam generated in the boiler house is distributed through an extensive network of steam pipelines to end users. The losses in steam distribution can be as high as 10–30% of fuel fired in boilers. Hence, the net boiler efficiency could be 10–30% lower from the user's point of view.

The losses do not necessarily attribute to a single cause, but are the result of a combination of various causes. It is common to observe a major steam loss caused by steam trap failure and condensate discharge problems. Steam loss could also occur due to poor insulation of steam pipes, leaks through flanges and valve seals, opened bypass and/or bleeder valves, and so on. For example, a 1/4 in. leak in a 600 psig line loses over 1,000 lb/h steam. In addition to noise and safety risks, this leak could cost $95,000 per year for the steam priced at $11.9/klb. Simple measures such as maintenance of steam traps and monitoring of steam distribution to determine if steam generated is in accordance with steam consumed can lead to significant cost-saving benefits.

Apart from distribution losses, other forms of energy losses could occur due to process cooling, poor insulation, condensate loss to drainage, pressure loss from steam letdown through valves, pump spill backs, and so on. To detect losses, you must know how much energy is generated versus how much is used in individual processes. The benchmarking method in Chapter 3 could be used to determine the overall gap of the energy performance, and individual losses are identified using different methods. Process energy losses can be detected using the energy loss assessment methods discussed in Chapter 8, while identification of steam losses are discussed in Chapter 18. Although these solutions can vary greatly, typical improvements to energy efficiency via minimizing losses are in the range of 3–5% and it can be even higher in some cases.

23.2.2   The Second Step: Effective Energy Use via Optimizing Operation

Early action should be taken in closing the overall energy performance gap to optimize operations. This is because operation improvement can be implemented quickly with little investment. Optimizing operations is to operate a process in the most energy-efficient manner via exploring synergetic interactions and running key equipment tight and close to their design limits. Simply put, optimizing operations is to get the most out of existing assets. At the same time, equipment must be operated within their reliability limits to avoid "equipment abuse." Otherwise, it could cause mechanical failure unpredictably.

By implementing operation opportunities at the early stage, it can boost engineers' confidence for implementing other improvement ideas. More importantly, the benefit captured creates a financial budget to pay off other investment projects for energy saving.

To determine how well a plant or a unit is doing, it is necessary to build an energy monitoring capability so that current energy use can be compared against a consumption target that reflects the current operations. Only then is it possible to do some analysis to determine the cause of deviations from the target and take appropriate remedial action. A good energy monitoring solution should allow the user to quickly *observe* the situation and assess the relative performance of multiple units; *orient* themselves by being able to drill down to get more details on key energy indicators of the most critical areas; *decide* on a set of possible actions based upon the determination of possible causes for deviation from target; and *act* quickly and decisively based upon well informed decisions. Actions can be best practices or control based depending on the resources and control tools available. The key to the success in optimizing operations is the concept of key indicators and details are provided in Chapter 4.

The key energy indicators (KEI) and targets can be readily implemented into an energy dashboard, which can show the performance gaps between current and targets on the computer screen as a snapshot summary. The level of a gap indicates the severity of deviations that form the basis to assign traffic lights for each KEI; that is, green implying the current performance is acceptable as it is within the target range; yellow, a warning sign indicating that a gap occurs and requires attention; red, an alarm sign urging for taking action at the earliest time possible. An example of monitoring for key indicators is Honeywell's Energy Dashboard (Sheehan and Zhu, 2009). This tool could be tremendously valuable to operators and engineers as to what to watch, what to focus on, and what actions to take.

For each key indicator, technical targets are established based on online data and simulation as the basis to compare with current performance. The difference between a target and the current performance for each key indicator defines the performance gap. Different gap levels indicate the severity and level of urgency.

Assessment methods (Chapters 5–7 for energy operation; Chapters 9–10 for heat recovery system; Chapters 12–14 for process operation; and Chapters 16–19 for utility system operation) are then applied to identify root causes—potential causes include inefficient process operation, insufficient maintenance, inadequate or lack of operating practices, procedures and control, inefficient energy system design, and outdated technology. Assessment results are translated into specific corrective actions to achieve targets via either manual adjustments, the best practices, or by automatic control systems. Finally, the results are tracked to measure the improvements and benefits achieved.

An example of a large multivariable control strategy was applied to an ethylene complex. This involves a total of 17 multivariable controllers that were linked together by an overarching optimization strategy that included the use of a nonlinear steam cracking model to predict product yields. The result of this control project was to enable the customer to increase feed rate by 3% over the previous best rate by being able to operate the process up against multiple constraints simultaneously. In addition, the application was able to reduce energy consumption by 3.25% by reducing steam consumption in the fractionators and minimizing excess O_2 in the furnaces. This resulted in a payback of less than 5 months.

Opportunities to operate process units more efficiently exist in most process plants. Little or no capital operational solutions can improve energy efficiency by 2–5%.

23.2.3   Third Step: Low Capital Investment via Increased Heat Recovery

Monitoring and controlling to improve process energy efficiency usually results in pushing the process up against multiple equipment constraints. To get to the next level of energy efficiency requires capital cost modifications to increase heat recovery within and across process units. Indeed, one of the key values of implementing operational solutions first, is that it highlights where the equipment constraints to the process are more clearly. Once specific units have been identified for improved heat integration, heat recovery opportunities can be identified via applying the methods as discussed in Chapter 9. Then the network pinch methodology (Chapter 10) is applied to determine the most effective energy retrofit projects. This method not only considers value and cost of improved heat recovery but also the impact in terms of product yields, operating flexibility, especially with respect to startup, shutdown, maintenance, and control.

A typical example of redesigning for improved heat recovery involved an older 1970s vintage diesel hydrotreating unit, which had a combined feed exchanger, charge heater, one reactor, and a stripper. After assessing this unit, the modifications of adding four heat exchangers were identified to recover more heat from the process and also generate steam. The capital cost for installing four exchangers was estimated to be $3 MM but resulted in energy savings of $4.5 MM/year.

Projects to improve process unit heat recovery can typically improve energy efficiency by 5–10%.

23.2.4   Fourth Step: High Capital Investment via Process Changes and Advanced Technology

Improved heat recovery is the most common type of capital project implemented to improve energy efficiency. However, recent work has identified other areas less commonly explored that may provide significant opportunities. These opportunities include process design changes and use of advanced process technology such as enhanced heat exchangers, high-capacity fractionator internals, new reactor internals, power recovery turbines, improved catalysts, and other design features, which are discussed in detail in Chapter 11.

One example of process changes comes from process flowsheet optimization for a hydrocracking process. This example was discussed in detail in Chapter 11. The original design features one common stripper, which receives two feeds containing very different compositions, one from the cold flash drum and the other from the hot flash drum; these two feeds are originated from the reaction effluent and go through a series of separations before mixing. The inefficiency caused by this process sequence can be described very briefly as: separation, mixing, and then separation again. This design feature can be observed in many other process plants.

To avoid this inefficient design feature, it is proposed to use two strippers, namely, a hot stripper that receives the hot flash drum liquid as the feed and a cold stripper that is used for the cold flash drum liquid. Furthermore, the cold stripper bottom does not pass through the main fractionator feed heater but goes directly to the main fractionators. Only the hot stripper bottom goes to the main fractionator feed heater. The net savings by the proposed two-stripper design is 110 MMBtu/h, which is a 42% reduction of the fractionator heater duty, or 23% of total energy reduction for the hydrocracking unit. This improvement is significant.

Another example is power recovery for an existing plant. Power recovery often represents a good opportunity for economic energy optimization. For an FCC (fluid catalytic cracking) unit with 60,000 bpd throughput, the FCC catalyst regeneration flue gas at high temperature was being used for steam generation alone via a waste heat boiler. A power recovery turbine (PRT) was quickly identified as a method for significant energy efficiency improvement, as the flue gas could be used for both steam and power generation simultaneously. Further improvement could be achieved by installing a power recovery turbine (PRT) combined with a steam turbine. The objective was to generate electricity from the regenerator flue gas, but also produce electricity from high-pressure steam letdown to make the medium-pressure and low-pressure steam required in the FCC unit. Compared to the base case that does not include a PRT and uses a condensing steam turbine to drive the main air blower, this new scheme could achieve a net energy benefit of $14 MM/year with payback in less than 3 years.

There is a variety of advanced technologies that can be applied, which vary in terms of cost to implement and return on investment. Careful evaluation of each of these solutions is required as capital is always limited; therefore, it is necessary to select only the best opportunities that can be best integrated with the existing processes that provide the highest return on capital employed. Although these solutions can vary greatly, typical improvements to energy efficiency are in the range of 3–7%.

23.2.5 Fifth Step: Utilities Optimization for Reduced Energy Generation Cost

In addition to using energy more efficiently in the process, another common strategy is to produce energy more efficiently. In this step, all the energy users and producers as well as commercial options are combined in a single system, and the interactions among them are optimized to maximize overall energy efficiency. Many manufacturing sites have their own on-site industrial power plants that primarily exist to provide steam and power to the process units, but may also supply electricity to the grid at times of excess capacity. Supply and demand optimization is achieved by a simulation of the utility plant that can take a set of process forecasted demands from the production schedule and determine the configuration and operating profile of the boilers and turbines to meet demand while taking into account the tiered pricing for natural gas and power contracts. Detailed discussions on utility assessment and optimization are provided in Chapters 16–19.

The opportunity for utility system optimization is evident when the electricity price varies from season to season, in some cases from daytime to evening. Optimizing the choice of on-site power generation and purchasing from the grid could reduce operating cost in the order of millions of dollars. Second, the opportunity is relevant when there is a large amount of steam going through the letdown valve. It could reduce generation cost by minimizing the letdown rates. Another key opportunity to reducing energy costs in utility plants is to balance the changing energy demands from the process with adequate supply from the utility plant without wasting energy by keeping spare capacity on hot standby.

Therefore, a utility optimizer could determine the most efficient and economic operation to supply energy to satisfy process demand. There are several major options to achieve this overall goal including driver selection, utility equipment loading allocation, steam head pressure control, steam letdown rate minimization, power generation versus purchasing, and contract management.

For load allocations of boilers and turbines, boiler and turbine efficiency curves are used as the basis to distribute the total heat requirement among all the boilers and turbines in the lowest cost manner. However, it also aims to maintain the effective steam production. Combined with the pressure control, this strategy allows for the fastest dynamic response while always trending to the most economic steady state position. Turbines are modeled based on performance curves, which can operate in either steam pressure control mode where the set of turbines aim to maximize power generated while maintaining steam pressure; or in power generation mode where steam consumption is minimized while maintaining a total generated power target.

When it comes to natural gas and electricity purchasing, the plant could face contract management issues and better contract management could save millions of dollars per year. This is because the plant could set up a contract with a choice of a fixed amount with a flat cost or a tied structure with different prices. For a tied structure, the base amount is fixed with a flat price while additional amounts can be provided with price(s) higher than the base price. With better forecasting, the plant will predict the amount of deficit in fuel and electricity and make the purchase contract well in advance. This could dramatically reduce purchasing costs depending on the contract terms.

An example of where the utility optimization solution was applied was at a petrochemical site in Korea. The plant had three oil-fired boilers and three back-pressure steam turbines that provide steam and power to the process units, and also supplied excess power to the national grid. The solution used the combustion

TABLE 23.1. Typical Energy Savings from Different Categories of Energy Savings

Saving Opportunities	Energy Improvement, %	Energy Saving, MM$/yr	CO_2 Reduction, kMt/yr
Step 1: Minimized losses	3–5	2.2–3.7	36–60
Step 2: Improved operation	2–5	1.5–3.7	24–60
Step 3: Improved heat recovery	5–10	3.7–7.5	60–120
Step 4: Advanced process technology	3–7	2.2–5.2	36–84
Step 5: Utilities optimization	2–3	1.5–2.2	24–36
Total	**15–30**	**11–22**	**180–360**

controls, the steam header pressure controls for three headers, and equipment load allocation across the boilers and the turbines. The results from the implementation led to significant reduction ($>10\%$) in CO_2 and NO_x emissions and improved boiler efficiency leading to overall benefits of more than $3 MM/year.

Optimization of the energy supply system can typically improve energy efficiency by 2–3%, while the main driver of utility optimization is in cost reduction.

23.2.6 Summary of Potential Savings from Five Steps

Table 23.1 combines all of the potential savings opportunities from the five steps mentioned to provide a perspective on the level of benefits for a typical refinery of 100,000 bpd (barrels per day) that could be achieved by adopting a comprehensive energy management program. Recent studies indicate that typical benefits of 15% energy reduction are expected, rising to 30% for refineries operating in the fourth quartile in energy efficiency. The carbon credits associated with the reduction in greenhouse gas emissions range from 180,000 to over 360,000 MT/year.

23.3 CASE STUDY OF THE SITE-WIDE ENERGY OPTIMIZATION STRATEGY

While many companies have looked at energy improvements, few have taken a holistic approach that examines changing process conditions and process design, as well as the technology required to achieve it. The five-step methodology discussed previously has shown that a total energy reduction of 15–30% is achievable. This case study demonstrates that this methodology can be applied for determining energy capital projects that can be used to substantially improve profitability by unlocking energy-saving opportunities. We will discuss how energy efficiency can be improved in existing facilities through relatively inexpensive capital projects. The savings from these improvements can then be used to fund more expensive capital projects that will provide additional savings.

TABLE 23.2. Heater Inlet Temperature and Pressure at Bubble Point (BP)

	Base Case	After Modifications
Heater inlet temperature (FIT), °F	430	461.7
Heater inlet pressure at BP, psia	185.9	216.1

23.3.1 Identify Inexpensive Improvement Options for Crude and Vacuum Unit

The process energy optimization starts with defining a basis in terms of feed and product slates, process configuration, and operating conditions. The base case is simulated to mimic existing operations. Pinch analysis is performed on the base case with the purpose of identifying improvement options. Then costs and benefits of each option are determined and the constraints are identified.

For the crude preheat train in the refinery plant, three beneficial modifications are identified. Modifications 1 and 2 add a new shell to each of the feed preheat exchangers serviced by the diesel pump-around (PA). Modification 3 adds surface area to the diesel product run down heat exchanger. Before implementing these changes, it is crucial to make sure that the existing pump can handle the increased diesel PA flow and the increase can be tolerated in product cut points. These modifications can save 47 million Btu/h of fuel in the crude heaters, equating to cost savings of $2.3 million per year. With only a $2.2 million installed cost, this project provides a fast payback.

However, these modifications cannot be implemented on a stand-alone as they would cause feed vaporization and flash before heater control valves (Table 23.2). To assess the effects of these modifications on feed vaporization, the bubble point (BP) as a function of feed temperature was calculated. This calculation is shown in Figure 23.1, which shows the actual feed temperature and pressure while the line indicates the bubble point as a function of feed temperature. The triangle above the line indicates severe feed vaporization could occur if three modifications

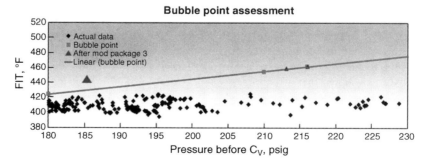

FIGURE 23.1. An extra 60 psi of pressure is needed to accommodate 462°F.

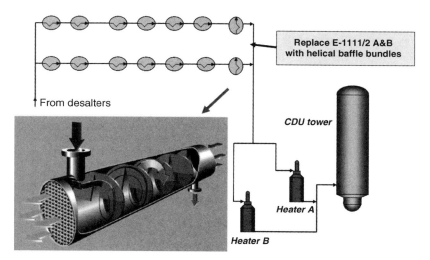

FIGURE 23.2. Use of helical baffle heat exchangers.

are implemented. This fact is also confirmed in Table 23.2, which shows the heater inlet temperature and the pressure at bubble point.

Several alternatives exist for avoiding flashing of the feed at the control valve. The cheapest is to reduce the pressure drop of the preheat train. Other options include replacing the impeller of the existing desalter boost pump or adding an additional pump after the desalter pump. An obvious first step would be to conduct regular heat exchanger cleaning to reduce pressure drop in the heat exchangers. Second, the design of each heat exchanger should be reviewed in terms of its velocity and pressure drop. In particular, the use of helical baffle heat exchangers (Figure 23.2) can significantly reduce pressure drop for the last two heat exchangers. Helical baffle heat exchangers provide a smooth flow pattern on the shell side that reduces pressure drop. In this case, a combination of helical baffle heat exchangers, regular cleaning of exchangers, and increasing the impeller size for the boost pump avoided the need for an expensive preflash drum.

23.3.2 Other Energy-Saving Opportunities

23.3.2.1 Reforming Unit Energy-Saving Opportunities Several energy-saving opportunities were identified for the reforming unit in the refinery. The first is to add a new shell for the feed preheat exchangers in the naphtha hydrotreating (NHT) section, thereby reducing the load on the NHT charge heater. The second is to add a new heat exchanger between the stabilizer feed and bottoms. The third is to replace inefficient reforming feed preheating exchangers with Packinox exchangers, to enable better heat balance between the NHT and reforming units and unload the existing reforming air coolers. Finally, redundant reforming heater economizers can be reinstalled. Together, these modifications would save 147 million Btu/h. These modifications can yield $7.1 million/year cost savings for a capital cost of $6.6 million.

23.3.2.2   Delayed Coker Unit Improvements   Two modifications were identified for the delayed coker unit. Additional reboiling serviced by the light cycle gas oil for the naphtha splitter and another reboiling serviced by the heavy cycle gas oil for the debutanizer. These modifications saved 12.7 million Btu/h or $0.6 million/year at a capital cost of $1 million. Another potential energy-saving project is the cross-unit heat integration. This involves generating medium-pressure steam at the vacuum unit to be used in the coker unit, saving 40,000 lb/h of high-pressure (HP) steam at the coker, thus reducing energy cost by $2.3 million/year. But the piping distance is long, which could be cost prohibitive.

23.3.2.3   Hydrocracking Unit (HGU) Improvements   Modifications to the hydrocracking unit were identified to save fired heat duty and increase throughput. Opportunities for reducing fuel to the reactor and fractionator heaters were identified. Installation of four heat exchangers that use reactor effluent heat for preheating the reactor feed and the fractionation feed can reduce total heater duty by 20%. With the above modifications, the energy-saving opportunity approached 108 MMBtu/h with a payback of less than 3 years.

23.3.3   Advanced Technology Projects

Once these quick-hit projects are implemented and start to provide cost savings, these can be reinvested in moderate- to high-capital projects that utilize the latest technology to provide even greater energy efficiency savings. Examples include enhanced heat exchangers, high-capacity fractionator internals, high-efficiency reaction internals, and high selectivity and activity catalysts. Specifically, process modification projects were also identified in order to achieve 20% capacity expansion and some of the major projects are listed here:

- Replacement of tower internals in CDU distillation using high-capacity trays
- Use new internals for the HCU main fractionators and debutanizer
- Replacement of reactor internals
- Added a new stripper in diesel hydrotreating
- Use of new HCU catalyst for better cold flow properties

23.3.4   Results Summary and Concluding Remarks

The synergy between energy savings and process technology know-how was critical to achieving these results. Twenty-four energy improvement projects were selected, which can reduce 400 MMBtu/h or save 17% of total energy use and provide operating cost savings of $19 MM/year with a total investment of $36 MM (Table 23.3). Furthermore, these projects can reduce CO_2 by 285,000 Mt/year creating CO_2 credit of over $5 MM/year for the site. More

TABLE 23.3. Cost and Benefit Analysis for Each Process Unit

	Number of Modifications	Fuel Savings, MMBtu/h	CO_2 Reduction, Mt/yr	Cost Savings, MM$/yr	Capital Cost, MM$	Simple Payback, years	NPV[a], MM$	Energy Reduction, %
Crude distillation unit	3	47.1	33,552	2.3	2.2	1.0	11.6	2
Hydrocracker	8	108.6	77,362	5.3	13.5	2.5	18.3	4
Delayed coker	3	12.7	9,047	0.6	1.0	1.7	2.6	1
Reforming unit	4	147.5	105,073	7.1	6.6	0.9	36.0	6
Distillated hydrotreating unit	2	23.3	16,598	1.1	0.6	0.5	6.0	1
Steam and power system	4	61.2	43,596	3.0	12.1	4.0	5.9	3
Grand total	24	400.4	285,228	19.4	36.0	1.9	80.4	17

[a]NPV $= 6 \times$ Cost Saving $-$ Capital Cost (equation (22.3)).

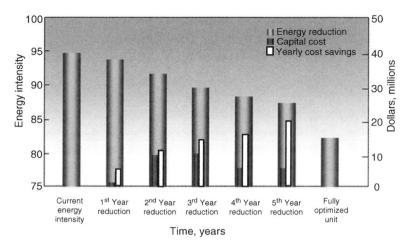

FIGURE 23.3. Energy reduction for 5-year implementation plan.

importantly these projects enable 20% capacity increase with minimum capital investment. Figure 23.3 shows a 5-year implementation. By the end of the fifth year, the site will become a first quartile energy performer in the oil refining industry after saving 400 million Btu/h of energy. The margin improvement from throughput increase is more than $80 MM/year.

This case study shows how effective return on investment can be achieved for energy capital projects and process/technology improvements by seeking integration opportunities between process and relatively inexpensive energy projects. The savings provided by these initiatives can be used to upgrade process design and technology to generate even larger savings from capital projects.

23.4 ESTABLISHING ENERGY MANAGEMENT SYSTEM

It is necessary to have an effective energy management system in place to drive measures for energy efficiency. It is not enough to just have good intentions from management and motivation from personnel. Some years ago, in a chemical plant, an energy program kicked off with enthusiasm, slogans, revival meetings, and pledges. An audit showed visible improvements from month to month. Management expected the path of improvement to continue. Instead, improvement ground to a halt. The trend leveled off after a few months and gradually turned upward.

What had happened? The rapid improvements seen at first came from correcting simple matters such as leaks, steam traps, insulation, and so on, which were detected by common sense. However, after the obvious improvements were accomplished, the improvement curve leveled off and became stable at an unacceptable level. After a while, despondence kicked in and the program arrived at a dead end since there was no preparation to keep digging deeper for hidden opportunities and thus no resources were assigned.

The lesson learned is that intention and motivation can produce good results only if right directions are developed, sufficient resources are given, and enablers are provided. Therefore, it is necessary to put into place an energy management system with clearly defined roles and accountability, metrics and measurement, tools, review, and actions. Simply put, an energy management system is about account-ability, work process, metrics, audit, and communications and so on. For sustain-ability, an energy management system must be integrated with existing technical management structures for continuous improvement, which is very important. Let us walk through these components one by one.

23.4.1 Accountability

The key element in the energy management system for a plant with multiple process units is to find a person as an energy coordinator or energy manager who has the passion for driving energy cost down, technical know-how, and communication skills. This person should come from within the plant so that he knows the people in the organization, processes, and existing management system. The role of the energy coordinator is to manage the energy management work process, facilitating imple-mentation of projects, handling interunit and offsite issues, and drive an optimization culture in the organization.

A large site may require an energy team consisting of a process engineer, an experienced operator, and/or a control engineer, all of whom are assigned to multiple responsibilities. The role of the operator is to go out to identify any inefficient operations visible on field such as furnace air register opening, bypass valve opening, flaring, steam leaks, hot flange, and so on. The role of the process engineer is to conduct opportunity evaluations and generate business cases for improvement options to push them through the capital gate review process. The control engineer is to develop a control strategy for improvement ideas and assist unit control engineers to implement in the distributed control system (DCS).

23.4.2 Work Processes

There are two major areas for an energy management system, which are energy capital projects and energy operation improvement. The work processes for these two areas are different due to the nature of the work. For example, energy projects should follow a techno-economic review process: define (process changes)–evaluate (cost/benefit assessment)–select (option prioritization)–execute (option implemen-tation). In contrast, for operation improvements, the work process should be based on the operation process: define (key indicators, targets, and limits)–monitoring–gap analysis (current versus targets)–actions/control.

The work process will define schedule, milestones, deliverables, and responsi-bilities. A well developed work process can avoid cracks under which good ideas could fall through while bad ideas are selected. The work process should be put on test in a trial for revision before it is integrated with the existing technical management structure.

With well-defined work processes for energy management, it is like setting the direction for a race. Often we observe situations in an energy management program driven by some strong personality without a well-defined work process in place. When this person leaves, the campaign suffers as this would cause a major disruption to the energy program.

23.4.3 Scorecard and Tracking

Success in operating a plant depends on awareness and alignment of objectives. The energy system must be developed as part of the overall business and operating plan. This starts with sharing the annual target for energy cost reduction including milestones and benchmarks. Tracking must be in place for reviewing progress toward goals, which could include a monthly review with plant management and an intranet site reporting scorecard of process energy performance.

Using a scorecard to show energy performance toward goals enhances alignment and awareness while tracking progress promotes motivation. Drilling the performance scorecard from plant level to process areas level until equipment level is the key for empowerment. This is similar to sports where the team will keep doing what they are doing well when winning. But when losing, the team will find root causes, fix them and win the next game. The essence of the scorecard is to tell if the team is winning or losing. Honeywell's Energy Dashboard (Sheehan and Zhu, 2009) is such a scorecard and tracking tool for process plants.

23.4.4 Documentation

Beyond the scorecard and tracking, it is critical to have a corporate-wide integrated documentation system for all management systems across the organization. When a new subsystem within a management system is created, it gets implemented throughout the corporation. The integrated document system, combined with the company intranet, can allow the company to integrate all management system processes throughout the company.

23.4.5 Technical Target System

Key indicators define the basis for continuous monitoring. Technical targets for key indicators must be developed and applied across the organization. Many industrial facilities have the potential to increase the efficiency of their systems, but have difficulty doing so because there are no metrics available to measure energy efficiency improvement and value created. The absence of energy performance metrics makes it difficult to establish a value structure for the improvements. Practically, a review of key energy indicators and targets should be part of the process performance review, which is a key part of integrating an energy management system with existing management structure. Since the development of key indicators and targets is a big ticket for the energy management system, it requires management support to acquire resources and technical capability.

23.4.6 Making Business Cases for Improvement Ideas

It is important to make good business cases for improvement ideas. No matter how good an idea is in technical merits, you need to do a lot of homework since you will have to answer many questions and deal with a lot of concerns for the effects that may occur if your idea is implemented. I have observed that many good ideas die in the infancy stage because they could not pass the evaluation gates. Such failure is commonly due to a lack of techno-economic assessment and communication. To avoid this, I list some important things you must to do.

First, you need to develop technical and economic merits to build a business case. Therefore, it is imperative that you determine the benefit of your ideas, that is, what does it worth, in the very early stages. Next, you should identify, with the help of process specialists, what it takes to implement the idea. You need to do the necessary evaluation to come up with rough estimates of capital cost required to deliver the benefit for your ideas.

Even if you made a superior technical and economic case, the doors of acceptance may not open automatically. An appropriate communication effort is required. This implies you need to elicit comments and feedback from technical specialists in the areas of operation, engineering, maintenance, and control. Their feedback will provide additional insights for the feasibility of implementing your ideas. Once you pass reviews based on technical merits, you need to sell your ideas to get buy-in from management. Although management expresses a strong voice for supporting energy efficiency improvement, management will not provide a blind check. You should remember the fact that the business objective of your plant is to produce desirable products and realize large economic margins. To successfully convince management, you need to connect your ideas with key business drivers and demonstrate the value of implementing your idea to the organization. Intelligent persistence wins in the end.

23.4.7 Corporate Energy Audit

The energy audit is intended to assist plant personnel in identifying both operational and capital projects that have short payback periods. The audit can be conducted in stages. The first phase is *benchmarking* of the overall site and each major process to determine where they stand. The benchmarking methodology was discussed in Chapter 3. The result of this assessment indicates the actual performance, which can be compared with either cooperate target or with peer performance to give the overall performance gaps. If the benchmarking is conducted from corporate, it can be done offsite with the plant sending required data.

If the overall gap is large enough, it warrants the second phase of audit or specific audit, which is designed to *identify root causes* and process improvement opportunities. The audit methods are explained in the book. The methods for energy audit can be put into two categories: equipment based and system based. The former includes furnace audit (Chapter 5), heat exchanger audit (Chapters 7 and 8), and fractionation column audit (Chapter 12). The latter consists of energy loss audit (Chapter 6), fractionation system audit (Chapter 13), heat recovery audit (Chapter 9), steam balance audit (Chapter 16), and steam system audit (Chapter 18). To estimate true values to be delivered by modifications, the methods in Chapters 20 and 21 can be employed. Last but not the least, one must make sure the modification options will be feasible and economically viable; feasibility assessment (Chapter 22) must be conducted to identify critical constraints and determine ways to overcome them with minimal impacts.

If the specific audits are conducted by a team organized by corporate, the company should provide audit standards and work processes that describe assessment topics and the requirements for conducting assessments, reporting, and documentation.

The third stage of the audit is to *review* the energy projects selected by the plant and recommend project prioritization although ultimate approval of projects will be made by plant management. Successful or failed results will be selected for sharing as experience and lessons learned throughout the company. Common tasks can be developed, which could include heat exchanger cleaning optimization, steam trap repairing and maintenance, control strategy for optimizing column energy and yield trade-off, utility contract management, flare gas reduction, key indicator system, and so on.

The key for a successful corporate audit is to have good preparation from plant personnel and from the lead auditor. The lead auditor needs to make plant engineers aware of what will be audited as well as the auditing process. At the same time, the lead auditor needs to get other auditors on the same page for the agenda and schedule. On the other hand, plant process engineers need to prepare and make the plant's performance data available in an organized manner. If plant personnel do not understand any part of the audit process, they should get clarification before auditors come to the plant.

23.5 ENERGY OPERATION MANAGEMENT

Operational success requires running the plant properly (safely, reliably, and economically). This is achieved by sufficient knowledge and experience, good training, and tools. One of the key elements for success is to provide operators with good monitoring capability for what to watch with limits and targets.

23.5.1 Operation Monitoring System—Key Indicators, Limits, and Targets

To keep process units in the desirable operating envelope, operating ranges such as critical limits, standard limits, and normal limits are developed for all process variables (Gillard, 2001). Critical limits are mainly reliability variables. An example is furnace tube skin temperature, which would cause furnace tube rupture if the temperature were to exceed a certain range. On the other hand, if the furnace were to be operated outside a standard limit, the results would not be economical. An example is furnace flue gas oxygen content. Normal limits are set statistically using typical process control algorithms. The limits are reviewed in each shift to identify operation outside of the ranges and determine actions to return to the operating ranges. This is a common practice in the process industry.

As a stretch from operating limits, major operating variables are defined as key indicators according to their impacts on profit, efficiency, and reliability. The intention of defining key indicators is to describe the process and energy perform-ance with a small number of operating parameters. A key indicator can be simply an operation parameter such as a desirable product rate, column overhead reflux ratio, column overflash, reaction temperature, spillback of a pump, heat exchanger U value, furnace heat flux, steam letdown rate through a letdown valve, and so on. By the name of key indicator, the parameter identified is important and has a significant effect on process and energy performance. The detailed discussions on key indicators can be seen in Chapter 4.

Application of the key indicators in reality follows a methodology based on three steps: defining key indicators, setting targets, and identifying actions to close gaps. The main objective of monitoring key indicators in operation allows operators to focus on important issues. Each indicator is correlated to a number of parameters—some are process-related and others are energy-related. In this way, energy optimi-zation is connected with process conditions and constraints.

For each key indicator, targets are established as the basis to compare with current performance. The difference between a target and the current performance for each key indicator defines the performance gap. Different gap levels indicate the severity and level of urgency.

(a) *The Best-in-Plant Target:* This target is determined based on the best performance achievable with existing facilities. This is determined using statistical methods based on historical data, which reflect the best performance in the past. Targets will be updated when performance is improved through operational improvement and upgrading of existing equipment.

(b) *The Best Potential Target:* This target represents optimized performance assuming equipment in the best condition with the process unit operated in an optimal manner. These targets help identify the potentials in improving process and control performance.

(c) *The Best-in-Class Target:* This target compares the operating results with the state-of-art technology—the difference is called the technology gap. Updates of technologies for the process and energy system can make big changes in energy performance, but usually with relatively high capital costs and longer implementation periods.

Gap analysis is then used to identify root causes—potential causes include inefficient process operation, insufficient maintenance, inadequate or lack of operating practices, procedures and control, inefficient energy system design, and outdated technology.

23.5.2 Operation Training—Key Indicators, Limits, and Targets

It is common practice that experienced operators train the inexperienced, which could be the easiest and lowest cost. This might be a good way for on-the-job training. However, it may not be effective for training with regard to more fundamental aspects of operations because the efficacy will depend on teaching ability and the interest of experienced operators. Operating methods evolve over time as improved understanding is developed, which is added to the existing training materials. In particular, inconsistent knowledge and experience from different operators may pass on to new operators and cause inconsistent operation for the same process and equipment.

Therefore, it is necessary to develop a more systematic approach. During the 5 years when I led a large energy project for a plant, the team developed a training model for the plant operators. Our overall goal was to equip operators with the knowledge and skills required for energy improvements in each process unit. The core of this training model is the key indicators and how to apply them properly. It is more like training people how to drive a car instead of designing and building a car. The experience shows that if key operating parameters can be operated close to targets, a process unit can achieve desirable yield and energy performance. Thus, the training model focuses on understanding of the key indicators, operating limits and targets, and relationships. At the same time, the Honeywell dashboard was used as the online tool to enable monitoring of the key indicators.

Thus, the training objectives were defined for this training:

- Understand key indicators (KIs) and their limits and targets
- Use of monitoring tools to measure the key indicators
- Hands-on training using DCS board
- Reach common understanding and achieve operation consistency

These objectives were communicated before starting the training of the unit engineers and operators. As a follow-up to in-class training, hands-on training was conducted at a control room under the supervision of experienced operators. The purpose was to help board operators to develop mental models for adjusting/ optimizing the KIs learned in class. Discussions and brainstorming were conducted and many improvement ideas were suggested by operators. After ideas were generated, they were screened and then a few top ideas were selected for implementation. To implement in DCS, experienced operators were chosen to make adjustments in the DCS board to show how the improvements could be implemented in DCS, while the energy dashboard tool displays the changes of key indicators and the trainer explains the fundamentals behind the adjustments.

The last step is testing. To ensure learning quality, online quizzes were used to verify understanding of training. If any operators failed any quiz, he/she was then required to go through the related training materials and do the quiz again until the questions are answered correctly. Overall quiz scores were recorded. Furthermore, a checklist for the required skills from training was developed. Operators were required to demonstrate proficiency via certain hands-on activities: for example, adjust column reflux ratio while observing overhead product specs—an experienced operator had to sign off on the competency using the checklist. This provides verification of field know-how and not just theoretical knowledge.

The KI training model above was provided to one process unit for trial. The results were found overwhelmingly positive. This specific training through key indicators helped to lay down the foundation for achieving high performance of process energy use and yields. The combination of training via class, DCS, and energy dashboard tool was proven effective in achieving the training objectives, which makes learning and sharing very focused and stimulating.

It was observed during training that operators have different ways of running the same process and equipment. This inconsistency in operation practices and procedures can cause poor performance. The KI training helps to determine best practices and procedures to be used for all shifts so that standard operating procedures are developed in place to operate the unit consistently. Through these training sessions, we also recognized that very often operators did communicate well with regard to business and technical issues. People tend to forget the fact that it is the board operators who are the end users and are ultimately in control of the units. Consequently, they have many ideas for improving performance due to the many years of work experience. It is very effective with the training outlined above to keep operators in the communication loop and make them feel they are valued for their opinions and performance, and it may well be the case that the company can gain valuable insights from these operators.

Due to the success of this KI training model in the process unit, the plant management decided to disseminate this training model for other process units and other operating areas. Reward mechanisms were also established to award operators who have made significant or valuable contributions to continuous energy improvement.

23.6   ENERGY PROJECT MANAGEMENT

Each company has well-defined work processes, procedures, and guidelines for capital project management. It is not the intention to repeat them here. Instead, certain important issues that may arise from capital project management are highlighted below.

- *A typical work process* for capital project management consists of four stages. The first stage is "'appraisal": technical people can give a consensus if an improvement idea is good or bad at a very high level. The second stage is "define": a cost–benefit analysis is conducted to give a rough idea for what is the benefit at what cost in ±50% accuracy. Once an idea goes through these two stages, it faces "selection," which is the third stage. In this stage, all ideas will be assessed in greater details and more accurate capital cost estimate in ±30% will be obtained. These ideas will be compared based on their NPV and return on investment together with other criteria, such as process turnaround schedule, risk factor, budget, and so on. For ideas that are selected, the fourth stage, "implementation," then starts. Engineering contractors are contacted to implement the capital projects, which will consist of changes to process equipment and to the control system architecture and associated software applications.

- Retrofit energy projects face tough challenges for being selected in competing with process related projects. The goal of the capital project evaluation is to *achieve minimum investment cost*. The key to achieving this goal is to explore alternative design options for each improvement idea and find solutions for overcoming process/equipment limits. Discovering these opportunities requires a good understanding of these interactions and the technical skills to exploit them for meaningful use.

- *Project Lookbacks* (Gillard, 2001): The lookbacks should be conducted for all projects to see if they deliver the promised results and address what went well and what did not and why. Care should be taken to avoid interdepartmental disputes. The emphasis should be on learning, avoiding mistakes, and doing better the next time. The review should answer the following questions:

  ○ What capital efficiency are the projects delivered on average?
  ○ Are projects delivering the economic return promised at the time of approval?
  ○ Are projects completed on schedule?
  ○ How does the cost estimates compare with the actual costs?
  ○ Is the deviation in project performance due to differences in economic assumptions used (feed costs, product values, capacity required), technical performance of the project, cost, or schedule?
  ○ Is the success of projects implemented during maintenance turnaround the same or different for projects outside turnaround?
  ○ For projects with performance problems, what are the root causes?

23.7 AN OVERALL WORK PROCESS FROM IDEA DISCOVERY TO IMPLEMENTATION

The potential benefits of these initiatives are substantial, but given the broad spectrum of solutions that have been described, what sort of capability would a company need to have to deliver them? That is where an effective work process comes in to play a critical role in accomplishing it.

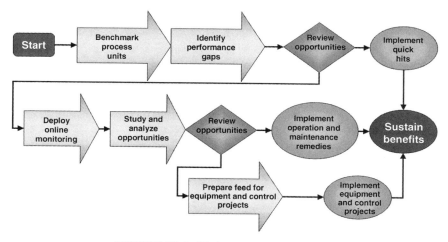

FIGURE 23.4. Work process methodology.

With the mentioned energy efficiency strategy, we can convert them into a solution methodology from idea discovery to evaluation and then eventually to implementation. Figure 23.4, which illustrates one example of how this can work, shows an energy solution methodology (Sheehan and Zhu, 2009).

The objectives of the new work process are to improve the integration between the in-house capability and the engineering vendors and provide a mechanism through which industrial plants can obtain guaranteed, sustainable benefits.

The first step is to benchmark the process units and identify performance gaps. This step helps focus the analysis on the best opportunities. It can also lead to operational improvements that can be implemented immediately. Energy monitoring application software is used to monitor the facility, identify performance gaps, and provide inputs to the detailed study and analysis step. Key indicator system and targets should form the basis for monitoring and control, which provides the means for continuous improvement.

Once the best opportunities are identified and analyzed, the plant chooses the projects to implement. At the same time, some advanced control and optimization applications along with other operational and maintenance solutions will be implemented where benefits that do not require capital investment can be realized. Operation improvements should be implemented at the earliest time possible because that can bring immediate cost savings. The savings could be credited to the energy program to fund capital projects in the latter stages. Operational improvements are often determined from online monitoring of actual performance against technical targets and the benefits are captured by optimizing process conditions. Many small savings are possible if the operators are given sufficient technical instructions to make changes.

The capital projects that the plant choose to implement usually go through a FEED phase to complete the engineering and confirm the capital costs. The next step is to choose a contractor to implement the capital projects, which will consist of changes to process equipment and to the control system architecture and associated software applications.

The last step in the process is to sustain the benefits from the energy projects that have been implemented. An energy management system must be established for continuously monitoring, maintaining, and improving the performance of the implemented energy projects.

REFERENCES

Gillard CF (2001) Creating plant operating excellence with measurable results, *Hydrocarbon Process*, April Issue, 57–70.

Sheehan BP, Zhu XX (2009) Improving energy efficiency, PTQ, Q2, 29–37.

INDEX

Advanced process control (APC)
  system, 45, 55–57, 306, 308, 319,
    324, 326
Aggregate energy intensity, defined, 13
APC. *See* Advanced process control (APC)
    system
Automated network pinch retrofit
    approach, 181–183
  mixed integer linear programming (MILP)
    model, 181
  nonlinear behavior, of stream heat
    capacity, 183
  utility modifications, 183

Benchmarking assessment, 16
  Process energy benchmark, 31
Benchmarking steam system, 386
  benchmark steam cost, 389
  costs of steam production, 388
  fuel selection, 388
  heating values, 389
  efficiency, 397–402
    cogeneration efficiency, 402
    R ratio analysis, 401

operation, 396–397
Benchmark steam, 387, 389
  and condensate losses, 389
    condensate discharge, 391
    condensate return, 392
    monitoring steam losses, 393
    recommendations, 392
    steam recovery, 392
    steam trap management, 389–391
  minimize generation cost, 387, 389
  process steam, usage and energy cost
    allocation, 394–396
Boilers, 329, 330
  blowdown, 332
  combustion efficiency, 387
  efficiency, 332, 370, 410
  feed water (BFW), 17, 403, 426
  fire-tube boiler, 330
  loading optimization, 405
  modeling, 332–333
  packaged boiler, 332
  specific fuel consumption, 370
  steam (*See* Boiler steam)
  water-tube boiler, 330–331

Energy and Process Optimization for the Process Industries, First Edition. Frank (Xin X.) Zhu.
© 2014 by the American Institute of Chemical Engineers, Inc. Published 2014 by John Wiley & Sons, Inc.